CHENEY

CHENEY

*The Untold Story of America's
Most Powerful and Controversial
Vice President*

S T E P H E N   F .   H A Y E S

HarperCollins*Publishers*

To Carrie
thanks for your help.
sfh

HarperCollins books may be purchased for educational, business, or sales
promotional use. For information, please write: Special Markets Department,
HarperCollins Publishers, 10 East 53rd Street, New York, NY 10022.

FIRST EDITION

Designed by Renato Stanisic

Library of Congress Cataloging-in-Publication Data is available upon request.

ISBN: 978-0-06-072346-0
ISBN-10: 0-06-072346-7

07 08 09 10 11 NMSG/RRD 10 9 8 7 6 5 4 3 2 1

CONTENTS

AUTHOR'S NOTE

On June 23, 2004, I waited in the foyer of the vice president's residence for my first one-on-one interview with Dick Cheney. Writing a book like this had crossed my mind, but I was at work on another project and busy with my duties at the *Weekly Standard*.

There was much to discuss. My first book, *The Connection*, had been released earlier in the month, and among top Bush administration officials, Cheney was one of the most receptive to the argument that Iraq had been part of a broad global terror network that included Al Qaeda. Leaks about the forthcoming 9/11 Commission Report suggested that it would downplay those ties. The Senate Select Intelligence Committee would soon release its own report on the intelligence about Iraq that guided policymakers before the war.

There was a lot of other news. In the morning papers were several stories about the Bush administration's detainee policy. Earlier that same month, CIA Director George Tenet had resigned, and congressional Democrats were pushing the White House to defer the selection of a successor until after the election in November.

I was to have thirty minutes, on the record. Shortly before we were to begin, Cheney's communications director, Kevin

Kellems, approached me with bad news. Scooter Libby, the vice president's chief of staff, had just changed the ground rules: No questions about Iraq and terrorism.

That's unacceptable, I protested, or words to that effect. He was sympathetic. If my questioning is restricted in that manner, I explained, then I'd write about the apparent unwillingness of Cheney to talk about the issue that was being discussed everywhere else. He took my complaint back to Libby and returned with good news.

"It's all on the record," he said. "Ask what you want."

We sat in the library. Kellems and Libby stayed in the room. For forty-five minutes, Cheney sipped a decaf latte as he answered questions or, in several cases, didn't.

There had been little actual news about the selection of a new CIA director, so advancing the story would require only a short comment from Cheney. Make the vice president talk about the ideal candidate, I naively thought, and perhaps he'd say something revealing.

In the search for a new CIA director, can you briefly describe what kinds of qualities you are looking for or what kind of personal characteristics you are looking for?

"Probably not."

I waited for him to continue, but he said nothing.

Is there a rough timeline for a decision?

"If there were, I wouldn't want to talk about it."

At another point, I asked him about new reports on prewar intelligence.

Russian president Vladimir Putin said recently that before the Iraq War began he gave the Bush administration information about potential Iraqi attacks on American soil. What kind of information was it and how serious was it?

"Yeah, I can't really give you anything on that."

Abu Musab al Zarqawi is the most dangerous man in Iraq. Can you talk about U.S. efforts to capture or kill him?

"There's nothing I can give you on that."

Not all of Cheney's answers were so brief, but he clearly knew where he wanted to draw a line. There were subjects he would discuss and others he would not.

Over the next thirty-four months, I would spend nearly thirty hours in one-on-one interviews with Cheney for this book: on the telephone from my home; on my cell phone from a dingy Wyoming motel; at the vice president's residence in Washington, D.C.; aboard Air Force Two somewhere over the Midwest, and again flying back from Afghanistan, and another time returning from Iraq.

He was considerably more forthcoming. Cheney spoke about his boyhood in Nebraska, his difficulties at Yale, and his two arrests for drunk driving. He talked about the psychological impact of having a heart attack at thirty-seven, the power struggles of the Ford administration, his frustrations in Congress, the high points and low points of his time at the Pentagon, and the cross-country driving trip he took alone as he considered a run for the White House. We discussed the dynamics of the Bush administration, his role as vice president, his reaction to 9/11. He surprised me with his candor about the mistakes of postwar Iraq and the personal and professional toll of the unrelenting criticism directed his way.

But he still knew where he wanted to draw a line. On August 9, 2006, I interviewed him at his home in Jackson, Wyoming, as I had the previous summer and would again in two weeks. Cheney was relaxed and actually seemed to be enjoying the interview. Throughout our conversation, now in its fifth hour, he had been remarkably open about his career in Congress and as secretary of defense.

There had been exceptions, though, as there had been two years earlier. To set up an amusing anecdote, he told me about an urgent call summoning him to the White House on his first day at the Pentagon. When he finished the story, I asked what seemed like an obvious follow-up.

Do you remember what the meeting at the White House was about?

Pause.

"I do."

There was a long silence. I looked at him expectantly, raised my eyebrows and gave him an encouraging, out-with-it nod of my head. He laughed a little, but said nothing.

"Umm," I said, "anything you can talk about?"

"It's classified still."

I told this story to another journalist traveling with me on one of Cheney's trips, a highly respected White House correspondent for one of the country's leading newspapers. We chatted a bit about covering the Bush White House and about writing books. Then he asked me a question that has stuck in my mind ever since.

"Don't you think your book will have to be a hatchet job in order to have any credibility?"

From his perspective it must have been a reasonable concern.

After all, I write for the *Weekly Standard,* a conservative magazine that has often supported the Bush administration, particularly in foreign and national security policy matters where Cheney's influence is greatest. One of my articles ran under the headline: "Dick Cheney Was Right." It was a lengthy defense of Cheney, pushing back against misreporting of his views by the mainstream press. The article offered lots of supporting documentation, and a heavy dose of outrage.

This book, a reported biography, has a different and much simpler purpose: to tell the story of Dick Cheney's life. And while I came to the book sympathetic to Cheney's views, I tried to go where the reporting took me. It will be up to others to determine whether I succeeded or failed.

At several points in the book, I describe a different Dick Cheney from the one who has emerged from more than six years of scrutiny in the White House. Given the access I had to him and those closest to him, that is not surprising. In an e-mail to a top Cheney aide, Bill Keller, executive editor of the *New York Times*, tried to persuade Cheney to cooperate more with reporters from his paper. Keller said that understanding political figures, particularly conservatives, requires access and journalists who possess "a deep quality of open-mindedness" and "an ability to listen without reflexive cynicism." He continued: "Our job is not to 'support' our leaders, not to buy in to any administration, Democrat or Republican, but our job should be to figure out what they believe and why, and how all of that shapes the

policies they make. We are obliged to get past the labels and slo-
gans. It's unbelievably hard to do."

Keller is right. But Cheney decided early in his vice presidency that his job would not include spending much time with
journalists. And much of the reporting about him lacked the
open-mindedness and rejection of cynicism that Keller thought
so important.

As the Bush administration began its final two years, caricature reigned. When President Bush spoke to the country on January 10, 2007, to describe his plan for more troops in Iraq, MSNBC's
Chris Matthews described the motivations of Bush and Cheney
to listeners of *Imus in the Morning* the following day.

"They're working through Cheney, of course, who always
wants to kill. And they drag us back into that same mentality of
'we're looking for any reason to strike.' I think that is the way he
is toward Iran right now."

Dick Cheney always wants to kill?

That's one way of looking at him. And while this book was
not meant to be the other, I sought to include that side in order
to tell a more complete story.

To do that, I conducted more than six hundred interviews of
people who have known Cheney over the course of his lifetime.
Many of these interviews were on the record and fill hundreds of
hours of audiotape. Many other interviews were conducted with
the understanding I could use the words without identifying the
source by name. And still others took place on deep background—
I could use the information, but never in quotes—and off the record. And while I made attempts to give each of those individuals
who has played a role in the story an opportunity to help shape
the story, not everyone agreed to speak to me for the book.

Those who did ranged from George W. Bush and Condoleezza
Rice to childhood friends Vic Larsen and Tom Fake. There were
numerous interviews with Cheney's notoriously tight-lipped
staff, including some who rarely, if ever, talk to the press: chief of
staff David Addington, staff secretary Neil Patel, staff director
of the Energy Task Force, Andrew Lundquist, and Cheney's on-
staff archivist, Jim Steen.

Many of the quotations in this book come directly from the individuals in those conversations or from documents that recorded them for history. Details of meetings often came from notes taken by one or more of the participants. As with every work of narrative nonfiction, I relied heavily on the memories of my sources, particularly for accounts of events that took place before Cheney entered public life.

This book is not the entire story of Dick Cheney's life or a comprehensive history of the Bush administration or even the definitive treatment of Dick Cheney's role in the Bush administration. Those stories, of course, are still unfolding. And I hope to keep telling them.

CHENEY

INTRODUCTION

Not long before Dick Cheney, America's forty-sixth vice president, agreed to become a candidate for that office, he offered a succinct evaluation of the merits of the job. It was a considered view, developed over more than two decades inside Washington's halls of power, two decades of firsthand exposure to the office and the men who occupied it.

The vice presidency, he said, is "a cruddy job."

This was not a novel opinion.

Thomas Marshall, vice president for two terms under Woodrow Wilson, spoke of the "utter uselessness and frivolity" of the position, and refused to go to cabinet meetings because no one listens to the vice president. "A great man may be vice president," said Marshall, "but he can't be a great vice president because the office itself is unimportant."[1] A century earlier, Daniel Webster chose to remain in the Senate rather than ascend to the obscurity of the vice presidency, saying, "I do not propose to be buried until I am dead."

John Nance Garner, the former Speaker of the House, who became Franklin D. Roosevelt's first vice president, famously said the position wasn't worth "a warm bucket of spit" and later declared that accepting it was "the worst damn fool mistake I ever made." And Harry S. Truman echoed those sentiments when Roosevelt approached him to be Garner's successor. "I bet I can go down the street and stop the first ten men I see and that they can't tell me the names of two of the last ten vice presidents of the United States."[2]

Cheney's own longtime political mentor, President Gerald R. Ford, "hated the job," says Cheney. At one point, Ford told Cheney that "the worst nine months of his life were the nine months he spent as vice president."

Indeed, the first vice president of the United States, John Adams, complained about the office to his wife. "My country has in its wisdom contrived for me the most insignificant office that ever the invention of man contrived or his imagination conceived."

But under Dick Cheney this traditionally inconsequential office has been transformed into a focal point of presidential power. In some ways Cheney's influence is obvious. He was the sole representative of the Bush administration on the nation's most-watched Sunday talk show five days after the attacks of September 11, 2001, at a time when the White House wanted to project calm and confidence; and he appeared in the same forum three days before the Iraq War. He had been the most aggressive proponent of that war and the most uncompromising critic of John Kerry in 2004.

But Cheney's real influence is unseen. It lies in his ability to work the levers of power at the highest levels of the U.S. government and, more directly, in his private conversations with George W. Bush. In the first week of his presidency, with blackouts washing over California and distortions in the energy market threatening catastrophic economic consequences, Bush tasked Cheney with a comprehensive interagency review of U.S. energy policy. Cheney brought to the job decades of experience in government and a deep understanding of the process that turns ideas into policy. It is no accident that the product of Cheney's Energy Task Force so closely reflected the views of its chairman.

"Mark my words," Bush had said in July 2000. "There will be a crisis in my administration, and Dick Cheney is exactly the man you want at your side in a crisis."[3]

Minutes after learning of the attacks on the World Trade Center on the morning of September 11, 2001, Bush gave his first order as a war president: Get me Dick Cheney. After Bush addressed the nation from the Oval Office later that evening, he met briefly in the conference room of the White House bunker with the full National Security Council. Moments later, he pared the group to a more select group of his principal national security and foreign policy advisers. And then, after that, he trimmed that group even further. Bush led his top adviser to the quiet of a small bedroom attached to the main room of the bunker for a private conversation. In those first hours of the global war on terror, the vice president provided his unfiltered views to a president eager to receive them.

In August 2006, Condoleezza Rice was asked to name an issue or a presidential decision in which Cheney has been particularly influential. "That's a long list," she responded.

After considering the question for a moment, she added: "I think the way that the vice president has had his biggest impact in many ways is just the intellectual contribution to the conceptualization of the war on terror."[4]

Shaping the national response to the defining issue of the early twenty-first century is a long way from the warm bucket of spit. Cheney, in fact, is arguably the second most powerful man in the world. There's a historical irony here: For more than 200 years, vice presidents, generally an ambitious lot, sought more power and craved more recognition. Cheney, who certainly does not want for influence, dislikes the attention that comes with it. So he does his best to avoid it.

"Am I the evil genius in the corner that nobody ever sees come out of his hole?" Cheney said in 2004. "It's a nice way to operate, actually."[5]

This reticence has a price. Where there is an information vacuum, people move to fill it, particularly in a town and a profession that operate on appearances. Because Cheney contributes so little to the public story line of his own career, there has been no authoritative version. Instead, everyone seems to have his own, and as a result, Cheney is controversial, polarizing, and perhaps misunderstood.

To many, Cheney is a man almost devoid of passion. His admirers find his demeanor reassuring: it says he is contemplative, unshakable, discreet. To his critics it registers as scheming, uncaring, and secretive. In public, he speaks of the most personal and emotional subjects—his daughter Mary's sexual orientation, for example, or the Iraq War—without so much as changing the inflection of his voice. He is no different in private. Even his closest friends and members of his family cannot recall hearing Cheney raise his voice or seeing him shed a tear.

He is in that sense a perfect complement to the man he serves— the man who campaigned for president as a compassionate conservative and who wears that compassion openly, the man who weeps publicly when discussing victims of tragedies that have happened under his watch. George W. Bush cares most deeply about the softer issues, the connective tissue of the body politic: promoting faith-based initiatives, leaving no child behind, promoting democracy and human freedom in those places least hospitable to such ideals. Bush's passion for these issues, say those who know him best, comes from his faith.

Cheney is Bush's opposite. He thinks government should do a few things well and otherwise leave its citizens alone. Defend the country. Cut taxes. Get out of the way. Cheney rarely gets carried away in the moment. His speeches to rally the troops in Iraq—judging from their emotive qualities—might well be formal addresses to economists in Chicago.

All this changes when Cheney talks about growing up in the West.

"Wyoming was different than a lot of places partly because it was an ideal place—the physical setting, the openness of it," Cheney says, his voice softened by a fondness approaching sentimentality. Sitting in his home at the base of the spectacular Grand Teton mountain in Jackson, Wyoming, wearing old New Balance sneakers, jeans, a denim shirt, and a green fleece vest, Cheney relaxes his shoulders as he waxes philosophical about growing up in the West. The tensions of his twenty-four-hour-a-day job seem to disappear. "Perhaps it lacked some of the sophistication you'd find back East, in big cities, both coasts. But there was a real sense of place here. Wyoming was a special place."[6]

Cheney arrived in Washington, D.C., in 1968. By the age of

thirty-five, he had raced from congressional intern to cabinet agency bureaucrat to low-level White House staffer to White House chief of staff under President Gerald Ford.

When it was over, he wanted to leave. "I've got a feeling that this town is too full of ghosts from previous administrations, and I'm not eager to become one," he said shortly after Ford lost the election of 1976. "Too many people, after they have once served in the White House, can't get it out of their system. They just sit around, maybe for the next thirty or forty years, waiting for lightning to strike again. And that's sad."[7]

So he headed back to Wyoming, but he didn't settle in for long. Within weeks he announced that he would run for Congress. In that campaign, Cheney's opponents tried to portray him as a Washington insider. Cheney countered by emphasizing his Wyoming roots and western values. It was a good strategy.

In January 1979, Cheney returned to Washington as the state's only member of the House of Representatives. "There's an independence that goes with the West, and a lack of phoniness,"[8] says Norma Fletcher, a longtime Wyoming resident who was a friend of Cheney's parents. For ten years in the House—first as a backbencher and, after just one term, as part of the Republican leadership—he brought those Wyoming sensibilities to Congress.

In 1989, Cheney, a ghost of previous administrations, joined the first Bush administration as secretary of defense. After a brief flirtation with a presidential run and five years in the private sector, lightning struck a third time and Cheney was back in the White House.

All this time Cheney has practiced politics as though he'd never left Wyoming, where blunt talk wins elections and substance matters above all else. He is in that sense an antipolitician. The rise of televised politics produced media-friendly leaders like Ronald Reagan and Bill Clinton, masters of political rhetoric who spoke extemporaneously as though reading from a teleprompter. The twenty-four-hour news cycle has led to a more intimate knowledge of American political leaders, a blurring of their public and private lives. The American public learned about the sexual proclivities of Bill Clinton and Jim McGreevey, the messy divorces of Newt Gingrich and Rudy Giuliani. Many public figures have volunteered their most private and painful moments in the furtherance of their political careers, as when Al Gore described his sister Nancy's slow death from cancer in a speech at the Democratic convention of 1996. As made-for-TV politics increasingly dominates public discourse, more elected officials have become made-for-TV politicians.

The nation's capital is a town full of talkers. If Los Angeles is a city built on movies and if Detroit is a city built on cars, then Washington is a town founded on words. The more one says, the thinking goes, the more people hear. So keep talking.

Not Cheney. Not his style. He still speaks in clipped English, drops unnecessary words—prepositions, conjunctions. The less you say, he believes, the more people listen. So when Cheney has nothing to contribute, he says nothing.

One of Cheney's congressional aides recalls having a breakfast with him at which only a few words were spoken. Another remembers driving for two hours with Cheney across the Wyoming wilderness in complete silence. It's not that Cheney dislikes being around people, exactly. But unlike most elected officials, he does not seek affirmation of his views from others, whether voters, staff members, or friends. He is uncomfortable with adulation and disdains sycophancy.

When Dick and Lynne Cheney moved from Casper, a blue-collar town in central Wyoming, to Jackson, they disagreed about where they should buy a home. According to their real estate agent, Jackie Montgomery, Lynne wanted a neighborhood near town and near other people, while Dick preferred a secluded lodge, tucked away in the woods at the base of the mountains. Lynne won that debate.

As secretary of defense, Cheney chose to delegate to Colin Powell many of the Pentagon's daily press briefings about the Gulf War. As a member of Congress, he gave only a handful of floor speeches. When he was running the White House for President Gerald R. Ford, Cheney would attend cabinet meetings that stretched on for hours without saying a word.

The bluntness of Cheney's manner can be disarming and, to reporters, frustrating. In November 2001, about two months after the attacks of 9/11, Cheney sat down for an interview with *Newsweek*'s Evan Thomas. Thomas and a group of other reporters for *Newsweek* were working on a detailed account of 9/11 that would weave together the experiences of four individuals—a terrorist, a firefighter, a survivor, and the vice president—on that tragic day.

Cheney's advisers thought it important that the vice president go beyond describing the events of the day and laying out the administration's policy response to terrorism. "We told him to give the American people some color,"[9] recalls Mary Matalin, Cheney's top communications adviser at the time. He didn't listen. Cheney gave Thomas an exhaustive report on his activities, but it was essentially a just-the-facts narrative.

Thomas, who has interviewed dozens of America's most powerful figures, artfully phrased his questions so as to elicit some emotion and capture some of the drama of the day. Despite these repeated efforts, Cheney kept his account simple and mostly understated. As the interview came to a close, Cheney seemed to understand that he wasn't providing what Thomas needed.

"Evan," he said, "you're asking me for all this good color stuff and that's just not the way my mind works."

This is as self-reflective as Cheney gets. Almost.

On Saturday, May 27, 2006, Cheney returned to Wyoming for an important speech. He had been invited to deliver the commencement address at Natrona County High School, the school that had given him his diploma, his best friends, and his wife. It was an unseasonably warm day in Casper, with the temperature reaching eighty-four degrees. Six thousand people attended; 340 students graduated.

It was an important speech to Cheney. His remarks are almost always prepared for him by his chief speechwriter, John McConnell, who in more than six years of service has come to know the vice president's voice as well as he knows his own. McConnell wrote the first draft of the commencement address, but en route to Wyoming on Air Force Two Cheney dismantled the draft and started to rebuild it.

Two days earlier, the front page of the *New York Times* reported that Cheney might be called to testify in the trial of his former chief of staff, Scooter Libby. The vice president had spent hours the previous week working out a compromise between the Justice Department and the angry leadership of the House of Representatives, after FBI officials raided the offices of Representative William Jefferson on a warrant related to bribery charges. He was on the phone with embittered congressional Republicans trying to broker a deal on immigration reform; a majority of the members of Cheney's party in the House of Representatives, a body he still calls his "political home," opposed the proposal backed by the White House. The newly elected prime minister of Iraq failed to name his top security advisers, as promised, and attacks on U.S. troops and Iraqi civilians continued unabated. The Iranian regime once again defied international nuclear inspectors, and its president issued belligerent statements about eradicating Israel. The Bush administration was in the midst of a far-reaching shake-up, and news reports indicated that Secretary of the Treasury John Snow would be the next to leave.

In Wyoming, Cheney was 2,000 miles from Washington, D.C., and its many troubles, and for just one afternoon he seemed to set them aside.

For those familiar with Cheney's life, the address was surprisingly personal and self-revelatory. On May 19, 2006, in a commencement address at Louisiana State University, he had spoken about hurricane Katrina, selflessness, and second chances in language recycled from previous addresses. On May 26, 2006, the day before he spoke in Casper, he had addressed the graduates of the United States Naval Academy. He spoke of honor, war, and sacrifice.

But in Wyoming he shared personal lessons.

Stay focused on the job you have, not the next job you might want. In your careers, people will give you more responsibility when they see that you take your present job seriously. Do the work in front of you. Try to find ways to make yourself indispensable. And I can almost guarantee that recognition, advancement, and other good things will follow.

I think there's also a lot of truth to the old wisdom that you should choose your friends carefully. They have a big influence on the kind of person you become. So when you see good qualities in people—things you admire, habits you'd like to pick up, principles you respect—keep those people close at hand in your life. In many ways, when you choose your friends you choose your future.[10]

There is little doubt that this advice sailed over the heads of many students in the crowd that day as they waited eagerly to hear their names announced as high school graduates. Like most of the words that pass the lips of politicians these nuggets of advice from Cheney could be dismissed as platitudes doled out without much thought in one of thousands of speeches given over the course of a public life.

They were not. For the graduates in attendance that day who paid attention, Cheney had done something extraordinary: in two short paragraphs, he had explained how a young man who sat among the Natrona County High School graduates forty-seven years earlier had become the most powerful vice president in the history of the United States.

CHAPTER ONE

The West

On January 30, 1945, Yeoman Richard Herbert Cheney was scheduled to return home to Sumner, Nebraska, on leave. That the break coincided with his son's birthday was serendipitous, but Marge Cheney told young Dick that his father had returned for the special celebration.

Cheney had grown up in Sumner, a rural, speck-on-the-map town 200 miles west of Lincoln, the state capital. For the first few years of his marriage, he supported his new family working as a bureaucrat in a program operated by the U.S. Department of Agriculture. But in 1944, shortly after the United States joined the fighting in World War II, Cheney joined the U.S. Navy; he was assigned first to the Great Lakes Naval Station and eventually to Naval Station San Diego in southern California. Visits back home to Nebraska were rare.

When Cheney left, his wife, Marjorie, and their two young sons, Dick and Bobby, moved into the basement of his parents' home. Young Dick asked the questions any four-year-old might ask: Where has my father been? And why was he gone? Yeoman Cheney explained to his son that he was in the U.S. Navy and pointed to the insignia on his shoulder. The most prominent feature of the yeoman's insignia is a large birdlike figure with its wings spread as if poised to fly. This only added to the confusion.

"I was convinced," Cheney recalls, "that when he was home he was my dad, but when he went back to the Navy, they put him through some kind of process and they turned him into one of those kind of deals I saw on his shoulder." Cheney thought the Navy turned his father into a bird. "I remember carrying that thought with me for years."[1]

DICK CHENEY LIVED his first thirteen years in Nebraska before moving farther west to Wyoming. His was an idyllic childhood and, for someone who would later be known for his gravity, remarkably carefree.

The family history of Richard Bruce Cheney, the forty-sixth vice president of the United States, is intertwined with the westward expansion of America. On his paternal side, Cheney's ancestors had virtually spanned the country in the course of the 1800s. His great-grandfather, Samuel Fletcher Cheney, was born in New Hampshire in 1829, and later moved with his family to Defiance, Ohio. There, he served as a captain in the Twenty-first Ohio Infantry and fought in the Civil War, becoming known as something of a Union hero; he served with distinction in many of the great battles of the war—at Stone's Ridge, the siege of Atlanta, and others.[2]

After the war, Samuel Cheney returned to Defiance, where he ran the family lumberyard and worked part-time as a cabinetmaker. Although he had managed to survive unscathed through thirty-four battles, including some of the bloodiest fighting of the Civil War, Cheney's good fortune ended once he returned to Ohio. He lost part of his left hand in an accident at the sawmill and was unable to continue his trade. In the 1880s, he joined a growing number of pioneer families moving west in search of land and a new life. He settled in Amherst, Nebraska, on a homestead that would remain in the family for more than a century. He had a son, Thomas Herbert Cheney, in Ohio in 1869, and moved to Nebraska with his family when the boy was in his teens. A few years later, Thomas struck off to Sumner, Nebraska, where he would take a job as a cashier at what was the only bank in town.

Millions of pioneering Americans preceded the Cheneys in the migration west, and millions more would follow, pouring into the lands opened by the growth of the transcontinental railroad. Among them were Dick Cheney's maternal grandparents, Dave and Clarice Dickey. In the years after the turn of the century, they owned and operated the town diner in Syracuse, Nebraska, known as Dickey's Cafe. They weren't wealthy by any means, but they made enough money to survive.

This work led directly to jobs working as cooks on the Union Pacific railroad. The Dickeys lived for months at a time in a railcar as it moved up and down the tracks supplying food to the "section gangs" of 100 men who repaired old tracks and laid new ones. (Later, as a board member of Union Pacific, Cheney would discover that although they both worked and lived on the railroad, Dave Dickey was not an official Union Pacific employee because he was partially deaf and failed his physical; his wife, however, shows up on the payroll records as a cook and part-time bookkeeper.)

The Dickeys enjoyed life. Dick Cheney recalls: "That whole side of the family loved to play cards, drink a little bourbon"—here he pauses for a slight smile—"and loved a good story."[3] In this environment, Marjorie Lorraine Dickey, Cheney's mother, developed a strong, energetic personality.

Richard Herbert Cheney's family was quite different. His father, Thomas Herbert Cheney, married young. His first wife died not long after their wedding, and several years later he married Margaret Ellen Tyler. Margaret, Richard Herbert's mother, was a devout Baptist who frowned on gambling and drinking and was determined to instill in her only child those same values.

Cheney was born in Lincoln, Nebraska, on January 30, 1941, ten months before the attacks on Pearl Harbor would draw the United States into combat. He had an upbringing typical of most kids raised in wartime America. His father was away in the Navy, his mother ran the household, and his extended family pitched in to ease the burdens—financial and other—of life during World War II. He moved to Sumner at the age of three and lived there until his father returned from California after the war in 1946.

Once reunited, the Cheney family returned to Lincoln so that Yeoman Cheney could reclaim his job with the Soil Conservation Service (SCS). The SCS was a bureau of the U.S. Department of Agriculture, created as part of Franklin Delano Roosevelt's New Deal. Its mission was to promote soil conservation and combat soil erosion, both tasks crucial to the development of farmland throughout the country, particularly in the West.

Housing after World War II was scarce in Lincoln, as it was throughout postwar America. So the Cheneys moved in with the Volkmans, old friends of the family, and lived for several months in the unfinished basement of their small house. Marge Cheney shared the kitchen upstairs with Mrs. Volkman.

In the fall of 1946, Dick Cheney started kindergarten in Lincoln, as his family awaited the completion of their new house in the College View neighborhood. The Cheneys moved into their new house in February 1947. College View was a pop-up community, like the famous Levittown and countless others sprouting up across the country: clusters of small, mass-produced homes affordable to recently returned veterans eager to begin or expand their families.

The house was small—800 square feet—with one bathroom, two bedrooms, a kitchen, and a living room. (The woman who bought the house from the Cheneys still lived there in 2004, when the vice president made a campaign stop in Lincoln and paid a visit to his boyhood home. "You can stand in the middle of the house and spit and hit all four corners," she told him.[4]) Cheney's father later finished off half of the basement, and in the summer the family often retreated to the coolness of the basement where they would eat dinner and listen to *The Lone Ranger* on the radio. After supper, Dick would return upstairs to the bedroom he shared with his younger brother, Bobby.

The Cheney boys became fast friends with another pair of brothers who moved in two doors down, Vic and Edsell Larson. Together they formed the core of what must have been southeast Lincoln's most fearsome band of rascals, the "Forty-fourth Street Gang." The gang members rode bikes and climbed trees;

they played Midget League baseball and Pop Warner football; they walked three-fourths of a mile to a small creek and spent hours fishing for crawdads; and, on Saturdays, they took the city bus to the local movie house to take in the latest Roy Rogers or Gene Autry movie for a dime. The boys spent hours making an elaborate slingshot to fight off the rival "Forty-sixth Street Gang," but the attacks they anticipated never came.[5]

There was higher adventure, too. Not far from home was a tunnel that led to the entrance of a large storm drain and beyond that an intricate maze of more tunnels that ran throughout the city of Lincoln. "You didn't want to get caught in there in a rainstorm, obviously, but you could go a long way under the streets through that tunnel," Cheney recalls.[6]

He began to explore intellectually, too. Cheney's paternal grandmother had been a schoolteacher, and his home was usually scattered with books. Young Dick grew into an avid reader.

As a fourth-grader at College View Elementary, Cheney spent his summer as a Junior Book Sleuth, devouring *Kit Carson* and *Myles Standish*, a biography of baseball player Lou Gehrig, and *John Quincy Adams, Boy Patriot*. His favorite author was Sanford Tousey, a well-known writer of children's books whose *Treasure Cave* and *Ned and the Rustlers* were among the books on Cheney's list.

Cheney started working early, at the age of nine, mowing lawns and delivering *The Lincoln Journal-Star*. Each day, Cheney would get a bundle of papers delivered to his house. He sorted the papers, folded them, and wrapped them with a rubber band. Then he grouped them in a bag, draped it over his bicycle handlebars, and pedaled throughout College View firing newsprint missiles at front porches.

The bike would serve him well. The Cheneys didn't have a car until 1949, two years after they moved into their College View home, when young Dick's great uncle died and left them a 1937 Buick coupe. It was a fine vehicle but for one detail: it had no backseat. Nonetheless, the Cheneys were frequent and enthusiastic travelers.

Even under the best of conditions (and seating arrangements) trips in those days were rough by modern standards. Interstate highways had not yet reached the Great Plains, let alone the Rocky Mountains or the California coast.

Dick and Marge Cheney would ride together in front; young Dick and his brother Bobby would ride in boxes on the floorboards where the backseat should have been. Short trips would take days; long ones a week or more. The Cheneys drove to visit a cousin who owned a dairy farm outside Denver, an uncle in Idaho Falls, and Grandma in California.

Marge Cheney kept detailed records of all the expenses on these trips, painstakingly noting the cost of gas in Broken Bow, Nebraska, or a sandwich in Barstow, California. (The vice president still keeps scrapbooks that his mother compiled with records of these trips.) The Cheneys usually packed lunches to avoid having to stop at restaurants and would stay only at motels with kitchenettes so that they could save money on dinner. Not infrequently, Marge brought a portable stove for picnic lunches along the side of the road.

Dick and Marge Cheney allowed themselves the occasional luxury, however. On one return trip from Whittier, California, the family endured a long drive through Death Valley in a car without air-conditioning. The Cheney boys were riding in their boxes behind the front seat when their father stopped at a roadside shop shortly after crossing into Idaho. He emerged with a six-pack of Coca-Cola and a bottle of bourbon. "Bob and I would drain the Coke down to a certain level," Cheney recalls, "and then they'd fix themselves a blend as we drove late at night across the desert coming back."[7] Cheney learned to like the backseat.

In 1953, the Cheneys prepared themselves for another, bigger trip. On January 20 of that year—a spectacular, springlike day—President Dwight D. Eisenhower delivered his first inaugural address in Washington, D.C. Dick Cheney was among those who saw Eisenhower's address. The Cheneys had saved enough money to buy a television and young Dick's entire sixth-grade class gathered at his house to watch the inauguration.

Eisenhower asked the crowd—and the nation—to join him in prayer and then began his speech. "The world and we have passed the midway point of a century of continuing challenge. We sense with all our faculties that forces of good and evil are massed and armed and opposed as rarely before in history."

Shortly after taking office, Eisenhower ordered the reorganization of the Soil Conservation Service (SCS), and Dick Cheney Sr. was to be reassigned. He was offered the choice of two towns farther west: Great Falls, Idaho; or Casper, Wyoming. He chose Casper.

Casper was booming when the Cheneys arrived in 1954. It was known as the "Oil Capital of the Rockies," and it was flooded with roughnecks and entrepreneurs and burgeoning with other businesses that had sprung up to harness some of the new money. The population of Natrona County exploded in the 1950s, experiencing a 63 percent increase.[8]

The growth brought attention and with it, tourists. One estimate of annual tourism revenues was $2 million,[9] an impressive total for a town of just 31,000. Frontier Airlines touted Casper as the "Gateway to All the West," and began direct flights from other western cities including Denver, Salt Lake City, San Francisco, Los Angeles, and Seattle. Airline ads boasted of Casper's shopping centers, water sports, and outdoor activities.[10] There was talk of bringing in a professional basketball team, and national politicians paid visits, including Harry S. Truman, Richard Nixon, and John F. Kennedy.[11]

Throughout the decade, Wyoming's politicians would struggle with many of the same issues Cheney would have to address as a legislator more than thirty years later: economic development, water distribution, agricultural productivity, soil conservation, and energy exploration.

Many of these issues were familiar to Cheney's father, who worked in Casper as a manager for the SCS, testing soil to determine which crops would be most suitable for the region. He worked a typical shift—forty hours a week—at the federal building downtown, now named for his son. As a government bureaucrat, Dick Cheney Sr. kept up with Wyoming politics but was by no means a news junkie. He was a lifelong Democrat who believed that he owed his job—his career—to Franklin Delano Roosevelt and the New Deal.

His wife, Marge, was also a Democrat, but her party affiliation, like that of so many of her contemporaries who had lived through the Depression, was more habitual than ideological. She was an artistic woman who taught her friends how to sew lingerie and taught herself how to paint by watching a local artist who gave weekly lessons on TV.

Although the Cheneys were Democrats, theirs was not a political household. Dinner conversations focused on everyday life and rarely touched on national politics.

Those conversations and others were marked by extended periods of silence. Cheney's father "offered his opinions reluctantly," says Norma Fletcher, "but people listened when he did."[12] Several family friends recall that among the Cheney men—Dick Sr.; the future vice president; and the younger brother, Bob— young Dick was the most talkative.

The youngest of the Cheney children, Sue, would follow that tradition.

The Cheneys were active in the First United Methodist Church of Casper. Dick served as treasurer, and Marge made pastries for the church staff and coordinated each Sunday's flower arrangements. "They were fairly religious folks," says Fletcher. "They were not expounding all the time, but they were fairly religious."[13]

These beliefs, however, didn't keep them from playing a friendly, low-stakes game of cards once a month. The Cheneys joined three other couples for a potluck supper to play "Nickels," a game Marge brought with her from Nebraska. Marge served as treasurer, and each person contributed $1 to play. The $8 each month stayed in a general kitty to be used for a dinner out when the pot was big enough to cover meals for the entire group. The games were spirited, in large part because of Marge's intense competitive drive.

Cheney's mother had been competitive all her life. As a young woman in the 1930s, she played on the Syracuse Bluebirds, a nationally ranked women's softball team that traveled in the region and beyond, playing top women's teams from across the West and Midwest. When young Dick wanted to practice baseball, it was his mother who often joined him in the front yard for a game of catch.

Both of Cheney's parents liked to fish, but his mother was the family's avid angler. Marge Cheney stole away for quick outings whenever she could and taught her sons—Dick and Bob—the art of fishing. These trips had a practical purpose, too; the results of her efforts frequently ended up on the table at mealtime. Her specialty was trout, which she cleaned and fixed herself, and most often served at breakfast.

The move to Casper in 1954 came as young Dick was just beginning adolescence, an awkward age for most kids. Cheney passed his first summer in Wyoming with activities that allowed him either to be alone or to excel: reading and baseball. He spent hours at the Casper public library—a fifteen-minute ride on the city bus—starting one book just as soon as he had finished another. Cheney remembers that his reading list included several military histories. When he wasn't reading, he could be found pitching and playing third base in Casper's Pony League. Cheney was selected to the all-star team, and he traveled with the squad to a Little League tournament in Richland, Washington. On the trip, he made friends with another all-star from Casper named Tom Fake, who would become one of his best friends. The pair would double-date together, play football together, graduate from Natrona County High School together, attend Yale University together, and later drop out together.

Cheney was reserved, but not shy. Playing sports almost year-round meant that he didn't have trouble making friends. In addition to baseball, he played sandlot football and spent much of his free time with his new friends fishing and hunting rabbits on the outskirts of town. (The boys sold the rabbit pelts to a local furrier for a quarter each.)

In his second year at Natrona County High School, he and a few other sophomores were invited to try out for the varsity football squad. Coach Swede Erickson asked the boys to partner up for some early hitting drills. At one session, Cheney warmed up with his friend Joe Meyer. The coach instructed the boys to stand ten feet apart, jog slowly toward their partners, and make light contact. The purpose of the drill was to get the players accustomed to hitting after a long layoff. Cheney and Joe Meyer, eager to make the varsity squad, ignored the coach's directions. On the whistle they sprinted full speed at one another, collided with the full force of their 130-pound frames, and tumbled to the ground. Meyer still remembers Coach Erickson's reaction: "God, these guys are nuts!" They both made the team.

High school football in Wyoming was a serious commitment. In addition to two-a-day preseason workouts and daily practices after school, the boys traveled across four states to play their games. Among their regular opponents were high schools in Rapid City, South Dakota (326 miles); and Scotts Bluff, Nebraska (175 miles). To get to Friday-night games in Grand Junction, Colorado, 392 miles away, the boys boarded buses on Thursday morning and returned to Casper late on Saturday.

Natrona County High School itself drew students from hundreds of miles away. Two of Cheney's football teammates—Dale and Dave Wight—lived on a ranch ninety miles west of town. The brothers came to Casper on Sunday night, stayed during the week with friends, and returned home after classes Friday or, during football season, on Saturday.

The Wights missed out on the weekend's evening activities in Casper. Students at Natrona spent hours cruising in their cars between the A&W restaurant on the east of town and the A&W restaurant on the west side of town. Some of the boys had saved up enough money to buy their own cars, and Saturday nights gave them an opportunity to show off the fruits of their hours of work in the garage. Cheney was not among them. Friends remember that Cheney, ever practical, didn't see the point of spending his free time driving back and forth along the same road. His family—like most families in Casper—had one car. Cheney was lucky to get permission to use it, and when he did, he preferred to use it to go somewhere.

One frequent destination was Alcova Lake, a giant reservoir thirty miles west of Casper. The lake winds along a craggy shoreline of red granite cliffs. The Alcova Dam collects the waters of the North Platte River to provide irrigation to Wyoming farmlands and power to the greater Casper area. For students at Natrona, it was a place for recreation: fishing, camping, water-skiing. Cheney's brother-in-law, Mark Vincent, remembers Cheney as a skilled water-skier in both the traditional method and a more idiomatic local version. One section of the Alcova complex, known as the Kendrick Project, featured dirt roads running parallel to sunken concrete irrigation canals that carried water from the dam. Cheney and his friends used to hitch a long rope to the back of Joe Meyer's Oldsmobile convertible and ski along the canals. "Of course you had to stay like you were skiing on the far right side of the boat all the time because if you got a straight line you were in the dirt," says Cheney. "And that was a real problem."[14]

Cheney and his friends frequently ventured beyond eastern Wyoming for overnight camping trips. At sixteen, Cheney took a four-day excursion with four football teammates to the Powder River basin, a rugged patch of terrain that stretches from northern Wyoming across the border into Montana. After fifty miles of highway, the group followed a seemingly endless dirt road to the top of a mountain where they set up camp. A long walk down a steep canyon delivered them to the middle fork of the Powder River, a pristine stretch of wilderness that offers some of Wyoming's best fly-fishing. Although Cheney had learned the secrets of traditional fishing from his parents, this trip marked his introduction to fly-fishing, a pastime that would take him on hundreds of trips around the world over the next fifty years.

During his junior year at Natrona County High School, Cheney took note of a classmate named Lynne Vincent. She was hard to miss. Lynne was a champion baton twirler and standout student who was involved in virtually every extracurricular activity sponsored by her school. As a majorette, she was a regular in parades in Casper and performed at many of the town's public events. She was the pride of her family.

"Her homecoming picture was the single largest object in our house," says Lynne's younger brother Mark.[15]

Lynne had varied taste in boys. She dated Jerry Moore, a rebel who wore a black leather jacket and cruised around Casper on his motorcycle; and Tom Fake, a devout Catholic who starred on the baseball team and was the quarterback of the Natrona football squad.

In January 1958, Cheney found himself without a date for an upcoming dance sponsored by ECLAT, one of several social clubs at the school. Lynne Vincent was in his chemistry class, and during a break Cheney asked her if she would go with him. He did not anticipate her response.

"Are you kidding me?"

Cheney was stunned. "My immediate reaction was, 'No,' she's not going to the dance." As it turned out, the exclamation was born of surprise, not distaste, and Lynne accepted his offer. "We got it worked out in short order," Cheney recalls.[16]

Dick and Lynne went to the dance on January 31, 1958, one day after Cheney's seventeenth birthday. Afterward, they went out with Tom Fake and his steady, Darla Howard. (Although Cheney and Darla had never dated, a note he had written in her yearbook suggests he would have been open to a courtship: "Dear Darla, It was really good getting to know you this year. I hope to get to know you A LOT better next year. Dick.")[17]

When the dance ended, the two couples piled into a 1950 Dodge owned by Fake's parents and, eschewing a popular make-out spot in Casper known as C-Hill, headed to the quieter parking lot of Garfield Elementary School. By all accounts the activity in the car was strictly PG-rated. ("Back in those days," Fake recalls, "girls wore panty girdles. They were tough to manage, so it had to be a big commitment.") The group had not been parked for long when they heard the hissing of air escaping from the tires. Jerry Moore, an old beau of Lynne's and a recent interest of Darla's, had let the air out of the two back tires as a prank. "The mood was gone," says Fake.[18] The car limped to a local filling station where Fake filled the tires; shortly thereafter the boys returned the girls home.

Days later, Cheney would learn that Lynne's mother, Edna Vincent, knew virtually every detail of the date. Mrs. Vincent worked as an assistant to Casper's police chief. She was, Cheney recalls, "a watchdog."

"Edna knew when you went out like that," he says. "She knew what filling station we'd been to. There were no secrets in a town like Casper, no secrets from the secretary to the chief of police."[19]

Despite this, Edna wasn't particularly strict with her kids. The Vincents were members of the First Presbyterian Church in Casper but were Christmas and Easter churchgoers.

In 1958, during the summer before his senior year, Cheney was nominated to participate in Wyoming Boys State, a week-long leadership seminar for high school students. After a series of lessons in civics and citizenship, the week culminated in a series of mock elections, with students running campaigns to be a state representative or governor of Wyoming. Cheney went to Boys State simply because he had been selected. His interest in government was no greater than his interest in a variety of other subjects.

Later that summer, Cheney spent five weeks in Evanston, Illinois, as part of the Cherubs program, an academic summer camp hosted by Northwestern University. He was one of ninety students across the nation selected for their interest in science and engineering. Several of Cheney's classmates were academic stars who would continue their studies the next year at Massachusetts Institute of Technology. That group included Cheney's roommate, John Castle, from Marion, Iowa, who was admitted to the program at Northwestern because of his advanced work on the binomial theorem. There were whispers among the students that someone in the group had registered an IQ of 180.

Castle had arrived before Cheney and checked into the room they would share on the first floor of McCulloch Hall, a three-year-old residence hall on the Northwestern campus. Castle was working busily at a desk in the room when his new roommate walked confidently through the door. "In walks this guy, who came right over to the desk and sticks out his hand. 'I'm Dick Cheney from Casper, Wyoming.' I remember him doing that to this day,"[20] says Castle, who has remained a friend of Cheney's.

"He was from a small town in Iowa and he was greener than I was," Cheney recalls.[21]

A typical day began with an early-morning breakfast followed by several hours of classroom work. After lunch, students would attend lectures on how to succeed in college—"How to study, how to manage your time, how it'll be different when you go to college than it was in high school, things like that," says

Castle—before adjourning for afternoon free time and athletic activities.[22]

The summer included several academic field trips—to the Argonne National Nuclear Laboratories in Argonne, Illinois; the International Harvester plant in Chicago; and a Standard Oil refinery in Whiting, Indiana. There were other, less academic outings such as a Cubs game at Wrigley Field and a trip to the Ravinia amphitheater north of Chicago for an outdoor concert by the Chicago Symphony Orchestra.

Although Cheney was a cherub for one summer, it was not a description that applied to his entire high school career. He was generally well liked by his friends' parents, but he and his friends managed to have some fun that would have not met with their parents' approval.

"Dick did his share of partying and having fun," confides Ron Lewis.[23]

Still, by today's standards, their behavior was tame. "These were the days before there were drug problems," says Cheney. "The most we did was probably drink a little malt liquor, smoke cigars."[24]

And get into a few scuffles with their classmates. Natrona County High School had a unique way of discouraging students from fighting in the hallways. The combatants were separated, scolded, and then told to get ready to fight again. The rematch came at the end of the year, in a boxing ring, with the proceeds going to the C Club, the booster club for the school's athletic teams.

Early in his senior year, Tom Fake decided to run for student body president. Fake's girlfriend at the time wore a sandwich board at school touting his candidacy and, naturally, inviting ridicule from fellow students. One day as she was wearing the sandwich board, Jerry Moore, the same prankster who earlier had let the air out of the car tires on Fake and Cheney's double date, stopped to tease her. Fake caught him in the act, and the two boys began throwing punches. Moments later, a football coach grabbed them both by their necks, offered a brief reprimand, and informed them that they would meet again in a grudge match.

Cheney, as it happens, had received the same punishment

for a tussle in the hallway. Cheney and his friends say they cannot recall the details of that skirmish, but everyone recalls that the "punishment" was a rematch at the C Club fights later that spring. With their reputations on the line Cheney and Fake decided to seek some help.

The two boys crossed the tracks to North Casper and paid a visit to a former boxer, by now a gruff old man with a pug nose and cauliflower ears. Though not enthusiastic, he reluctantly agreed to give them one lesson in his living room—for a price. "How much money do you have?" he inquired. Together they had $5—not a small sum in the late 1950s. Their would-be coach handed them two sets of boxing gloves. Cheney was smaller and quicker; Fake was bigger and more powerful. They pounded on one another as if their friendly sparring were a prizefight. "We got a little carried away punching each other," Fake recalls.[25] The veteran pugilist was so amused that he agreed to coach them free. Fake says they beat each other up regularly that spring, sparring primarily in Cheney's garage. Both Cheney and Fake won their grudge matches.

They also won their bids to run the Natrona student government. It was Cheney's first bid for elective office, and it seems to have been inspired less by any political ambition than by the fact that two of his best friends—Fake and Dave Nicholas—had decided to run for student leadership jobs. Cheney simply thought it sounded like fun. "There really wasn't a lot of heavy lifting," he recalls. "It was more a social thing than anything else."[26]

Still, it added to his already impressive résumé. And if Cheney didn't think much about his high school accomplishments, others did. A local businessman by the name of Thomas Stroock approached Cheney and Fake with an extraordinary offer: full scholarships to Yale University.

Before coming to Casper and making millions in the town's oil boom, Stroock had graduated from Yale, where he'd been a classmate and friend of George H. W. Bush. He was in a position to provide the same opportunity to Cheney and Fake.

"In those days, you could do things you can't do now," Stroock would later recall, "so I called Yale and told 'em to take this guy" and his friend Tom Fake.[27]

Stroock took an interest in Cheney for two reasons. First, the oilman had employed Cheney's girlfriend, Lynne, for several summers during high school. Second, and more important, Stroock was a regional recruiter for Yale. Although Lynne was the truly outstanding senior at Natrona County High School, she wasn't eligible to apply to Yale, in those days still an all-male institution. So Stroock turned his attention to Cheney and Fake.

Cheney hadn't given much thought to college. His father had attended college only briefly, before his family's financial difficulties forced him to drop out and get a job. The younger Dick Cheney always assumed that he would go to college, but had made no real plans to do so. Still, he recognized that he'd been presented with an excellent opportunity. Assured of admission by Stroock, Cheney didn't bother applying to another school.

Although the recommendation from Stroock was decisive, Cheney's high school record made its own case. His grades were solid; he had been selected for several off-campus leadership programs; he had served as captain of the varsity football team and made the all-state team as a senior; and he had been elected senior class president. Judging from the index of the Natrona County High School yearbook for 1959, few students were as active and popular as Richard Bruce Cheney. Most students had one or two citations. Cheney and his girlfriend Lynne each had eight.

On one page of the yearbook was a picture of Cheney and his best friend smiling broadly: "Ivy League bound Tom Fake and Dick Cheney show their happiness upon receiving notice of their acceptance by Yale University."[28]

But after their first year in New Haven, neither young man would be smiling.

CHAPTER TWO

To Yale and Back

In early September 1959, Dick Cheney and Tom Fake drove 120 miles southwest to Rawlins, Wyoming, where they boarded a Union Pacific passenger train from California and settled in for a long journey. The trip would take them from the open spaces of the North Platte River valley, over the plains of the west and the cornfields of the Midwest, to New Haven, Connecticut, on the shore of the Atlantic Ocean. They changed trains in Chicago and again in Manhattan—the first time either had stepped foot in New York City.

Cheney and Fake spent two days and two uncomfortable nights sitting upright in narrow coach seats. Of the many new sights they'd taken in along the way, Fake was most struck by the dramatic contrast between the dry moonscape of eastern Wyoming and the foliage of the northeast. "We'd never seen so many trees in our life."[1]

But cultural differences, as it turned out, mattered more than geography. And in that sense, too, Yale was a world away.

To a boy in Casper, Wyoming, local history didn't stretch back very far. The past was the Oregon Trail, pioneers, and cowboys. Fort Caspar, the site of a battle between members of the Sioux and Cheyenne tribes on one side and a volunteer cavalry company protecting the local Pony Express office on the other, is the region's oldest landmark. It was built in 1865. Yale University, the nation's third oldest private institution of higher education, was founded in 1701. The college cemetery contains headstones of men who fought in the Revolutionary War.

To that same boy, prosperity was defined by the doctor who lived down the street in a modest house with a television. Although Casper grew rapidly throughout the 1950s, few of its residents were wealthy, and those who had money had gotten it recently, most through hard labor.

Yale, on the other hand, had for centuries been a way station for the scions of America's richest and most powerful families. It counted among its students the sons of senators and of industry titans. The campus was thick with privilege and expectations. Cheney would go to class in the same Gothic buildings as Cole Porter and James Fenimore Cooper; political leaders like Vice President John Calhoun, President William Howard Taft and Secretary of State Dean Acheson; inventors such as Eli Whitney and Samuel Morse; and numerous Supreme Court justices.

William Samuel Johnson, a delegate to the U.S. Constitutional Convention in 1787 and an early proponent of executive power, graduated from Yale in 1744. The father of American football, Walter Camp, created the game at Yale in the 1880s. Prescott Bush II, Connecticut's Republican senator when Cheney first arrived in New Haven, received his BA from Yale in 1917. And Gerald R. Ford, an up-and-coming politician from Michigan often mentioned as a potential running mate for Richard Nixon in the presidential campaign of 1960, had coached boxing at Yale and graduated from its law school in 1933.

The physical settings, too, were dramatically different. Casper was a young town growing from the arid and mountainous terrain of the new west. New Haven was an established city on the heavily populated eastern seaboard. In 1960, during Cheney's first year at Yale, New Haven County had a population twice that of the entire state of Wyoming.

When Dick Cheney arrived at Yale he brought the west with him. "Everybody arrived with the pride of geography," explains a classmate, Dennis Landa. "Cheney and Fake were known as 'the cowboys.' They came with a ride 'em cowboy kind of thing about them."[2]

Jacob Plotkin, one of Cheney's roommates during his freshman year at Yale, remembers that as a native of Wyoming, Cheney stood out. "He wore cowboy boots and drawled a little when he spoke—very Western."[3]

Not surprisingly, the freshmen typically associated with classmates from similar backgrounds. At the time, the student body at Yale was all male, and until the mid-1960s the students came predominantly from private schools. The prep school graduates hung out with other "preppies," and the public school kids stuck with other "high school Harrys," as they were known.

Even among other public school graduates, Cheney and Fake had very distinct interests. "He and Tom talked about hunting and fishing in a way that none of us understood," says Jim Little, a close friend who was a product of public schools in New Jersey.[4]

Yale assigned Cheney to Room 321 of Wright Hall, where he lived with Plotkin and Stephen Billings. Cheney and his roommates got along, but from their first days at Yale it was clear that they would not become close friends. For both of Cheney's new roommates, Yale was the first step in long academic careers. Billings studied theology in Cambridge, Massachusetts, before getting his doctorate in ministry; Plotkin received a PhD in mathematics from Cornell University and went on to teach for more than three decades at Michigan State University.

By contrast, Cheney spent most of his orientation week with Fake and two other teammates from the freshman football squad: Wolf Dietrich and Dennis Landa. "He was quiet," recalls Landa, whose football locker was next to Cheney's. "He was always a quiet guy, though not quite taciturn."[5]

Cheney had excelled at Natrona County High School with very little effort, but his study habits served him poorly at Yale. Classes started at eight AM on September 23, 1959, and his difficulties began a short time later.

"I was not a diligent student—didn't work very hard at it, was not motivated," says Cheney. "Seemed like a good idea at the time, but I hadn't given any serious thought to why I was there, why I wanted to go to Yale."[6]

For most of the other students, Yale was a long-sought goal. Some had even taken a fifth year at prep school as further preparation for the rigors of an Ivy League education. "They were all smarter, taller, faster, and richer than we were," says Pete Cressy, a classmate of Cheney's who was also a public school graduate.[7]

Yale did not give athletic scholarships, and those attending Yale with need-based assistance were required to work throughout their first year. Cheney and Cressy were employed in the freshman commons during mealtime. The jobs were assigned by the bursar's office and those who filled them were known derisively as "bursary boys." Some weeks they worked on the food line, serving their classmates milk, punch, and the meal of the day. Other weeks, they simply bused tables. Some of the bursary boys thought the work seemed to separate them even more from their wealthier classmates. According to those who knew him well, Cheney, who had worked at various jobs in high school, never complained.

Even sports didn't provide the outlet they once did. Cheney played right field and third base on the freshman baseball squad and halfback and linebacker on the freshman football team. His gridiron teammates remember that what he lacked in size he made up for in quickness and tenacity.

Playing on the freshman football team required both physical ability and mental acuity. Like Yale's Walter Camp a century earlier, the freshman coach, James "Gib" Holgate, was something of a football intellectual. In an indication that the academic maxim "publish or perish" may have extended from the classroom to the stadium, Holgate had released a book called *Fundamental Football* a year before Cheney and Fake arrived.

In Wyoming, Cheney and Fake had been standout athletes, but at Yale they were merely average. According to Cheney's teammates, Holgate pushed Cheney harder than other players. Holgate barked, "God damn, Cheney!" so many times on the Yale practice field that Cheney's teammates still laugh when they think of the scolding forty years later.

Freshmen at Yale were not allowed to have cars, a rule that severely restricted their free-time activities. "We were totally isolated as freshmen," says Rees Jones, a friend of Cheney's who is now a well-known golf course architect. Cheney spent weekends that fall watching football—usually the New York Giants—and otherwise goofing around. Every Sunday night, a large group gathered to watch *Maverick*, a popular television show set in the old West. "We had no outlet for our social life," says Jones, "so we were involved in a lot of shenanigans. That's why we were all pranksters. He was a jokester and a prankster just like all of us."[8]

Regular diversions included water balloon ambushes and, after a shared bottle of Thunderbird wine, disrupting the artsy films shown by the Yale Film Society. Cheney seemed particularly fond of "short-sheeting" the bed of his roommate Stephen Billings.

As he had been while growing up in Wyoming, Cheney was an avid reader. But at Yale he read what interested him, which did not often include books for his classes, and his grades suffered as a result. He read military histories, in addition to periodicals such as *Sports Illustrated, Time* magazine, and *The New York Times.* But his reading seldom provoked conversation among his friends about politics and world affairs. "I'm hard-pressed to remember more than two such discussions," Little recalls.[9]

One class, however, did arouse Cheney's intellectual curiosity: Introduction to International Relations (PS-13A), taught by H. Bradford Westerfield, a noted expert on U.S. foreign policy and intelligence. Covering U.S. diplomacy and recent history, from the end of World War II through the early stages of the cold war, it was the closest thing to a current events course offered by Yale. It fascinated Cheney, who not only churned through the required reading but also began studying contemporary histories in his free time.

Several members of Yale's class of 1963—the former senator David Boren; the longtime White House adviser David Gergen; and the administrator in Iraq, L. Paul Bremer—seemed to their classmates to be destined for the world of Washington politics. Cheney, however, was not among them. Most of his friends assumed he would return to Wyoming, perhaps to a career in business.

To his friends, it was obvious that Cheney missed the West. Unlike their classmates from the East Coast and the Midwest, Cheney and Fake couldn't simply head home for a long weekend to get a break from Yale. They couldn't get to holiday gatherings back in Casper; instead, they spent Thanksgiving and Easter of their freshman year with Jim Little and his family in New Jersey.

On the Saturday night of Thanksgiving weekend, Little tried to set Cheney up with a young woman he had known from high school. Cheney had no interest. As his friends would quickly come to understand, he wasn't interested in anyone other than Lynne Vincent. He passed many nights in his dorm room in Wright Hall composing long letters to his girlfriend, who had left Casper for Colorado College at the same time he had departed for Yale.

When Cheney returned to Casper in 1960, the summer after his freshman year, the mail brought unwelcome news. Because of his poor academic performance, his full scholarship to Yale had been revoked. He was free to return, the letter advised, but he would have to finance his own education. Cheney knew that his family would have to borrow money to pay the tuition, and this knowledge added to his uncertainty about going "back east."

Even worse, Tom Fake had gotten a similar letter and decided to stay in Wyoming. Since they first met on the baseball road trip they had taken six years earlier, Cheney and Fake had played sports together, dated together, boxed together, and gone to Yale together. Now, Fake would stay behind in Casper.

But Cheney's family and friends pushed him to return to Yale. Most vocal among them was Lynne, who didn't want to see him throw away an opportunity that she—despite her academic excellence—never had. Cheney accepted the school's offer to provide loans.

In New Haven, Fake's absence was palpable right away. At the end of the previous term, Cheney, Fake, and ten friends had arranged to live in a set of adjoining four-man suites in the Berkeley College living quarters. Now there would be only eleven young men in the rooms.

They were at the end of a long hallway on the third floor.

Each suite had two bedrooms and a common room; Cheney and his friends had agreed that they would share all three common rooms, assigning each one a distinct purpose: a "salon" for studying, a television room, and naturally a party room.

Cheney's suite mates were drawn largely from the freshman football team. Of the eight young men who had played on that squad, only one played football in his sophomore year. The Berkeley suite crowd came up with a variety of ways to spend their newfound free time: playing cards, drinking beer, and throwing parties. "The master of the college," says Jim Little with a roguish grin, "knew us."[10]

One weekend they decided the throw a beach party. The boys hauled in sand to cover the floors in the rooms. They planted fake palm trees. Most ambitious, they deliberately stopped up the drains of the two showers in the bathroom across the vestibule, so that the water flowed into the hallway and down the stairs to the second floor.

Except for the beer—Schaefer and Teal's—the Berkeley boys spared no expense. And they bought cheap beer that night only because they never drank anything other than cheap beer. The party even featured "surfing," or the closest thing to it that could be done indoors in New Haven: a bike launched down the water-soaked stairs. "It was a bicycle," remembers one suite mate, Ned Mason, "with Mr. Cheney on it."[11]

(Forty years later, several of the suite mates gathered in Jackson Hole, Wyoming, for an informal reunion. Cheney was presented with a mangled child's bike in recognition of the beach party surfing.)

It didn't take long before Dr. Charles A. Walker, the master of Berkeley College, paid a visit to Cheney and his friends early one weekend morning. Walker, a professor of engineering who spent his days at work on heterogeneous catalysis and other serious concerns, had been the master of Berkeley College for a little more than a year. He had been relatively tolerant of the tomfoolery, but his patience was waning.

The young men were groggy when Dr. Walker summoned them for their scolding. He lined them up and began by calmly listing their offenses. He worked himself into a controlled rage and punctuated his lecture with a direct threat.

"If you don't clean up your act," he bellowed, "you will be rusticated!"

The young men looked at their shoes. No one dared to speak. After several uncomfortable moments, a muffled snicker broke the silence. And then, a short time later, another. The boys could no longer contain themselves, and laughter cascaded down the row of miscreants.

"Nobody knew what the hell he was talking about," says Mason.[12] So after Walker's harangue they returned to their rooms and consulted the dictionary. They remained confused. Could it really be the case that more shenanigans would result in their banishment to the countryside?

Cheney and his teammates from the freshman football squad weren't good enough to play intercollegiate ball, but they didn't abandon the game altogether. Yale's intramural football league was perfect for young men who lived by only the latter half of the Ivy League aphorism: "Work hard, play hard." Practice was infrequent, but the games were intense. Each squad lined up in full pads and in uniforms. Injuries were common, but playing hurt was a point of pride, and winning the championship meant bragging rights for an entire year.

If the games were intense, the pregame warm-ups were not. Typically, they involved little more than some light stretching and a few full-bodied beers. Cheney and the rest of the "Berkeley Birds" drank several beers before one particularly hard-hitting game during their sophomore year. Cheney, as always, played halfback, and his good friend Bob Tomain was quarterback. At one point, Tomain rolled out and pitched the ball to Cheney. The future vice president darted toward the sidelines and began to let up as he saw the white chalk line indicating that he would soon be out of bounds. Just as he relaxed, an opposing player launched himself at Cheney. The violent collision left Cheney on his hands and knees, dazed. But he was determined to keep playing. "He blew lunch on the sidelines, wiped his mouth off, and ran back to the huddle," says Little.[13]

....

THE BERKELEY BOYS soon decided they needed to clean up their image. So they did what seemed most sensible: they threw a party. Cheney and his friends put on their best shoes and slacks, pressed shirts, coats, and ties. They traded their cheap beer and Thunderbird wine for a drink of more sophistication (sherry) and invited the distinguished English professor Richard Purdy, an expert in Victorian literature, to join them for a gathering of the enlightened.

The get-together marginally improved their public standing, but Cheney's academic performance remained a problem. Much of his learning came outside the classroom. During sophomore year he had been assigned to a work-study job in the Yale language lab under the direction of a German who had fought in World War II. The former German soldier regaled Cheney with tales from the front lines about how his unit had fought its way to the outskirts of Moscow only to retreat a short time later. Cheney found the stories fascinating.

In December 1960, after the first semester of his sophomore year, Cheney received word from Yale that his continued academic struggles made him no longer eligible for the school's loans. It was a severe blow. Cheney had already lost his scholarship; and his family, getting by on the salary of a mid-level bureaucrat, now had to come up with enough money to pay for school.

There was even more bad news. Yale made Cheney a nonnegotiable offer. "They said, 'Why don't you take a year off?' I said, 'OK.' Not like I had a choice in the matter."[14]

So it was back to Wyoming. After graduating from high school, Cheney had had a summer job loading trains with 100-pound bags of bentonite—clay used in drilling oil and gas. After a short time he was offered a better opportunity, a job building power lines across Wyoming. The work paid well and Cheney was able to put in enough time to earn his union card with the International Brotherhood of Electrical Workers. He had returned to the job after his freshman year at Yale, and now, during this involuntary hiatus in the middle of his sophomore year, he would go back again. For nearly all of 1961—what should have been the end of his sophomore year and the beginning of his junior year in college—Cheney worked on a crew that laid power lines in Wyoming and neighboring states.

If Cheney was ambivalent about college, his girlfriend was not. Lynne excelled at Colorado College, where she studied literature. And she made clear her disappointment in Cheney. Several friends say that without Lynne's nagging Cheney would not have returned to Yale.

But by working throughout 1961, Cheney was able to save enough money to pay for one final shot at the Ivy League in the spring of 1962. It didn't work.

"At the end of that semester they basically said, 'Well, thank you. We're glad we got to know you,'" Cheney recalls with a laugh. "And they sent me on my way." The problem was "basically grades at that point. I'd worked hard that last semester, that first half of the semester. But I basically just didn't like it. I didn't like the setting. I was doing it because everybody said that's what you're supposed to do. But it held no real appeal for me."[15]

So it was over.

Nearly forty years later, it would all seem funny. After all, Cheney was the vice president of the United States, working for a president who, despite his reputation as an intellectual lightweight, had actually managed to graduate from Yale.

On May 21, 2001, George W. Bush was invited back to New Haven to address its graduating students. "To those of you who received honors, awards, and distinctions, I say, well done. And to the C students I say, you too can be president of the United States. A Yale degree is worth a lot, as I often remind Dick Cheney, who studied here but left a little early. So now we know—if you graduate from Yale, you become president. If you drop out, you get to be vice president."[16]

Bush warned Cheney in advance that the gibe would be coming. "He laughed like hell when he told me," says Cheney.[17]

No one was laughing when Cheney departed from Yale for the final time. He returned to the decidedly unglamorous work of constructing Wyoming's transmission lines. Cheney romanticized his return to the West, telling himself that he preferred

Wyoming to Yale, that this job was a step toward fulfilling his dreams of traveling throughout the West. There was some small measure of truth in those thoughts, but a much greater measure of self-delusion. In reality, Cheney had slid to a new low.

Cheney labored as a groundman, supporting the more experienced journeymen and equipment operators. He worked with the "powder monkeys," who transported explosives to blast holes in the rock to accommodate the creosote-treated poles.

The job took him across Wyoming and the region. He was on one crew that constructed the power transmission lines for Warren Air Force Base near Cheyenne and another that increased the capacity of the new Dave Johnston Coal Plant near Glenrock. Other projects included an eight-mile stretch of lines between Kemmerer, Wyoming (elevation 6,958 feet) and Ogden, Utah, dropping down 2,700 vertical feet over the Wasatch mountain range. Another job involved running lines for the radio tower atop Laramie Peak, a dangerous project in a place so remote that helicopters were needed to access sites leading to the mountaintop.

It was a hard life. Cheney and his crewmates lived for weeks at a time in cheap motels in towns near the work sites. The men worked from morning's first light until early evening, when they would often haul themselves to a local tavern for a night of hard drinking.

Cheney occasionally joined them, and sometimes the drinking led to trouble. In October 1962, he was arrested in Cheyenne for "operating a motor vehicle while intoxicated and drunkenness."[18] He forfeited the $125 he paid in bond, and his driver's license was suspended for thirty days. Nine months later, in June 1963, Cheney was arrested again for the same offense[19] in Rock Springs, a rough town that later earned national attention for its nightly bar fights and rampant lawlessness.

The same month that he was arrested for the second time, Cheney's friends and former classmates received their diplomas from Yale University. As he sat in the jail cell in Rock Springs, the contrast struck him hard. For eighteen years, Cheney had a carefree life marked by a series of seemingly effortless accomplishments. His admission to Yale, on a full scholarship, appeared to continue this promising trajectory. But after he received his acceptance letter, his life had taken a different course. He moved away, quit sports, struggled academically, and eventually dropped out of college. Now, almost four years after the excitement and anticipation of that first cross-country train trip to New Haven, Cheney found himself alone in jail, left to contemplate everything that had gone wrong. Even for someone who had been—and would be—known for his equanimity, it was another disturbing new low.

If Cheney had any remaining illusions, they were soon punctured by his high school sweetheart. "He had quite a serious discussion with Lynne," says Cheney's longtime friend Joe Meyer.[20] In December 1962, Lynne had graduated from Colorado College a semester early, and she spent the spring of that year in Europe. She planned to attend the University of Colorado to pursue a master's degree in literature. She did not, she made clear, plan to marry an electrical worker who was in trouble with the law.

Cheney had to get his life in order. "One way to do that was to move out on the job, away from the bars in Rock Springs, Wyoming," Cheney admits. He decided to stay with a crotchety veteran of World War II who remained overnight at the various job sites instead of frequenting local bars. Bob Leibrantz was a difficult man, a loner whose best friend was a mangy, ill-tempered mutt. He and Cheney hit it off.

One reason was that Leibrantz liked to tell stories about his wartime service, and like the veteran who ran the language lab at Yale, he had in Cheney an eager listener.

To Cheney's new roommate, home was an old Army truck with a trailer that he had rigged up as a makeshift RV. It wasn't fancy, but it kept Leibrantz and his dog warm and dry. Cheney pitched a tent alongside and slept on a cot. During the week, the men cooked their meals on an old Coleman stove. Each Saturday, they stayed in whatever town was closest to the job site to do laundry, drink beers at the local tavern, and take their weekly showers.

While Cheney was living with Leibrantz, he began reading John Keegan and Winston Churchill's six-volume history, *The Second World War*, and once again at the urging of Lynne, made plans to enroll at the University of Wyoming in the fall of 1963.

"She made it clear she wasn't interested in marrying a lineman for the county. That was really when I went back to school in Laramie. I buckled down and applied myself. Decided it was time to make something of myself."[21]

Cheney came to Laramie with two years of credits from Yale and a determination to succeed. He moved in with his high school friend Joe Meyer, studied hard, and worked evenings to help pay for school. Through the Veterans Administration he found a job reading education textbooks to a World War II veteran who had lost his sight. The man, an officer in the Air Force, was in school to receive his certification to teach the blind. For three hours a night, Monday through Thursday, Cheney read aloud to the airman.

Cheney changed his behavior in the aftermath of his two arrests, but he avoided the dramatic lifestyle changes that often follow problems caused by alcohol. He did not turn to religion or quit drinking altogether, as George W. Bush later did when he worried about his own alcohol consumption. Cheney simply toned it down a bit. "We still drank a lot of beer," says Joe Meyer, Cheney's roommate in Laramie. "We still had a lot of fun. We weren't Goody Two-shoes by a wide margin."[22]

If Cheney had cleaned up his act figuratively, he had not done so literally. Cheney and Meyer lived in the basement of a house off-campus. Cleaning was not a priority. It was common practice for Cheney and Meyer to throw their trash on the floor, often under the kitchen table to keep it out of sight. "I was the head cook for most of that stuff," Cheney says. "And we'd whip up a big batch of chili and leave it on the back of the stove. When you wanted some you'd turn it on."[23]

These culinary habits were a result of both laziness and thriftiness. To save money, Cheney and Meyer bought cases of tomato rice soup on sale at the local grocery and ate it for consecutive meals until their supply was exhausted. "God, I still can't eat it to this day," whispers Cheney, with a shudder of disgust.[24]

To Cheney's good fortune and improved health, Lynne received her master's degree from the University of Colorado in June 1964 and they were married in Casper on August 29, 1964. There had been no formal proposal. They just decided to get married.

The modest wedding was held at the First Presbyterian

Church of Casper. It was small—just family members and close friends—and rather low-key. Afterward, the gathering moved to the Henning Hotel for a coffee-and-cake reception. And that was it: Mrs. and Mrs. Richard Bruce Cheney.

Three weeks before their wedding, President Lyndon B. Johnson signed the Gulf of Tonkin Resolution, drawing America deeper into the growing conflict in Vietnam. To Cheney, the war seemed remote. "Like most Americans, at the outset certainly I supported it. And initially it wasn't that big a deal. In the early 1960s we had advisers there, but we had advisers a lot of places. . . . We were focused on a lot of other things besides what was going on in southeast Asia."[25]

At the University of Wyoming, Cheney's focus, for the first time, was on his classes. "Dick just got hooked by political science," remembers Joe Meyer. Cheney's approach to the subject was analytical, not partisan or ideological; he cared about the ideas that animated political fights, but not the fights themselves. "I don't think either one of us thought, cared, or talked about politics," says Meyer. "I don't think I ever knew what his politics were."[26] Neither, for that matter, did Cheney. And this ambivalence would have its benefits.

In December 1964, Cheney was finally awarded his bachelor's degree and immediately began coursework toward a master of arts in political science. Along the way he participated in several extracurricular fellowship programs that did far more for his future than his academic training would. The first of these came in January 1965: an internship program sponsored by the National Center for Education in Politics (NCEP) and funded by the Democratic and Republican parties of Wyoming. Along with another young man, Cheney was selected to serve as an intern in the Wyoming state legislature for its forty-day session, which ran through the end of February.

The other intern was active in Young Democrats and naturally wanted to intern with his party, which then controlled the legislature's lower chamber. The Republicans, who controlled the state senate, agreed to take Cheney, the last man standing, as their intern.

Cheney, by his own admission, "didn't have a political identity."[27] His parents had been Democrats and if the other intern had been a strong Republican, Cheney would gladly have worked for the Democrats. In essence, Cheney became a Republican by accident.

Following his internship, Cheney wrote an essay on the experience that won a national prize—the Borden Award—from the group that sponsored the internship. Although he had done well in the classroom, this outside recognition caught the attention of his professors. Ralph Wade, the chairman of Wyoming's political science department, approached Cheney about another opportunity offered by the NCEP. Each year, twelve graduate students in political science from throughout the country were chosen to intern with governors or big-city mayors. With Wade's encouragement, Cheney applied. Maureen Drummy, an executive from the organization, recognized Cheney as the recipient of the Borden Award and presented him with an opportunity in Wisconsin: to work directly for Governor Warren Knowles. Cheney quickly accepted.

In January 1966, the Cheneys relocated from Laramie to Madison, Wisconsin, where both Dick and Lynne would continue graduate school. The move wasn't the only major development. Lynne was pregnant with their first child. The Cheneys settled into an apartment in a modest living unit for graduate students in Park Village. Their apartment had one bedroom and one large closet, which would soon double as a nursery. As the new year began, Lynne Cheney, who had recently completed her master's thesis on Yeats, began a PhD program in literature and set to work researching her doctoral thesis on the Victorian poet Matthew Arnold. Her husband prepared his one suit—"I think it glowed in the dark," he would later admit—to begin his internship.[28]

Warren Knowles had been elected to his first two-year term in November 1964, bucking a national trend of difficulties for the GOP that included Lyndon Johnson's rout of the Republican presidential nominee Barry Goldwater. Knowles had served three terms as the state's lieutenant governor before he moved into the governor's mansion. A moderate Republican, he was a strong supporter of civil rights, and an avid outdoorsman who pushed the state to purchase land for public use. Although many state leaders at the time believed that the federal government was responsible for protecting the environment, Knowles carved out an active role for Wisconsin's policy makers. (In 1999, an environmental stewardship program of the Wisconsin Department of National Resources was named after Knowles and another longtime Wisconsin politician, Gaylord Nelson, the founder of Earth Day.)

Cheney came to Knowles's staff as so many young people enter politics: He was a grunt, an intern without any defined responsibility who reported to work each day to do the jobs that others in the office didn't want to perform. His six-month fellowship ran from January through June 1966. When it was over Knowles hired him full-time.

Knowles was up for reelection that year, and as the campaign intensified, he picked Cheney as his personal assistant. The work wasn't glamorous, but it gave Cheney his first taste of retail politics. Cheney called himself "the bagman" because he carried the briefcase. At campaign events—parades, county fairs, fundraisers, whenever Knowles stopped to chat with a group of potential voters—Cheney would snap a photo of the group with a Polaroid instant camera, a novelty at the time. He pulled the cartridge from the camera, waited for the photograph to develop, tore off the protective cover, left the picture with the governor's new friends, and ran to catch up with Knowles, by this time glad-handing another group and posing for the camera. It was a smart way to campaign. Not only were the voters delighted to meet the governor—a minor celebrity in Wisconsin—but they had an instant memento of their encounter to show their friends.

Knowles believed in this kind of bank-shot campaigning, using interactions with small groups of voters to reach a broader audience. He insisted on at least two stops each time he and Cheney visited one of Wisconsin's small towns: at the local newspaper, which made immediate sense to Cheney; and at the barbershop, which baffled him.

"I couldn't figure out why in the hell he'd go to the barbershop," says Cheney.

Cheney still remembers Knowles's explanation. "He said,

'Look, every guy in town has to get a haircut. They all go to the barbershop. And that's the one place they all talk politics. You don't ever want to hit one of those small towns without stopping by to say hello to the barber. He'll talk about it to every guy who gets his haircut there for the next two months.'"[29]

Knowles won with nearly 54 percent of the vote. That same year, voters in Elroy, Wisconsin, elected Tommy Thompson, a recent graduate of the University of Wisconsin law school, to the state assembly.

Before coming to Madison for the internship, Cheney had made arrangements with the political science department at the University of Wyoming to continue working toward his MA. He received it in June 1966. In July, his daughter Elizabeth was born.

The birth came as the conflict in Vietnam was intensifying. Young men across the country were being drafted into service, and there was a growing schism in the country over the ethics of conscription. Cheney would later say that he would have been "happy to serve" had his number been called.[30] And he registered with the Selective Service in 1959, as he was required, when he turned eighteen. But Cheney plainly did everything he could to avoid service. He had twice been classified as draft-eligible (1-A), once two years before the Gulf of Tonkin Resolution, and again just as the public debate over the draft was heating up. The first time, in the spring of 1962, Cheney was still at Yale and the military was enlisting only older men. The second time came in 1965 after Cheney was married, making him less likely than his unmarried contemporaries to be drafted. He had received four student deferments while bouncing from Yale to Casper College to the University of Wyoming. In January 1966, soon after learning of his wife's pregnancy, Cheney applied for and received his fifth and final draft deferment. On January 30, 1967, Cheney turned twenty-six and was no longer eligible for the draft.

His lengthy academic career meant that he was eligible for student deferments for six years, rather than the four years of a typical potential draftee. He filed for a deferment each time he might have been eligible for the draft. As he famously told the *Washington Post* in 1989 during his confirmation hearings to become secretary of defense, "I had other priorities in the 1960s than military service."[31]

Asked about the comment and the Vietnam War, Cheney says: "It didn't figure as prominently in my life. I think a lot of people who look back on it and feel like it was the only thing going on in the country at the time. It wasn't. There were an awful lot of people who were living their normal lives and the war was something that was off there, unless it affected you personally, as it actually did when you were on campus and got all of these protests and demonstrations. . . . We were struggling graduate students living in student housing trying to get our PhDs, and everything else was sort of irrelevant to that basic thrust of what we were trying to do."[32]

The fighting in Vietnam was relatively light during the first four years when Cheney was of draft age, with two soldiers killed in 1959, five in 1960, sixteen in 1961, and fifty-three in 1962. The casualties grew over the next four years, rising from 118 in 1963 to 6,144 in 1966. The deadliest time came immediately after Cheney turned twenty-six and was no longer eligible for the draft, with more than 39,000 of the war's 51,000 deaths coming from 1967 through 1969.

With the rising death toll came dissent, and in the late 1960s, the campus at the University of Wisconsin became, in effect, a regional headquarters for the antiwar movement and its legions of protesters.

"In Wisconsin, they were an aggravation, if I can put it in those terms," Cheney says. "We had protests over napalm, over the draft, over the war in general. . . . So you were aware of what was going on, but it was an aggravation if you couldn't get to class because the protesters were blocking you or there was tear gas released that morning. And we were serious about doing what we wanted to do in terms of pursuing our educations."[33]

Cheney spent 1967 and 1968 completing the classroom requirements for a PhD in political science. Every day he spent hours buried in his books and at the computer lab. Fellow students spotted him walking around campus carrying boxes of IBM computer punch cards that cataloged the votes of members of

Congress from years past.[34] Cheney was using the data in an expansive study of congressional voting records that he conducted with one of his professors, Dr. Aage Clausen. They used multiple regression analyses to find patterns of behavior in congressional voting. The resulting paper was clearly the product of a more serious Dick Cheney.

The manifest purpose of the roll call analysis described in this paper is that of demonstrating the existence of two policy dimensions in Congressional voting: economic and welfare. Support is sought for two propositions:
> *I. Each of the two dimensions appears in both the House and the Senate in each of six Congresses, the 83rd through the 88th, 1953–1964;*
> *II. Roll call voting on the economic policy dimension is more heavily influenced by partisan differences while welfare policy voting is more subject to constituency constraints.*

The second proposition is significant as an attempt to distinguish between a policy dimension on which partisan differences appear to be responsible for the greater part of the voting variation, and a policy dimension on which constituency factors have a substantial impact. This bears upon the more general concern with distinguishing those party differences in voting behavior which are a function of an independent partisan factor from those which may be attributed to any number of factors correlated with partisan affiliation. This problem will be viewed from different analytic perspectives, including an analysis of the effects of intra-party and inter-party personnel turnover on the policy positions taken by representatives of the same constituency.[35]

Those were the stated goals. But the two scholars had something more ambitious in mind, too. By studying these recent congressional votes, Cheney and his mentor hoped to create a computer model that might predict the outcomes of similar votes in the future.

The paper was published as an article in the spring 1970 issue of the *American Political Science Review*. In the publishing-mad world of higher education, writing an article that appears in an academic journal was a tremendous accomplishment. For a graduate student to share full writing credit on an article in the *American Political Science Review*, the preeminent political science journal, was a very high achievement. (Lynne Cheney, meanwhile, was published as a graduate student in *Modern Fiction Studies*.)

Even as his academic career flourished, Cheney felt the pull of politics as it was practiced in precincts and legislatures across the country. In early 1968, a group of well-heeled Republicans in Wisconsin approached Cheney about running a congressional campaign. Cheney was torn. He was interested in the position. In Richard D. Murray, the party had recruited a strong candidate to challenge the incumbent Democrat, Bob Kastenmeier. The money was good, too—$1,000 a month—and came at a time when the family was strapped for cash. Lynne Cheney had had a teaching fellowship that came with a small stipend, but it was barely enough to support a growing family.

But there was a drawback. The preliminary exams for Cheney's doctorate were scheduled for August, and running a competitive congressional campaign would leave him with little time to prepare. His only good option, he figured, was to ask to postpone the tests until after the campaign. After much deliberation, Cheney took his problem to the chairman of the political science department.

The chairman was unsympathetic. "You have to decide whether you want to be a politician or a political scientist."

"That was a little off-putting," Cheney later recalled. He rejected the notion that there was conflict between the theoretical world of academe and the handshake-and-a-Polaroid world he had seen during Knowles's campaign. "I thought, gee, I'd be a hell of a lot better professor if I'd actually run a campaign," he says.[36] But Cheney had spent two years completing the coursework toward his PhD and accepted the advice of his department chairman. He reluctantly turned down the campaign job.

Within weeks, though, he would have a new opportunity, one

that would allow him to continue his studies while working in practical politics: the Joe Davies Fellowship with the American Political Science Association. The dean of the graduate school had recommended Cheney for the position, endowed by the U.S. ambassador to the Soviet Union under Franklin Delano Roosevelt. The fund would cover his expenses for a full year living and working in Washington, D.C.

In early April, Davies's grandson, a Democratic senator from Maryland, Joseph Tydings, was in Wisconsin campaigning for the presidential candidate Robert F. Kennedy. During a stop at the campus of the University of Wisconsin to recruit students to help Kennedy in the upcoming Indiana primary, the dean suggested a get-to-know-you session with Cheney.

Following the speech, Tydings and Cheney met for beers at the student union. Cheney was nervous. Although he had already been chosen as the recipient of the fellowship, he thought it was an interview. Tydings found Cheney "a personable young man, the type we wanted."[37]

The congressional fellowship was a good match. It wouldn't start until September 1968, after Cheney's preliminary doctoral exams, and since he planned to write his dissertation on Congress, he could participate in the fellowship at the same time as he wrote his thesis. Pushed along as much by circumstance as by desire, Dick Cheney was on his way to Washington.

Years later, after Cheney had been elected vice president, he attended a dinner in his honor at the University of Maryland. Joseph Tydings, a member of the university's board of regents, was among the guests. He had long forgotten that the first Joseph Davies Fellowship recipient was Richard Bruce Cheney. The vice president had not.

When Cheney arrived, he and Lynne walked directly to Tydings and his wife. The vice president thanked Tydings for his crucial role in Cheney's career.

"If it hadn't been for Joe Tydings," Cheney said to the senator's wife, "I wouldn't have come to Washington."

Tydings was bewildered.

"You know how politicians, if they don't know something, you just fake it?" asks Tydings. "I told him, 'Sure, I remember.

I was happy to help.' I was faking it. I was blinking—my wife knew it."[38]

The minute the Cheneys walked away, Tydings's wife called him on it. "I had no idea what he was talking about," says Tydings.[39] He remained confused until Cheney, from the podium, told the entire gathering about the good work of the Joseph E. Davies Foundation and his beers with Senator Tydings.

Of Cheney's many fellowships and internships, none would be as important. For the second time in ten years—a decade interrupted by failures, both academic and personal—Cheney would find himself thrust into a new world of prestige and power. He was off to Washington.

CHAPTER THREE

Choosing Government

The Cheney family—Dick, Lynne, and their daughter Elizabeth—arrived in Washington, D.C., in September 1968. It was a busy time. Elizabeth, two years old, was precocious, impressing her parents' friends with an extensive vocabulary. Lynne was three months pregnant. She and Dick had both taken their "prelims"—preliminary doctoral examinations—in August. For Lynne, it was the culmination of months of exhaustive preparation, and for her husband the test ended a short week of intense cramming. They both passed. After packing up their possessions, the Cheneys drove from Madison, the capital of Wisconsin, to Washington, D.C.

The family landed in a town in turmoil. After sweeping to several primary victories in the spring, Robert F. Kennedy was assassinated on June 5, 1968. Lyndon B. Johnson, with his much-ballyhooed War on Poverty winding down and the war in Vietnam escalating, had abandoned his short-lived bid for reelection. In late August, Hubert Humphrey had been nominated as the Democrats' presidential candidate at their convention in Chicago. For many Democrats, Humphrey was their third choice, and his nomination was overshadowed by other news from the convention. Local police and national guardsmen had confronted angry hordes of antiwar protesters in violent clashes that were covered live on national television.

Richard M. Nixon, meanwhile, was making a dramatic political comeback—Nixon had lost the presidential election of 1960 to John F. Kennedy and only two years later had lost the race for governor in California. After having been written off as a failed politician, Nixon was riding the promise of an honorable end to the Vietnam War and gaining a healthy advantage in the polls.

Capitol Hill, too, was bustling with activity. Chief Justice Earl Warren had been looking at the polls, too, and decided to retire in time to give President Johnson an opportunity to appoint his replacement. Johnson chose Abe Fortas, already an associate justice, to take Warren's seat. But Fortas, a reliable liberal vote, had been a close political ally of Johnson, and had provided the president with political advice even while serving on the Supreme Court. Senate Republicans, exasperated by the activism of the Warren court and concerned that Fortas would accelerate the court's shift to the left, decided to filibuster the nomination. Johnson's legendary arm-twisting was not enough to carry Fortas to victory, and in early October the president withdrew the nomination.

Cheney was just six years removed from his ignominious departure from Yale University and the troubles that followed. And yet in many ways the next six years would be even more remarkable, as he would rise from a part-time political science fellow working for a neophyte congressman to one of the most powerful men in the nation's capital. It was in these critical years that Cheney got his first exposure to the vast federal bureaucracy—an experience that quickly started shifting his political beliefs to the right.

Although he had the temperament of a scholar—quiet and contemplative—Cheney also had the confidence that came from experience with policy and campaigning. His appearance, too, differed from that of a stereotypical academic. He kept his hair relatively trim and wore suits and boots to work, doing his best to look respectable even as a penniless congressional intern. At five feet nine inches Cheney hardly cut an imposing figure, but he was slim; and though he was a frequent smoker, he kept himself in relatively good shape.

In early September 1968, Cheney and forty-six other congressional fellows from the American Political Science Association gathered for an orientation. ("Fellows" was an appropriate term; all but two of them were men.) Members of the group spent the early part of their fellowship listening to guest speakers, arranging internships, and otherwise getting acclimated to life in Washington, D.C. Their internships didn't start until January. Each fellow was required to split his time between the House and Senate, and whenever possible, between Republicans and Democrats; a congressional fellow who worked for a Republican in the House during the first half of the year was expected to intern with a Senate Democrat for the second half.

Among the speakers at the orientation was a young representative from Illinois, a rising star in the Republican Party named Donald Rumsfeld. Rumsfeld, a former naval aviator first elected in 1962 from a wealthy district in suburban Chicago, had been marked early on as a savvy and ambitious legislator. In 1964, during his second term, he'd helped engineer the surprising election as House minority leader of another midwestern Republican, Gerald R. Ford, from Michigan. Rumsfeld was not known as a gifted orator, but young Dick Cheney was impressed with his speech and decided to explore the possibility of an internship in his office. Rumsfeld agreed to see him for an interview.

It did not go well.

"It lasted about fifteen minutes and I found myself back out in the hallway," Cheney recalls. "He didn't like me. I didn't like him. I thought he was an arrogant, abrasive young congressman from the North Shore of Chicago, which he was. And he thought I was some airy-headed academic. . . . He'd never really met one he liked. And he was right. I probably was giving him a lot of my PhD dissertation. . . . He quickly perceived that I couldn't do anything for him."

"It was," says Cheney, "probably the worst interview of my life."[1]

There was another option. Maureen Drummy, the woman who had read Cheney's essay about the Wyoming state legislature and who later arranged for his internship with Governor Warren Knowles of Wisconsin, presented him with yet another opportunity. Drummy had taken a new job in Washington, working for Representative Bill Steiger, a young Republican from Wisconsin. Steiger, who had met Cheney several times during his tenure with Governor Knowles, agreed to hire Cheney to spend the next six months in his Capitol Hill office. In a nod to Cheney's academic background and his experience in state politics in both Wyoming and Wisconsin, Steiger promised his new hire a meaty role in substantive legislative work—and a desk in the congressman's personal office.

Steiger was first elected in 1966, one of a remarkable class of forty-seven first-term Republicans elected that year. The new members formed the "Ninetieth Club," so named because their first term came in the Ninetieth Congress. At twenty-eight, Steiger was not only the youngest member of this club but also the youngest member of Congress. He exuded youthful enthusiasm about even the most mundane obligations of public service and was well regarded by members of both parties—a distinct advantage for a newcomer in the minority.

On January 3, 1969, three weeks before the inauguration of President Richard Nixon, Cheney's internship officially began. On that day, Steiger was sworn in for his second term and invited his new aide to the ceremony. Among the representatives inducted that day were four men who would play major roles in Cheney's political future.

A panoramic photograph of the ceremony captured a boyish-looking Steiger standing at the far left side of the podium, behind the last row of seats on the House floor, staring straight ahead without expression. Four members to his right was another member of the Ninetieth Club, Representative George H. W. Bush, Republican from Texas. Closer to the front was Bob Michel, a veteran legislator from Illinois. In the center of the photograph, just below the speaker's rostrum, stood the House minority leader, Gerald R. Ford. And in the upper right-hand corner—standing tall to see over the heads in front of him, but craning his neck to avoid a protruding lamp—was Dick Cheney, a new congressional fellow.

Cheney was involved in most aspects of Steiger's operation from the beginning of the fellowship, and clearly made an impression quickly. In a letter to Dr. Evron Kirkpatrick, the executive director of the American Political Science Association, Steiger wrote: "Either I have been the luckiest member of the House of Representatives, or the American Political Science Association has the best group of professionally competent aides-to-be in Washington. I have been exceptionally pleased with the help of Dick Cheney."[2]

Steiger, a moderate Republican, had responsibilities no different from those of other young representatives: constituent service, fund-raising, floor votes, and service on committees. The most important of these was the committee for Health, Education, and Welfare—HEW—which provided oversight for many of the programs at the heart of President Johnson's War on Poverty. Cheney spent much of his time on Steiger's staff working with the HEW committee and on its issues. It was experience that would later come in handy.

In the spring of 1969, as he approached the halfway point of his fellowship, Cheney was due to move to a new internship in the Senate. He landed a coveted position in one of its highest-profile offices. At the time, such an appointment must have seemed only natural for a young man of Cheney's abilities. In retrospect, given his politics and his views on the media, it is hard to imagine a more unlikely job: deputy press secretary for Senator Edward Kennedy.

"I don't think Kennedy realizes it to this day," Cheney says.[3]

He had the job on paper, anyway. Cheney had a friend, Bob Bates, working in Kennedy's office as a congressional fellow of the American Political Science Association on loan from the Office of Economic Opportunity at the White House. After working for Kennedy during the first half of his fellowship, Bates wanted to continue there; Cheney, for his part, wanted to remain on Steiger's staff. So the two fellows pretended to swap assignments.

"We got together and worked up a scheme where we would do a paper switch. So I actually went over and spent a few minutes with Kennedy so I could say I had. He did the same thing with Steiger," Cheney recalls. It was a deal they brokered over martinis with an official from the American Political Science Association. "And we told headquarters that we made the switch, but we'd never made the switch. Someplace in the annals of the American Political Science Association there's paperwork showing that I worked for Ted Kennedy."[4]

Although Kennedy had a reputation as one of the Senate's most liberal members, Cheney had no ideological objection to working for him. Even at twenty-eight, Cheney did not yet have a strong political philosophy. His work in graduate school, as evidenced by the paper he cowrote with Aage Clausen, had been focused sharply on analytical issues like congressional voting patterns.

Cheney stayed with Steiger for practical reasons. The two men were only three years apart in age, and over the course of the fellowship they had become close friends. Steiger had given Cheney tremendous substantive responsibility for a mere congressional fellow, responsibility that Cheney did not want to trade even for a more glamorous position in the office of a senator from America's most prominent political family. It was a telling choice.

In April 1969, President Nixon tapped Donald Rumsfeld to run the Office of Economic Opportunity (OEO). Created as part of President Johnson's War on Poverty, the OEO was a cabinet-level agency that operated out of the White House. The idea behind the agency was simple. State and municipal programs designed to help the poor had proved ineffective. With the OEO, the federal government would provide grants directly to community-based programs, which would ensure that the money went to those who needed it most. Its proximity to the president would provide built-in accountability.

Rumsfeld knew very little about these antipoverty programs, so when he agreed to take the job he turned to Steiger for advice on his confirmation hearings. Steiger was familiar with OEO programs from his time on the HEW committee. Cheney was perhaps even more familiar with them.

When Cheney learned that Rumsfeld had been appointed to run the OEO, he drafted an unsolicited twelve-page strategy memo on the upcoming confirmation hearings. He gave the memo to Steiger, who then passed it to Rumsfeld. Cheney's memo focused on accountability, and—not coincidentally—so did Rumsfeld's testimony:

> *If we are to strengthen the innovative role of the Office of Economic Opportunity, we must also improve its program evaluation. It is meaningless to talk about innovation and to experiment if you cannot—or do not—measure the results.... We may set high goals—we must set them high—but we must insist on knowing whether we achieve them, how close we come and why we miss.*[5]

It seemed to Cheney that Rumsfeld had relied heavily on his memo. And yet Cheney heard nothing at all about it after he handed it to Steiger.

In late May 1969, shortly after Rumsfeld's confirmation, Cheney received a call from a former classmate of Rumsfeld's at Princeton named Frank Carlucci. (Although the OEO and the Pentagon have little more in common than a reputation for bureaucratic inefficiency, Carlucci, like both Cheney and Rumsfeld, would one day serve as secretary of defense.) Carlucci invited Cheney to attend a meeting of Rumsfeld's transition team. Rumsfeld had asked Steiger if he could borrow Cheney for ten days while he set up his new office, and Steiger had agreed.

Cheney made the short trip from Capitol Hill to the OEO headquarters, located on the top floor of a run-down office building at the corner of Nineteenth and M streets in downtown Washington. Rumsfeld appeared briefly in a large OEO conference room, where he spoke to the three dozen people who would be working on the transition. When he left, his secretary, Barbara Dorsey, appeared in the doorway.

"Is there someone in here named Cheney?"

When Cheney raised his hand, she beckoned him into Rumsfeld's office.

Inside, Cheney took note of the rusty buckets arrayed throughout the room to catch rainwater as it dripped through the leaky ceiling. Rumsfeld glanced up at Cheney from behind an old desk.

"You, you're congressional relations," he said. "Now get the hell out of here."

Nearly forty years later, Cheney remembers the brief meeting with disbelief:

> He didn't say: "Hey, I liked your memo," or "Would you like to come to work for me?" or "I'm sorry I threw you out of my office six months ago." He said: "You're congressional relations, now get the hell out of here." So I turned around, walked out. In the other office, I said, "Where's congressional relations?" And they told me, "Halfway down the hall."

After ten days, Rumsfeld asked Cheney to stay at the OEO until the fellowship ended in September. Then, in September, he asked Cheney to join his staff full-time.

For the third time in as many years, Cheney faced what he calls a "decision point"—a moment that required a choice between politics and academe. The first time he faced such a decision point came after he finished working for Governor Knowles, and Cheney decided to pursue a doctorate in political science. Then, after he had completed his coursework, he was offered a position running a congressional campaign. But accepting the job would have required him to postpone his exams, so Cheney reluctantly turned it down.

Rumsfeld's offer was the third and final decision point. Cheney could reject the offer and return as planned to Wisconsin to complete his dissertation. Or he could accept it and effectively end his academic career. It was a good job that paid reasonably well, working on issues he enjoyed and for a man he admired. But did it hold enough promise to justify abandoning his hopes of becoming a professor? After twice choosing academe, he opted for government. Cheney accepted the job.

Thus was born one of the strongest and most mutually beneficial relationships in recent political history. The years Cheney spent at Rumsfeld's side would provide the young political scientist with a political education that wasn't available at Yale, Wyoming, or Wisconsin. And what he saw would shape his views for the rest of his political career.

When Rumsfeld joined the Nixon administration in 1969, he was given two titles: director of the Office of Economic Opportunity and special assistant to the president. He and Cheney had offices at both the OEO and the White House, in the space occupied by the White House counsel during the second Bush administration. They began each day attending staff meetings at the White House, spent the middle of the day up the street at the OEO, and returned to the White House in the evening.

As a member of Congress, Rumsfeld had voted against the creation of the OEO, and he now brought a deep skepticism to its leadership. Cheney, despite having worked for Republicans since his time in the Wyoming state senate, came without concerns about the agency's mission.

News reports about the programs funded by the OEO were filled with stories of inefficiency, political favoritism, and corruption. Its short existence had done little to suggest to Rumsfeld that his original judgment had been wrong. Rumsfeld recalled that by the time he started with the agency:

> *OEO had gotten to the point where a reasonable amount of the money had ended up going into radical organizations that were antisociety, and they were doing things that were, in some cases, illegal, and in other cases very irritating to the rest of society. A great many of the mayors and elected officials around the country were up in arms; people felt that the program had run amok. It wasn't just President Nixon who was sensitive to the problems in these programs; an awful lot of people were about ready to throw the baby out with the bath water because of anger over the way it had been administered.*[6]

Rumsfeld sought to have many OEO programs reassigned to other departments, a move that many observers interpreted as the first step in a plan to dismantle the agency. "The president sent Rumsfeld there to close it down," recalls Christine Todd Whitman, future governor of New Jersey and administrator of the Environmental Protection Agency, who began her political career at the OEO as a special assistant. "Some of us thought the programs were worth saving, but we were all aware that the agency's time was limited."[7]

Rumsfeld believed that many of these programs simply duplicated less effective projects funded by other parts of the federal government. The solution, in his view, was to consolidate these programs in the agencies that had responsibility for them in the first place.

In October 1969, one of the elected officials up in arms about an OEO program was Louie Nunn, the Republican governor of Kentucky. Nunn wanted Nixon to cancel OEO funding of an antipoverty project in eastern Kentucky called the Middle Kentucky River Area Development Council. Nunn reported that the program was rife with corruption and double-dealing and that its administrators were redirecting money intended for his state's poor to their political allies, all Democrats. The White House wanted to be responsive. Nunn had served as Nixon's campaign chairman for the southern United States in the tumultuous presidential race of 1968, in which Nixon defeated Hubert Humphrey and George Wallace, in part owing to the success of his notorious "southern strategy." Under normal circumstances, the White House would have simply granted Nunn's wish.

But these circumstances were far from normal. The program in question was located in the congressional district of Carl Perkins, an influential Democrat. At a campaign event in 1968, two months before he was assassinated, the presidential candidate Robert F. Kennedy appeared with Perkins and joked about his extensive power.

> *One day [Perkins] said to my brother, "Why don't you run for president?" And my brother said, "I can't run for president." And Carl Perkins said, "You should be president of the United States." President Kennedy said, "I don't have any of that kind of following around the country." And Carl Perkins said, "I'll get everybody together." And the next thing we know, without hardly doing anything John Kennedy was elected president of the United States.*[8]

Kennedy's account was exaggerated, but Perkins might have seemed that powerful to Donald Rumsfeld and his colleagues at the OEO. Perkins was chairman of the Committee on Education and Labor. For Rumsfeld, that was a serious complication. Perkins's committee controlled the budget for the OEO: he was being asked to cut funding for a pet project of the man who, in turn, controlled his budget.

Rumsfeld met with Nixon and John Ehrlichman to discuss the sensitive politics. The president suggested sending a team from the White House to investigate, a suggestion that struck Rumsfeld as unwise.

"You don't want to do that," he told Nixon and Ehrlichman. "That's like embracing an asp. You just do not want to get near this. You can always disown me or disown Louie Nunn or something, but you can't just take it right into the White House and stick it right in the president's plate."[9]

Rumsfeld proposed an alternative: he would dispatch his best man, Dick Cheney. And so Cheney went to Kentucky to meet with the woman who was running the controversial program, Treva Turner Howell.

Howell came from a family of powerful Kentucky Democrats. Her father had been a judge, a prominent state senator, and a school superintendent; her mother had recently resigned as a school superintendent; at the time, her brother was in the state senate; and her husband was chairman of the Breathitt County Democrats.[10] Howell did not take kindly to the accusations from Governor Nunn. She lashed out at the people in his administration, calling them "character assassins, purveyors of perjury, crybabies and cowards."[11]

Howell turned on the charm for Cheney. After leading him on a tour of the facilities funded by the Middle Kentucky River Community Action program, she invited him to her house and, once he arrived, poured him three fingers of bourbon, neat. Howell did her best to persuade Cheney that the program was run according to the highest principles.

Cheney left Kentucky unconvinced. He believed that there were irregularities in the way Howell administered the OEO funds, but he also thought that it would be nearly impossible to prove any malfeasance. Back in Washington, he recommended to Rumsfeld that the OEO continue to fund the program, and Rumsfeld concurred.

Louie Nunn was furious. He threatened to hold public hearings that would make the "case of the taxpayers versus OEO."[12] Haranguing the White House, Nunn managed to extract a promise that the program would be reviewed a second time. Again, Nixon proposed that the White House investigate, and again Rumsfeld objected. "Let's get the FBI to do it," he suggested. "If there's something wrong, somebody ought to know that something's wrong."[13] Rumsfeld met with the FBI director, J. Edgar Hoover, to explain the situation; the FBI investigated the matter and confirmed Cheney's initial assessment that while there were questions about the program and its administrators, there was no provable criminal activity to justify cutting the funding. Throughout the episode Rumsfeld backed his young aide against pressure from one of Nixon's staunchest supporters.

The difficulties with the OEO continued. Some of these problems were serious. Others, such as a dispute in California in early 1970, were farcical. In late 1969, Governor Ronald Reagan of California vetoed funding for an OEO program in his state, the East Oakland–Fruitville Planning Council. This planning council had clashed with Mayor John Reading of Oakland, a Republican and staunch ally of Reagan's; an executive from the organization had challenged him in the mayoral election the year before, and since then the two had carried on a running dispute. When Reagan vetoed further funding, supporters of the program argued that the governor was playing political games by punishing the political opponents of a strong supporter. The Republicans, meanwhile, claimed that the planning council, using federal government money, had become something of a "shadow government" whose main objective was creating trouble for Mayor Reading.

Rumsfeld agreed with Reagan, and he overruled the OEO's own Office of Program Development (OPD), which had recommended that the funding continue. Rumsfeld had Cheney explain his decision to the media. "If the OPD can override a veto, then you don't need a director," Cheney told the *Washington Post*.[14]

The Planning Council sued to continue the funding and hired the Legal Aid Society of Alameda County to represent it. That was a comical twist: the Legal Aid Society was itself funded by the OEO. So one OEO-supported agency was suing the OEO on behalf of another OEO-supported agency to restore its funding.[15]

Incidents like those in Kentucky and California began to shape Cheney's views on the size and scope of government. He spent several hours each day at the OEO grappling with the inefficiencies of large-scale federal programs. His next job would add to this growing skepticism.

In the summer of 1971, the U.S. economy was in trouble. Inflation was on the rise and so were the anxieties of American voters. And when the voters were nervous, so was Richard Nixon.

On Friday, August 13, President Nixon assembled a small team of economic and political advisers for a weekend at Camp David. It was soon clear to all involved that this would be no ordinary meeting. John Connally, Nixon's big-thinking secretary of the treasury, presented a new economic plan, audacious in design and breathtaking in scope. It included tax cuts at home, tax hikes on imports, and, for the first time, an abandonment of the gold standard for the dollar. But the most dramatic change—and the one least debated at Camp David[16]—was an across-the-board freeze on all wages and prices in the U.S. economy. The freeze would be monitored by a new agency called the Cost of Living Council (CLC).

Several officials at Camp David, including Nixon, had concerns about the economic impact of wage and price controls. George Shultz, director of the Office of Management and Budget, who had a PhD in industrial economics from MIT, thought the plan disastrous. But with inflation rising sharply, and the election of 1972 a little more than a year away, Nixon was less concerned with long-term recovery than with reelection. He wanted to be seen as a decisive leader, willing to take action to assuage the nation's increasing economic worries. The freeze would be temporary, lasting only three months, and little thought was given to how it would end and what, if anything, would follow. Participants in the session at Camp David were struck by how much time was devoted to selling the package and how little time to actually gauging its likely results.

The new policy would affect every aspect of the U.S. economy, from the price of a meal in Seattle to the rent of a one-bedroom apartment in Tuscaloosa to the salary of a laborer in Hanover, New Hampshire. If a half-gallon of milk cost $0.58 on August 15, 1971, it would cost $0.58 on November 13, 1971, and on every day in between.

Nixon announced the plan to the nation that Sunday in a prime-time address from the White House. The speech won rave reviews even though it marked a significant policy reversal for an administration already on record as opposing such wage and price controls. Democrats who might have been expected to criticize the plan were caught flat-footed. Congress was out of session at the time of the announcement, so whatever their complaints, they would have to make them from their districts across the country. More important, Nixon had been able to pull off this audacious move only because congressional Democrats had given him the authority to freeze wages and prices the year before, when they passed the Economic Stabilization Act of 1970.

The vote had been, in a sense, a game of political chicken. Democratic leaders had signed off on the bill never imagining that a White House generally opposed to governmental intrusion in the free market would use the authority it granted. They were wrong, and at least in the short term, Nixon's gamble paid off politically. The controls proved popular with the public and provided Nixon with a boost as he headed into the final year of his first term.

Still, the plan had its share of skeptics. The economist Murray Rothbard wrote in the *New York Times,* "On Aug. 15, 1971, fascism came to America. And everyone cheered, hailing the fact that a 'strong President' was once again at the helm." He asked, "By what right do you use coercion to tell buyers and sellers what prices they may or may not agree upon, or employers and workers what wages they may pay? What possible stretch of the Constitution gives the President the right to freeze rents in a Sioux City boarding house?"[17]

Complicating matters further were the unresolved questions of what to do when the initial freeze ended and who would supervise the implementation of that still undetermined plan. The executive director of the Cost of Living Council, an economist from the University of Chicago named Arnold Weber, had made clear to Nixon that he wanted to return to academe following the initial ninety-day freeze, which was scheduled to expire on November 13, 1971. Nixon picked Rumsfeld to replace Weber, and he asked George Shultz to make the phone call.

"Don, the president and I want you to run phase two of the economic stabilization program," Shultz told Rumsfeld.

"Well, I don't agree with it," said Rumsfeld. "I don't believe in wage or price controls; I'm a market man."

"Well, we know that, Don. That's why I want you to do it."[18]

Rumsfeld took the job at the Cost of Living Council and asked Cheney to come with him as director of operations. Cheney, though, already had another offer. Over the course of his time at the OEO and the White House, he had quietly begun working with a small group that would become the heart of Nixon's reelection team in 1972. Along with another young White House staffer, he had set up the campaign's surrogate speaker program, an operation that would dispatch campaign representatives to events around the country: while Nixon was stumping in New York, Cheney's operation might send the campaign chairman to address the Rotary Club of Topeka, Kansas, or send a prominent senator to attend the annual Lincoln Day dinner in Orange County, California. Cheney's reward for laying the groundwork was a position on the Committee to Re-Elect the President, which was soon to be announced.

It seemed like an obvious decision: on one hand was the fast-paced excitement of a presidential election, with a high profile and potentially big rewards; on the other, the anonymous drudgery of a bureaucracy regulating technical aspects of the U.S. economy. As he had when he passed up the internship in Senator Kennedy's office to remain on the staff of Representative Steiger, Cheney opted for substance over flash. He chose to toil in obscurity at the Cost of Living Council rather than work on Nixon's reelection campaign.

On October 7, 1971, in a televised speech to the country, Nixon announced that wage and price controls would continue beyond phase 1, the initial ninety-day freeze. In his speech, he provided an overview of phase 2, but the lack of planning at Camp David meant that the details of those regulations were left to his new team. The plan would go into effect at midnight on November 13. That left the team—in the words of the *Washington Post*—"the luxury of a month" to devise the rules.[19]

Two bodies were set up under the Cost of Living Council to govern phase 2: the Price Commission and the Pay Board. But over the next weeks, both were racked with internal politics, and their deliberations were often reduced to bickering over minutiae and tactical maneuvering among the board members. Time ticked away, and little was accomplished.

Finally, on Thursday, November 11, 1971, two days before the deadline, Rumsfeld called an emergency meeting of his senior staff. There was one item on the agenda: to write the regulations that would run the U.S. economy for the foreseeable future. Working through the night, economists and noneconomists alike tossed out ideas about how to regulate the prices that American businesses could charge their customers and the wages they could pay their employees. Cheney, as the typist, had a front-row seat.

The results of this session were published in the *Federal Register* on November 13, 1971. The entire U.S. economy would be governed by rules that took up just twelve pages. Even in their rudimentary form, though, the regulations included several pages of mind-numbing detail.

Raw honeycomb honey was exempt from the new price controls; "processed and blended honeybutter product" was not. Unpopped popcorn was exempt; popped popcorn was not. Garden plants and cut flowers were exempt; floral wreaths were not. Wigs and toupees were exempt, and so was taxidermy.

If most Americans understood what their employers meant by a raise, this was not so simple to the bureaucrats at the Cost of Living Council.

"Pay adjustment" means a change in wages and salaries which includes all forms of direct or indirect remuneration or inducement to employees by their employers for personal services, which are reasonably subject to valuation, includ-

ing but not limited to: Vacation and holiday payments; bonus, layoff and severance pay plans; supplemental unemployment benefits; night shift, overtime, production and incentive pay; employer contributions for insurance plans (but not including public plans, e.g., old age, survivors, health, and disability insurance under the Social Security system, Railroad Retirement Acts, Federal Insurance Contributions Act, Federal Unemployment Tax Acts, Civil Service Retirement Acts and the Carriers and Employees Tax Act); savings, pension, profit sharing, annuity funds and other deferred compensation and welfare benefits; payments in kind, job perquisites, housing allowances, uniform and other work clothing allowances (but not including employer-required uniforms and work clothing whether or not for safety purposes); cost-of-living allowances, commission rates, stock options and other fringe benefits; and benefits which result in more pay per hour or other unit of work or production (e.g., by shortening the workday without a proportionate decrease in pay).

Although Rumsfeld objected to the government's massive intrusion into the free market, Cheney did not. "At the time I didn't think much about it," he says. "Nixon had done it so we were going to make it work."[20]

On the day after the new regulations went into effect, Cheney wrote talking points for Nixon for an afternoon meeting with the Cost of Living Council. "We have now moved to a system of flexible, equitable controls and the mechanism is in place," Cheney wrote. Despite the chaos that had characterized the work of the CLC, he suggested that the president commend the two men running the Pay Board and the Price Commission. "They have wrestled with tough issues and made decisions which are essentially sound. Their standards give us a good chance of reaching the goal of reducing inflation to 2 to 3 percent by the end of next year. The task ahead is to see that the effort is fair and equitable for all Americans—that the mechanism is responsive to the needs of individual citizens—and that we continue to move toward our goal of prosperity with peace."[21]

Nixon's remarks hewed closely to Cheney's suggested language. When he departed from the talking points, it was to offer assurances that were as obviously false as they were optimistic. Phase 2, he said, "will succeed because it has been so carefully worked out, not simply in terms of setting goals that may sound good at the moment but that are totally unachievable, but in setting realizable goals, goals that can be achieved, and then making up our minds that we are going to follow through and see that those goals are achieved."[22]

Three weeks after the implementation of phase 2, the Pay Board and Price Commission were lost in a thicket of claims and counterclaims. The Pay Board, in particular, was devolving into chaos. "The board has yet to devise some of the basic procedures under which it must operate," wrote James L. Rowe Jr. of the *Washington Post* in a news analysis. "Meetings are often so confused that members are not always aware of what they are voting on. Some members have failed to be recorded on an issue because they were unaware a vote was being taken."[23]

With phase 2 under way, Cheney's responsibilities shifted. As director of operations for the Cost of Living Council, he monitored the enforcement of the wage and price controls, and directed a small army of IRS agents—some 3,000—who were charged with finding and punishing violators. They investigated such claims as those brought against nine grain workers in Chicago who were set to receive pay increases in excess of the 5.5 percent permitted by the government.

Cheney soon grew frustrated with the politics of enforcement. In a sarcastic memo of December 27, 1971, to Ken Cole, Nixon's deputy assistant for domestic affairs, he protested the unrealistic procedures: "The AFL-CIO in its usual helpful attitude has run off several hundred thousand copies of the form which the public is to use to report suspected violations to the Internal Revenue Service. Any day now we expect to be inundated under a blizzard of alleged violations."[24]

In the summer of 1972, less than six months before the presidential election, the White House was getting reports of mounting public concerns over the rising cost of groceries. In fact, the food component of the consumer price index had not risen for the previous six months; groceries weren't getting more expensive. In short, there was no problem, a fact that did not deter Nixon from offering a solution.

On June 29, 1972, Nixon called a meeting of the Cost of Living Council in the Cabinet Room at the White House and listened to his economic advisers debate the wisdom of refreezing food prices. When they were done, he offered a story that made a lasting impression on Cheney.

Nikita Khrushchev, Nixon said, had once given him a sage bit of advice: in order to be a statesman, it is sometimes necessary to be a politician.

"If the people believe there's an imaginary river out there, you don't tell them there's no river there," Cheney remembers Nixon saying. "You build an imaginary bridge over the imaginary river."[25] In late June 1972, the CLC froze food prices again.

"I became a great skeptic about the whole notion of wage-price controls," says Cheney. "The idea that you could write detailed regulations that were going to govern all aspects of an economy as big as the U.S. economy is loopy. But that's what we were trying to do. . . . You know, all of a sudden the price of hamburger was our problem. The American people out there go to the store and have to pay eighty-nine cents for a pound of hamburger. And last week it was seventy-nine and all of a sudden it became a political problem."[26]

The entire experience, he says, "created strong feelings that I have to this day about the government trying to interfere in the economy—moved me pretty radically in the free-market direction, the importance of limited government."[27]

If Cheney's views came to resemble Rumsfeld's more closely, he didn't always agree with his mentor's management techniques. Throughout the time Cheney worked for Rumsfeld in the Nixon administration, he had to deal with what Rumsfeld called "snowflakes." These were short, off-the-cuff memos that Rumsfeld dictated to his secretary for distribution among the staff. Some of the snowflakes required action; others simply laid out a series of questions for the staff to consider as they formulated one policy or another. Occasionally, a snowflake would include biting personal criticism of a member of Rumsfeld's staff, the kind of critique that might embarrass the employee and affect staff morale.

Rumsfeld's secretary brought the snowflakes to Cheney before distributing them to the staff. Without ever telling Rumsfeld, Cheney unilaterally decided which snowflakes would be sent to the staff and which ones—usually those that included personal criticism—would be quietly withheld.

One day near the end of their time at the Cost of Living Council, the phone rang in Cheney's office. It was Rumsfeld.

"Get your ass down here!" he barked into the receiver.

"Yes, sir."

Cheney arrived to find Rumsfeld's secretary standing sheepishly next to Rumsfeld. On his desk were two piles of carbon copies. In one pile sat copies of the snowflakes that had been sent to the staff; in the other were the snowflakes that had gone undistributed.

Cheney was nervous. It occurred to him that such insubordination could get him fired. Rumsfeld did not go that far.

"I want you to know I know what you're doing," he said, pausing to let Cheney think about his transgressions. "Now get out of here."[28]

Nixon won reelection on November 7, 1972. The next day, he met with his cabinet and instructed the officials to obtain letters of resignation from all political appointees, including those of relatively low rank. Not all the employees would be let go. Those whom Nixon wished to retain would get a call informing them that their letter had been discarded. Those who did not receive a call were expected to clear out of their offices. Rumsfeld regarded the move as cowardly and refused to carry out the order. Not long afterward, he was eased out of the White House.

Rumsfeld accepted an offer to serve as U.S. ambassador to NATO and moved to Brussels, Belgium. He asked Cheney to join him; but the Cheneys had two young children and Lynne was teaching at George Washington University, so it made more sense to stay in Washington.

Instead, Cheney took a job with Bradley Woods, a consulting firm in Washington, where he watched the turmoil of Nixon's second term with great interest. He thought frequently of his decision to forgo working on Nixon's campaign to join Rumsfeld at the Cost of Living Council.

As Cheney watched, transfixed, from a distance, two of the men with whom he had worked most closely on the small White House campaign team were swept up in the Watergate conspiracy and the subsequent cover-up. Bart Porter, who worked alongside Cheney on the surrogate speaker program, spent a month in prison for lying to the FBI. Jeb Magruder, who had run Rumsfeld's first congressional race and supervised Cheney's campaign activities in the White House, served seven months for conspiracy to obstruct justice.

Asked about his decision to pass up the job at the Nixon reelection campaign, Cheney says, "I made the right choice."[29]

CHAPTER FOUR

The Ford Years

On August 8, 1974, Dick Cheney took a phone call from overseas. Don Rumsfeld's personal secretary was calling from Brussels, where Cheney's former boss had remained as U.S. ambassador to NATO throughout the early part of Nixon's second term. The secretary was calling to ask Cheney to meet Rumsfeld's flight at Dulles Airport in suburban Washington at one o'clock the next afternoon. She didn't give him a reason. He didn't need one.

That summer had been tumultuous. On July 24, the Supreme Court ruled unanimously that Richard Nixon must turn over to prosecutors the tapes of secretly recorded conversations in the Oval Office. Three days later, the House Judiciary Committee concluded that the president had obstructed justice and passed the first article of impeachment. On August 5, the White House disclosed that one of Nixon's tapes included a conversation in which the president discussed a plan to block the work of Watergate investigators.

Then, at a cabinet meeting the next day, one of Nixon's most stalwart defenders formally withdrew his support. Vice President Gerald R. Ford told his colleagues that he could no longer defend Nixon against claims that the president was guilty of impeachable offenses. It was a decisive blow. Two days later, on

August 8, 1974, the same day that Cheney received the call from Brussels, Nixon announced to America and the world that he would resign the presidency the following day.

Cheney was one of nearly 100 million Americans who spent the next morning watching the surreal events on television. Nixon delivered a rambling farewell address to his staff in the East Room of the White House:

We want you to be proud of what you have done. We want you to continue to serve in government, if that is your wish. Always give your best; never get discouraged; never be petty; always remember, others may hate you, but those who hate you don't win unless you hate them, and then you destroy yourself.

Moments later, after climbing the five stairs to board Marine One for the final time, he turned and bade the nation farewell with a defiant wave that would be played thousands of times in the years that followed. It was a strange end to a strange time.

With Rumsfeld's arrival fast approaching Cheney left his spot in front of the TV and drove his 1965 Volkswagen to the airport. At the gate, he was surprised to see that he was not the only one waiting for Rumsfeld. A White House messenger was there, too, waiting to hand Rumsfeld an envelope with a letter from the man who had an hour earlier been sworn in as the thirty-eighth president of the United States. Rumsfeld had been called back to the United States to serve on Ford's transition team. The letter from Ford asked him to lead it.

As the two former Nixon administration officials got into Cheney's car for the drive back to Washington, they tried to make sense of the momentous events of the past forty-eight hours. Just as he had done when he made the transition from Congress to the Nixon administration, Rumsfeld asked Cheney to help him for ten days as he coordinated the presidential succession. Once again that short-term assignment would be extended for several years.

Cheney had come to Washington without a definite political philosophy. Though he had worked for Republicans in state politics, he had not been a partisan. Most young people came to Washington to end the Vietnam War, or to help Lyndon B. Johnson win the War on Poverty, or because they had walked precincts for Senator Barry Goldwater or passed out leaflets for Richard Nixon. They came with plans to change the world or at least to help their side.

By contrast, Cheney's initial interest in national politics was procedural and methodological, almost technical. He was fascinated by how things were done in Washington, why some programs worked and others didn't, why some policies made sense and others seemed doomed to fail. It was the political science professor in him, detached and almost aloof.

His experience in the Nixon administration began to change that. He saw well-intentioned government programs that solved one problem and created a dozen others. A plan by the Office of Economic Opportunity to train migrant workers to grow azaleas in South Carolina would have provided jobs for the workers but destroyed the market for azaleas in the process. Need-based assistance in the poorest parts of the country was diverted to "community-action programs" that did little more than line the pockets of local politicians. Through the Cost of Living Council, the IRS targeted small businesses because their owners wanted to give employees a raise. Grocery stores had to fight with the federal government to raise the price of a dozen eggs. To protect the American public, the Price Commission directed McDonald's to reduce the price of Quarter-Pounders.[1] To Cheney, these experiences not only demonstrated the inherent inefficiencies of big government but seemed to confirm the wisdom of individualism and self-reliance, the cardinal virtues of his home state.

Cheney had long been conservative by disposition and temperament, but he was not yet a political conservative. Twenty-seven months at the highest levels of the Ford administration would make him one. In that short time he would develop strong views on issues of war and peace, on intelligence and the media, on the mechanisms of policy making and the importance of discretion, on the manipulation of the levers of power and the function of the vice presidency, and—importantly—on the proper role of the executive branch in the American constitutional system.

Cheney worked daily with men whose thinking would shape the conservative movement for years. He drafted language for the president with the economist turned speechwriter Milton Friedman and discussed monetary policy with Alan Greenspan, a devotee of Ayn Rand serving on Ford's economic team. Cheney refereed foreign-policy fights between UN Ambassador Daniel Patrick Moynihan and Secretary of State Henry Kissinger. He regularly attended briefings put together by the administration's in-house academic, Robert Goldwin, which brought leading intellectuals in for long, freewheeling discussions with the president and his senior staff. It was an experiential education without parallel.

The challenges facing the Ford administration were monumental. Ford was the first president in history who had never been elected either president or vice president. Nixon selected Ford to replace Vice President Spiro T. Agnew, who resigned in October 1973 after pleading nolo contendere to charges that he had accepted bribes and filed false tax returns. Then, not even nine months later, Nixon resigned and Ford was elevated to the presidency, stepping into a breach of unprecedented proportion. It was, as Ford later remarked, a government in "chaos."[2] As the nation approached its bicentennial, these betrayals of trust put immense pressure on the constitutional order that had served America so well for nearly 200 years.

As if those fundamental questions were not enough, Ford also faced countless practical problems. Most significantly, the new president had to convince Americans that his White House would be nothing like Nixon's while convincing the rest of the world that his conduct of foreign affairs would not differ markedly from that of his predecessor.

Ford spent much of his first afternoon in office reassuring foreign governments—allies and adversaries alike—that the United States would make good on its commitments and that the foreign policy team would remain intact. Fulfilling that second promise meant primarily one thing: keeping Henry Kissinger.

As secretary of state and national security adviser, Kissinger had been the focal point of foreign policy under Nixon, and one of Ford's earliest priorities was to affirm publicly that he would remain so. Ford announced his decision to keep Kissinger even before he formally became president. At a press conference in front of his home in Alexandria, Virginia, on August 8, 1974, the night before he took the oath of office, Ford announced that Kissinger would remain a key decision maker.

Cheney thought that this announcement was a shrewd move; it reassured allies abroad, and also demonstrated Ford's understanding that the flamboyant diplomat Kissinger had more international credibility than a man who had until recently been a congressman from Michigan's fifth district, even if that man now occupied the Oval Office.[3] Ford's pledge also meant that James Schlesinger, a favorite of congressional and administration conservatives, would remain secretary of defense; and that William Colby would be retained as director of the CIA.

Hours after he took the oath of office, President Ford met with his transition team. The new president instructed the small group—Rumsfeld, Cheney, the former Democratic congressman Jack Marsh, and Secretary of the Interior Rogers Morton among them—to examine all aspects of the White House operation except for the making of foreign policy.

As part of his effort to restore confidence in government and to contrast his administration with the ultrasecretive Nixon White House, Ford wanted an executive branch that was accessible to the public and responsive to the press. He also wanted a White House that allowed him to receive advice from a broad array of top officials. The transition team proposed a system that would be known as the "spokes of the wheel." In theory, President Ford was the hub of the wheel and nearly a dozen advisers were the spokes. Each of these advisers was to have equal access to Ford. Such a system seemed natural and sensible to Ford, who had spent twenty-five years in the House of Representatives. But it would prove more appealing in theory than in practice.

After ten days working on the transition, Rumsfeld went to Ford and recommended that his team be disbanded. Ford had decided to retain—at least temporarily—Nixon's last chief of staff, General Alexander Haig; Rumsfeld believed that the two teams working side by side would create more confusion than clarity. The transition team filed a report for President Ford, and its members left the White House. Rumsfeld flew back to Belgium, and Cheney rejoined his consulting firm. They wouldn't be gone for long.

Shortly after the transition team completed its work, Ford made an announcement that stunned the nation. On September 8, 1974, the new president granted Nixon an unconditional pardon.

As we are a nation under God, so I am sworn to uphold our laws with the help of God. And I have sought such guidance and searched my own conscience with special diligence to determine the right thing for me to do with respect to my predecessor in this place, Richard Nixon, and his loyal wife and family. Theirs is an American tragedy in which we all have played a part. It could go on and on and on, or someone must write the end to it. I have concluded that only I can do that, and if I can, I must.

With the benefit of hindsight, there is widespread agreement that Ford's decision was the right one. Those who felt that way at the time, however, were in a distinct minority. Ford's press secretary resigned, and his former colleagues in Congress roundly criticized the move.

Worse, the pardon ended the short honeymoon that the Washington press corps had given Ford. Editorialists condemned the pardon with virtual unanimity. Many critics went beyond the decision itself and questioned Ford's motives. One respected journalist in Washington wrote a book heaping scorn on Ford, called simply *The Man Who Pardoned Nixon*. The president was in trouble, and he once again called on Rumsfeld. Ford and Rumsfeld had grown close during their years together in the House. As House minority leader, Ford had come to trust Rumsfeld's judgment and often consulted him in difficult times. This was one of those times and, as it turned out, Ford would call on Rumsfeld for more than short-term consultation.

While he was in Brussels, Rumsfeld had been approached informally about serving as Ford's chief of staff. He was interested, and shortly after he returned to Washington, he met with Cheney to discuss the job. They met at the Key Bridge Marriott, across the Potomac River from Georgetown, on Saturday, September 14. Rumsfeld, who was scheduled to meet with Ford later that same day, told Cheney about the developments and asked him to return to the White House. The discussion was, at this point, hypothetical, so the two men agreed to speak again the next day after Rumsfeld's meeting with Ford.

As expected, Ford offered the job and Rumsfeld accepted it. When the new chief of staff met with Cheney the following day, he presented his former deputy with an offer that included a unique power-sharing arrangement. "The understanding when I took the job was that I would get access to the president and that I would be Rumsfeld's surrogate," says Cheney. "When he wasn't around, I would be in a position to make decisions for him and to operate as though I had his job. We agreed at the outset that we would alternate trips. He would take one presidential trip and I would stay in the White House and run it; the next time, I'd go on the trip and he'd stay at the White House."[4]

Cheney was hired as a consultant on October 1, 1974, earning $138 a day. The expectation was that he would officially join the White House staff as Rumsfeld's deputy on completion of a full field investigation by the FBI.

This review is in most cases a formality. But when the FBI learned that Cheney had been arrested twice for drunk driving in Wyoming, it nearly put an end to his work with the Ford administration—and his young political career.

Rumsfeld, who had not known about the incidents, called Cheney into his office and grilled his protégé about the arrests. In particular, Rumsfeld wanted to know whether Cheney had disclosed the violations, as required, on the paperwork he completed at the outset of the background check.

"You put it down on the form, didn't you?"

"Yeah," said Cheney. "I did."[5]

Asked if he remembers the incident, Rumsfeld says, "Do I ever."

Rumsfeld was satisfied. "My attitude was, that was history," Rumsfeld recalls. "I knew Dick well." In his mind, the arrests themselves were less important than the fact that Cheney had been honest about them even with the knowledge that such a disclosure could jeopardize his job. And Rumsfeld wasn't worried about how Cheney would conduct himself in the White House.

But he was concerned that any public disclosure of the arrests would embarrass the new president at a time when he could ill afford any additional problems. Ford's new chief of staff also understood that the decision about whether Cheney would serve in the administration was not his to make. So he took the issue to the president. It was not a short discussion.

"I spent a good deal of time with the president, who of course did not know Cheney at all, and convinced him that this was a good guy."[6]

Cheney did not learn until later that Rumsfeld had personally intervened with the president. "It impressed the hell out of me that he was willing to stand by me," Cheney recalls.[7] It was loyalty Cheney would reciprocate thirty years later, when Rumsfeld, as secretary of defense, was under pressure to resign from critics across the political spectrum.

Cheney was thirty-three years old when he began his work for the Ford administration. He had sat in on meetings with Ford when they were both working in the House. But Cheney didn't meet his new boss until after starting full-time at the White House. Ford was a trusting soul, a rarity in the cutthroat politics of Washington, and he immediately saw Cheney as part of his inner circle. "He is as comfortable with Cheney as he is with Rumsfeld," one senior aide of Ford's said. "He doesn't hesitate to say 'Get me Cheney,' if something comes up and Dick is the one close at hand."[8]

This did not go over well among some on the White House staff, particularly the holdovers from the Nixon administration.

Early each weekday morning, a messenger hand-delivered a copy of the president's daily schedule to the senior White House staff. On a typical day, Ford's first meeting was at eight-thirty AM with his chief of staff, who briefed the president about the day ahead. Because Rumsfeld had agreed to share his responsibilities with Cheney, one morning shortly after they came to the White House, the first entry on Ford's schedule read, Eight-thirty: Cheney, Oval Office—thirty minutes.

On this morning, the copy of the schedule delivered to

Cheney's office came with an additional notation. The entry indicating Cheney's meeting with President Ford was circled and a note in the margin registered the disbelief of its author.

"Can you believe this crap?"[9]

The author of the note was Tom Korologos, a former aide of Nixon's who kept his job under Ford. He intended the message for George Joulwan, a friend who served as the military aide to the former chief of staff, Alexander Haig. But days earlier, Joulwan had vacated his old office to make way for its new occupant, Dick Cheney.

It was the kind of mistake that almost certainly would have cost Korologos his job had it occurred in the Nixon White House. But Cheney, mindful of how Rumsfeld had handled his own unauthorized withholding of "snowflakes" at the Cost of Living Council, had a different plan. He waited. And later, rather than provoke a confrontation, he simply let the military aide know that he had seen the snide comment.

"I saved that for about six months and then I went to see him to let him know I had received it," said Cheney. "He was cooperative from then on."[10]

Rumsfeld's title was staff coordinator; Cheney was deputy staff coordinator. The bureaucratic designations were an attempt to dispel the image of the imperious chief of staff personified by the man who held the title in the Nixon White House, H. R. Haldeman. A veteran of World War II, known for his close-cropped hair and his hyper-disciplined operation, Haldeman zealously guarded access to Nixon and proudly referred to himself as "the president's son-of-a-bitch." Many on the White House staff called him much worse.

In his new position, Cheney immediately became involved with the administration's most sensitive decisions about policy and personnel. At eleven AM on October 24, Cheney sat next to Ford as the president informed John Sawhill that he would be replaced as director of the Federal Energy Administration.

Sawhill had long advocated raising gasoline taxes as a solution to the nation's energy woes; Ford strongly disagreed.[11] Cheney had prepared talking points for Ford, and made it clear that Sawhill was being forced out and that the administration hoped to portray the dismissal as a resignation. Cheney encouraged Ford to be blunt: "It is in his best interests as well as your own that everyone leave the meeting with the firm understanding that he is leaving. It should not be left vague, nor should he be given the impression that the decision to replace him can be reversed." The suggested language made clear that Ford was to waste little time on small talk. "I want you to know I appreciate the fine job you have done under very difficult circumstances. I know it has not been easy. I have decided, nonetheless, to replace you as director of the Federal Energy Administration." Cheney proposed that Ford offer Sawhill another, unnamed position in the administration and volunteered to make the arrangements for his departure: "The details of how we handle your resignation and the announcement of a replacement should be worked out with Dick Cheney."

Little more than a decade earlier, Cheney had been a college dropout living in a tent and working as a grunt laying power lines in rural Wyoming. Now, he was working directly for the leader of the free world, coordinating the unceremonious dismissal of the man in charge of energy policy for the United States. It's hard to imagine a more dramatic change in direction, but Cheney doesn't remember spending much time reflecting on where he had been and where he was headed.

"I was too busy to worry about it, I guess. We had an awful lot to do. . . . You didn't have time to sit around and scratch your head and say, 'Gee, isn't this significant?' We were busy."[12]

One of Cheney's first responsibilities as Rumsfeld's deputy was to handle external politics for the White House, including outreach to the congressional campaign committees and the national party. Before he started full-time, Cheney had picked Bob Teeter to design and execute a polling strategy. Teeter had done polling for Ford back in Michigan and, more recently, had conducted national surveys for the Republican National Committee.

In these early days, the White House political operation consisted of five main players. In addition to Cheney and Teeter, there were Jack Marsh, a former Democratic congressman from Virginia, who had been a good friend of Ford; Dean Burch, a political adviser who started at the White House under Nixon; and

Foster Channock, a brilliant, hardworking recent college graduate who started as a gofer and came to be one of Cheney's most
trusted advisers.

Despite efforts to distance the Ford White House from Watergate and the Nixon administration, the midterm congressional elections, coming just three months after Nixon's resignation,
were a disaster for the Republicans. Already a minority in both
houses of Congress, Republicans lost forty-eight seats in the
House and five in the Senate.

The results were devastating for the White House. In a very
practical way, the postelection makeup of Congress would make it
extraordinarily difficult to govern. Democrats not only controlled
the congressional agenda; they had large majorities—in many
cases, by a proportion of two to one—on legislative committees.
Equally worrisome, the election outcome would worsen the already abysmal morale of the White House staff. In a meeting with
the senior White House staff on November 15, Ford implored
his team not to lose heart. The message, for a relatively conservative politician, was somewhat ironic. Ford told his staff that the
political fortunes of the GOP—even then a party that sought to
constrain government by noting its many failures—would be enhanced only by restoring confidence in the federal government.[13]

Three factors made that task nearly impossible. Ford had inherited an increasingly unpopular war in Vietnam; the economy
was in recession, racked by record-high inflation; and the White
House press corps was eager to find the next Watergate.

"Ford became the first victim of that new and deep journalistic cynicism toward government officials," said Ron Nessen,
Ford's press secretary. "And the press's contempt was increased
by the embarrassment of those correspondents assigned to cover
the White House on a regular basis. They were acutely embarrassed because they had failed to expose Watergate. It was un-
covered by Bob Woodward, Carl Bernstein and other reporters
who never went inside the White House gate."[14]

This was in many ways an understandable inclination, but it
was also deeply destructive. The most obvious manifestation of
these views was the daily White House briefing. John Hersey, a
novelist and contributor to the *New York Times Magazine,* cap-

tured the mood of those briefings in his short book, *The President,* a first-person account of a week he spent in the Ford White House with nearly full access to the staff and President Ford. "Ever since Watergate days, reporters' questions in these briefings have been searching, prolonged, often fierce—the sum of all the questions being: Does a President ever tell the truth?"[15]

The deep cynicism of the press corps meant that Ford's unfiltered public statements and speeches took on added importance. Shortly after Ford took office his staff began preparations for the State of the Union speech for 1975. Cheney had a leading role in preparing the remarks, which would focus on energy and economic issues. Ford was to meet his Economic Policy Board on December 21 to discuss the upcoming address, and Cheney sent him a three-page strategy memo prior to the meeting.

"The group needs to be encouraged to pull together a coherent set of policies," he wrote. "But they also need to devote as much time to finding ways to package the policies and sell them to the American people and the Congress as they spend developing the policies in the first place." Cheney was brimming with ideas: Ask Vice President Nelson Rockefeller to serve as the administration's lead spokesman; use cabinet and subcabinet officials as surrogates; reach out to the business community and interest groups to gain their support for the program; hold presidential press conferences and editorial board meetings; and place op-eds in newspapers and magazines throughout the country. "Finally," he advised, "you might want to remind the group not to discuss any of these matters with the press."[16]

Ford had compiled a relatively moderate voting record in the House, and Cheney thought he should seek opportunities to reach out to conservatives. This State of the Union message provided one such opportunity. The White House, Cheney thought, should reach out to leading conservative politicians and public figures to enlist them in the effort to win backing for Ford's proposals. He recommended that Alan Greenspan have lunch with Senator Barry Goldwater and that the White House let reporters know that prominent economists endorsed Ford's economic proposals. And Cheney believed that another prominent conservative, Governor Ronald Reagan of California, would be pleased that Ford had dropped plans for several proposals that would have given more power to the federal government. "Reagan was opposed to welfare reform, the new income supplement program, and national health insurance," Cheney wrote in his strategy memo. "We ought to get some points from him for having decided not to go ahead with those programs."[17]

Cheney also saw the speech as one of the first major opportunities for Ford to show presidential leadership in a way that would further separate him from his predecessor and continue the gradual process of restoring confidence in the federal government.

Those plans were dealt a severe blow on Sunday, December 22, 1974. The *New York Times* published an article that would not only prolong the steady erosion of trust in government but would also lead to the eventual reorganization of the U.S. intelligence community. Written by Seymour Hersh, one of the leading opponents of the Nixon White House, the story ran under the headline: "Huge C.I.A. Operation Reported in U.S. Against Antiwar Forces, Other Dissidents in Nixon Years." Ford read it aboard Air Force One as he flew from Washington to Vail, Colorado, for his annual Christmas skiing trip.

The allegations were explosive.

> *The Central Intelligence Agency, directly violating its charter, conducted a massive, illegal domestic intelligence operation during the Nixon Administration against the antiwar movement and other dissident groups in the United States, according to well-placed government sources. An extensive investigation by The New York Times has established that intelligence files on at least 10,000 American citizens were maintained by a special unit of the CIA that was reporting directly to Richard Helms, then the Director of Central Intelligence and now Ambassador to Iran.*

In addition, the article continued, a review of the CIA's domestic activities "produced evidence of dozens of other illegal activities by members of the CIA inside the United States, beginning in the nineteen-fifties, including break-ins, wiretapping and the surreptitious inspection of mail."[18]

Within hours of publication Henry Kissinger recommended that Ford require the CIA's director, William Colby, to provide the White House with a full report on the allegations. The CIA had forty-eight hours to respond.

Ron Nessen, on Air Force One with Ford, told the traveling press corps that Ford had spoken to Colby and that a report was expected within days. The specter of Watergate hovered over the exchange. One reporter said to Nessen: "We remember President Nixon saying, 'Nobody will tell me anything.' He ordered this investigation and that investigation and said never did the facts come to him."[19]

According to Colby, the CIA had studied American antiwar groups, beginning under President Johnson in 1967, in order to determine their links to foreign organizations and governments. In the process, the CIA had developed 9,994 files on U.S. citizens. Colby acknowledged that the CIA had "overstepped proper bounds" by penetrating these groups in order to determine their foreign ties, but his report disputed Hersh's characterization of the CIA's activities as "massive domestic intelligence activity." The error in Hersh's article, Colby wrote, was linking the study of antiwar groups with other, unrelated domestic activities of the CIA.

Those other operations, according to Colby's report and a subsequent presentation he provided to the Justice Department, included a wide array of questionable operations by the CIA: the CIA had conducted wiretapping and "personal surveillance" operations on reporters, including the columnist Jack Anderson and several members of his staff (such as future Fox News Channel anchor Brit Hume); the CIA detained a Russian defector at its facilities in the United States for nearly two years; the CIA's Office of Security broke into a business operated jointly by an employee of the CIA and a Cuban national, and—in a separate incident—an office of a former defector who was still on contract with the CIA; for nearly twenty years CIA officials intercepted mail at the Kennedy Airport mail depot as it was routed between individuals in the United States and the Soviet Union; CIA officials also intercepted mail between the United States and China for three years. Colby also disclosed that the CIA had developed plans to assassinate foreign leaders, too, although he insisted that the plans had not been carried out.

Cheney, as a congressional staffer and then a midlevel bureaucrat in the executive branch, had not previously worked on intelligence issues. Hersh's story provided quite an introduction. As Cheney's notes at the time anticipate, the article would have long-term implications for the future of American intelligence and for the relationship between the executive and legislative branches of the U.S. government. These changes would figure prominently in public debates thirty years later when Cheney, as vice president, worked with—and against—the intelligence community.

In handwritten notes dated December 27, 1974, Cheney reviewed the situation: "CIA engaged in domestic intelligence activities, contrary to its charter, illegal entry, compiled a data bank on U.S. citizens. Extensive congressional investigation anticipated. Expect review of Nixon tapes and papers to determine W.H. involvement. Debate will take place w/in the context of Watergate, the Ellsberg case, the Huston Plan, etc."

Cheney rarely wrote memos, but when he did his thoughts reflected an orderly mind. He defined the problem, reviewed possible solutions, identified related issues, and finally recommended a course of action. In this case, Cheney wrote, the White House response should have four objectives:

(1) Ascertain validity of charge that CIA has violated its charter and/or engaged in criminal activity.
(2) Ensure proper presidential posture; avoid being tarnished by controversy.
(3) Adopt adequate safeguards at CIA.
(4) Protect CIA from overreaction by Congress, which could inhibit their ability to perform their primary function.

That last point was a particular concern. Congress was likely to call for public hearings, he wrote, with potentially damaging consequences to the intelligence community, Ford's political standing, and the separation of powers. "This could in turn lead to serious harm to the CIA, charges the Pres. is weak and refuses to take the initiative, and to a serious legislative encroachment on executive power."

Cheney recommended that President Ford take decisive action by releasing the unclassified portions of the report prepared by the CIA—which were most of it—and by creating a special commission, backed by the White House, to investigate the alleged abuses. Doing this, Cheney argued, "offers the best prospect for heading off Congressional efforts to further encroach on the executive branch. It clearly demonstrates presidential leadership and a willingness to accept responsibility for putting our own house in order. It offers the best opportunity for convincing the nation that gov't does have integrity, that our institutions are sound, that the Administration is capable of governing."[20]

On January 14, 1975, Ford named Vice President Nelson Rockefeller to head a special commission to investigate the charges. But Congress was not placated. Since the creation of the CIA with the National Security Act of 1947, what had passed for congressional oversight of intelligence were informal meetings, often at a local watering hole, between senior intelligence officials and members of Congress with authority over the community's budget. That was about to change.

On January 27, the Senate created what would become known as the Church Committee to investigate the CIA's misdeeds. The scope of the fifteen-month investigation was wide, and the recommendations were comprehensive. The committee conducted more than 800 interviews, held hundreds of hearings, and collected more than 100,000 pages of documentation. "In carrying out its Senate mandate," wrote one observer, "the Church Committee conducted one of the most sweeping and intensive investigations in the history of the United States Senate."[21]

News reports on the activities and findings of the Church Committee meant that the operations of the U.S. government's most secret agencies were splashed across the front pages almost every day. The political battles over the CIA—which pitted the Ford administration against congressional Democrats and the antiwar left—heightened the tensions the new administration had wanted so badly to ease.

Efforts to discredit the CIA were numerous, and in some cases the tactics were as contemptible as the CIA's own transgressions. A story published in November 1974 in the *Washington Monthly* laid out for the world "How to Spot a Spook." And in 1975, Phillip Agee wrote a book, *Inside the Company,* in which he outed dozens of covert CIA operatives with the active assistance of some former CIA officials. A magazine called *Counterspy* published the name and identity of Richard Welch, the former CIA station chief in Lima, Peru. And though some of this information was publicly available before the article's publication, Welch's subsequent assassination in Greece caused a national uproar.[22]

The Ford White House sought to use the resulting controversy against the most aggressive congressional critics of the CIA. Administration officials, including the president, hinted that such outrages were a predictable consequence of the public airing of America's dirty laundry. The administration claimed that leaks from Congress were imperiling the work of the intelligence community.

Then in late May, Hersh published an article in which he disclosed the existence of a fleet of U.S. submarines located off the coast of the Soviet Union. The U.S. submarines were at the heart of a highly classified operation to gather intelligence on Soviet capabilities:

> *For nearly 15 years, the Navy has been using specially equipped submarines to spy at times inside the three-mile limit of the Soviet Union and other nations. The highly classified missions, code-named Holystone, have been credited by supporters with supplying vital information on the configuration, capabilities, noise patterns and missile-firing abilities of the Soviet submarine fleet.*

Critics of the program, Hersh wrote, thought such intelligence collection was too "provocative" and inconsistent with the goals of détente. "Many of the critics acknowledged that they agreed to discuss the operation in the hope of forcing changes in how intelligence was collected and utilized by the government."[23]

Top White House officials were furious. As he had in December, Cheney drafted a strategy memo outlining possible White House responses to the disclosures. Cheney's reaction to Hersh's "domestic surveillance" story had focused on its substance, not its origins; but this second story by Hersh triggered quite a different response. "Did anyone on the Hill have access to this information?" Cheney wondered.

Cheney once again organized his memo in outline form. The five-page handwritten memo had seven headings: "Goals," "Issues," "Alternatives," "Timing," "Questions," "Political Considerations," and "Options." He opened the document by asking, "What, if any, action should we take as a result of publication?"

He listed six items under "Goals":

(1) To enforce the law which prohibits such disclosures.
(2) To discourage the NYT and other publications from similar actions.
(3) To find and prosecute the individual in government who provided the information.
(4) To discourage others from leaking such information in the future.
(5) To demonstrate the dangers to nat'l security which develop when investigations exceed the bounds of propriety.
(6) To create an environment in which the ongoing investigations of the intelligence community are conducted w/o harming our intelligence capabilities.

He set forth five options the White House might take in response to the article:

(1) Do nothing—ignore the Hersh story and hope it doesn't happen again.
(2) Go quietly to the NYT—tell them we could prosecute, but would prefer a simple commitment from them that they would cease and desist.
(3) Start FBI investigation—with or w/o public announcement. As targets include NYT, Sy Hersh, potential gov't sources, Marchetti, et al.

(4) Seek search warrant to go after Hersh and remaining materials.
(5) Seek criminal indictments of one or more parties based on information now in hand.
(6) Seek contempt citation against ex-CIA employees for violating court orders on release of classified info.

Cheney also wondered about using the disclosure to fight off inquiries from Congress. "Can we take advantage of it to bolster our position on the Church Committee investigations to point out the need for limits on the scope of the investigations?"[24]

Ironically, the second story by Hersh wasn't news. It provoked very little public discussion, and an article published in the *Washington Post* months earlier had included almost all the same information.[25] An investigation of the leaks, Ford officials concluded, would bring unwanted attention to the intelligence operations. Plus, Cheney wrote in a follow-up memo to Rumsfeld on May 29, despite the breach in security, "the Navy believes operations *can* continue, repeat *can* continue."[26]

So they did nothing.

The work of the Church Committee culminated in six final reports making nearly 200 specific recommendations, many of which would limit the scope and freedoms of the U.S. intelligence community. The two committees that conducted investigations—the Church Committee in the Senate and the troubled Pike Committee in the House—became permanent oversight committees, giving Congress a new, formal role in the world of intelligence.

Ford assented, at least in part. He endorsed widespread changes in the U.S. intelligence community that included restrictions on domestic activities, a ban on assassinations of foreign leaders, and a reorganization of the intelligence agencies that was to enhance interagency coordination.

The controversies over intelligence during the early 1970s, says Cheney, "ultimately resulted in the passage of legislation that set up the current committee structure and reporting arrangements. But it also was, I thought, a further infringement on what up till then had been the president's prerogative in the intelligence arena."[27]

The result was a U.S. intelligence community subject to the political whims—and grandstanding—of Congress. And although even defenders of the CIA and related agencies recognized that some congressional oversight was necessary, the reforms produced lawyering and legislative second-guessing that would lead to near-paralysis of the intelligence community in the future. When, in the months before the first Gulf War, an Air Force general discussed the possibility of targeting leaders of the Iraqi regime in a war in Iraq, Secretary of Defense Cheney's strong rebuke included a warning that such "decapitation" strikes might violate the ban on political assassinations that dated from the Ford administration. By the late 1990s, the intelligence agencies were so cautious about their activities—and about taking blame for operations gone wrong—that they passed up opportunities to kill Osama bin Laden, a man who had declared war on the United States and made good on his threats by repeatedly attacking U.S. interests. There were other factors, to be sure, but the reforms of the mid-1970s would lead directly to intelligence failures twenty-five years later.

"I think they undermined our capabilities in some important respects," says Cheney. "The intelligence community is a lot like the military. It takes a very long time to train somebody who's capable, for example, of being a command master sergeant. A really first-rate NCO is the work of several years. You just don't do it overnight. And that's not the top of the heap, but a very important part of the process. And the military you would go to war with today is the one that was created many, many years in the past. And I think the intelligence community is a lot like that. It takes time for those people to gain the experience and move up and do what they're doing. If you look back at that period of the seventies—there was a time when we paid a significant price, we lost a lot of good people."

In response to the reforms, the intelligence community began to shift its priorities from human data collection, and its inherent errors of fact and judgment, to a reliance on national technical means. The result of the reform movement, Cheney believes, was a weaker intelligence community.

"I think it downplayed the value of human intelligence, the

importance of working with oftentimes seedy characters around the world, from an intelligence standpoint. The ability to get next to the world's worst—we needed to lie down with fleas sometimes. There was a sort of holier-than-thou attitude; I think part of it was tied to Vietnam, part of it was tied to alleged abuses that had occurred previously. We ended up reducing our ability to collect human intelligence, spies."[28]

If there was an area of the White House operation where Cheney's influence was relatively small, it was foreign policy. "Kissinger was such a huge figure," recalls Nessen. "He so dominated foreign policy that there wasn't room for many others. Cheney focused on the White House organization and on domestic and political stuff."[29]

Still, Cheney's role as Rumsfeld's right-hand man and a top adviser to Ford meant that he was in the room for some of the most significant foreign policy decisions of the Ford administration. In the fall of 1974, the Vietnam War had been a central issue in American politics for almost a decade. The war had brought an end to many political careers and given rise to many more. And more important, it had taken a ghastly toll on the lives and families of many young Americans. Nearly 60,000 died there, and those who returned were scarred for life, physically and emotionally.

Ford's arrival in the Oval Office ended, as he memorably put it, the "long national nightmare" of Watergate. In the country, and among the White House staff, there was a growing consensus that by bringing U.S. troops home from Vietnam he could end another nightmare. In an address he gave at Tulane University on April 23, 1975, Ford showed that he agreed: "Today America can regain the sense of pride that existed before Vietnam, but it cannot be achieved by refighting a war that is finished as far as America is concerned."[30] Five days later, Ford signed off on Operation Frequent Wind, the order that would bring the last American soldiers home from Saigon. Nessen described the scene in his White House memoir, *It Sure Looks Different from the Inside:*

On the way back to his residence, the president paused outside Rumsfeld's office. Ford, Rumsfeld, Cheney and I stood there silently, staring at the carpet, alone with our thoughts, unable to say anything appropriate. A spring rain pinged against the windows. Finally, the president started away.[31]

The war was over. Decades later it would not yet be history.

There was little need for Cheney to offer advice on foreign policy. When he agreed with Kissinger, his advice was redundant. And when he disagreed with Kissinger he was spitting into the wind—a young White House staffer who had never traveled abroad contradicting the oracle of U.S. foreign policy. These were fights Cheney was unlikely to win, so he rarely offered his views on international affairs.

One exception was the case of Alexander Solzhenitsyn. Solzhenitsyn had spent most of his adult life in the harsh concentration camps of the Soviet Union for criticizing Joseph Stalin in personal correspondence with a friend. In prison, he wrote works he thought would never be published. When they were, Solzhenitsyn's writings revealed to the West the full horror and depravity of Soviet-style communism. When *The Gulag Archipelago*, Solzhenitsyn's gripping account of Soviet prisons, was published in the West in 1973, he became a symbol of the struggle against totalitarianism. And in 1974, he was stripped of his Soviet citizenship and expelled from the Soviet Union.

The United States, meanwhile, wrestled with the proper approach to the Soviet Union. Both the Nixon administration and the Ford administration, under the guiding hand of Henry Kissinger, had pursued a policy of détente toward the Soviets. The United States' interests were best served, they argued, by working with the Soviets to reduce arms and decrease tension. At a series of summits, leaders of the two nations negotiated details of arms reduction treaties and worked out arrangements for the United States to sell massive quantities of discounted grain to the Soviets. Ford explicitly rejected a return to a "cold war relationship" with the Soviet Union.

But the tales of torture and cruelty in Solzhenitzyn's book made détente seem less like reconciliation than appeasement. Prominent conservatives outside the Ford administration argued that détente as practiced by the administration was tantamount to accommodation. By cooperating with the Soviet Union, they argued, the Ford administration was providing legitimacy to a corrupt and inhumane regime. The leading exponent of this view was Governor Ronald Reagan of California, an ardent anticommunist, who framed the conflict with the Soviet Union in moral terms. He became a tough critic of détente, Kissinger, and President Ford.

The criticism put Ford in a difficult spot. Ford wanted to answer this criticism, but Kissinger and his acolytes told him that embracing Solzhenitsyn could set back relations between the United States and the Soviet Union. Conservatives saw Solzhenitsyn as a symbol of everything wrong with Soviet-style totalitarianism and believed Ford could send a powerful message—to the Soviet Union and to the world—by receiving the dissident at the White House.

Cheney had strong views of his own and took the unusual step of recording them in a memo dated July 8, 1975, that he sent to Ford through Rumsfeld:

> My own strong feeling is that the President should see Solzhenitsyn for any one of the following reasons. . . . I think the decision not to see him is based on a misreading of détente. Détente means nothing more and nothing less than a lessening of tension. Over the last several years it has been sold as a much broader concept to the American people. At most, détente should consist of agreements wherever possible to reduce the possibility of conflict, but it does not mean that all of a sudden our relationship with the Soviets is all sweetness and light. . . .
>
> I can't think of a better way to demonstrate for the American people and for the world that détente with the Soviet Union . . . in no way means that we've given up our fundamental principles concerning individual liberty and democracy. Solzhenitsyn, as the symbol of resistance to oppression in the Soviet Union, whatever else he may be, can help us communicate that message simply by having him in to see the President. Seeing him is a nice counterbalance to all of the publicity and coverage that's given to meetings

between American presidents and Soviet leaders. Meetings
with Soviet leaders are very important, but it is also impor-
tant that we not contribute any more to the illusion that all
of a sudden we're bosom-buddies with the Russians.

[The Soviets] have been perfectly free to criticize us for
our actions and policies in Southeast Asia over the years, to
call us imperialists, war-mongers, and various and sundry
other endearing terms, and I can't believe they don't under-
stand why the President might want to see Solzhenitsyn.

Ford scrawled "GRF" in the upper right-hand corner, an
indication that he had seen it, but he was not persuaded. Solzhenitsyn, and all that he came to represent, would remain a
hotly contested political issue at the 1976 Republican National
Convention, two months before the general election.

Despite his ability to offer advice directly to the president,
Cheney kept a low profile. The Secret Service gave him the appropriate code name "Backseat." He refused the government car
made available to him as the deputy chief of staff, opting instead
to drive his old Volkswagen to the White House each day.

In a sense, the car was an apt symbol of the humbler parts of
his job, the day-to-day aspects of running the White House operation. In January 1975, he instructed the White House staff to
think of ways to get cabinet secretaries to the White House for
meetings so that they would feel more "plugged in."[32] Later, he
rewrote the travel rules for White House officials to cut back on
the overall travel budget, restricting the travel of low-level officials and insisting that anyone making less than $42,500 fly coach
class. When a White House fellow asked for a raise, Cheney
denied the request, saying, "If they're in it for the money, they
should be doing something else."[33]

Among his broader responsibilities was preparing for unexpected developments. On July 14, 1975, Cheney wrote a memo
to Rod Hills, the White House counsel: "We want to review
existing arrangements concerning President succession and in-
capacitation. You will remember the 25th Amendment of the
Constitution, I believe, provides special provisions for what happens in the event of an incapacity on the part of the President.

You should quietly dig into what currently exists and develop a paper on the subject which can go to the President."[34] Seven weeks later, the memo would seem prophetic.

On September 5, 1975, Lynette "Squeaky" Fromme, a follower of the cult leader Charles Manson, confronted Ford with a loaded handgun in Sacramento, California. She pointed the gun at the president, but before she could fire a shot she was restrained by the Secret Service. "An agent behind the President saw her pull a gun and disarmed her before she could fire. He jammed the web of skin between his thumb and forefinger down in between the hammer and firing pin so the pistol couldn't be fired," said Cheney. "The other agents swarmed all over President Ford, pushing him into a crouched position and literally carrying him away."[35] Fromme was eventually imprisoned for attempting to assassinate President Ford.

Less than three weeks later, on September 22, 1975, Sara Jane Moore attempted to shoot Ford during a visit to San Francisco. Oliver Sipple, a bystander who saw Moore point her gun at the president, wrestled her to the ground as she began firing. Cheney's memo to Hills began what would become an intense interest in the presidential succession and continuity of government.

The Cheneys gradually eased into life in Washington. They lived with their two daughters, now aged six and nine, in a modest house on Swords Way in Bethesda, Maryland. They had an active, if not robust, social life. As a senior White House official, Dick Cheney received invitations to parties and receptions that could keep him busy seven nights a week. He declined most of them, believing that White House officials who were serving the president well could not afford to spend time on the cocktail party circuit.

Lynne Cheney was teaching classes part-time at George Washington University and had begun writing for several local publications. She and Phyllis C. Richman wrote a guide to Washington's popular culture, which was published in the *Washington Post*. (Richman, then a freelance writer, would go on to a remarkable career as the *Post*'s food critic.)

Washington, they argued, has a distinctive set of preferences in pop culture. Washingtonians, according to the article, read

TV Guide, Playboy, and *Penthouse.* Their coffee was either Taster's Choice or Maxwell House. Beer preferences were broken down into subregional preferences: Georgetown residents drank Heineken; in southeast and southwest Washington it was Miller "ponies"; in northeast Washington, it was Pabst Blue Ribbon; and in northwest D.C. and suburban northern Virginia, Budweiser was most popular. The average weekly dining-out expenditure for a family with an income under $6,000 was $6.92; for those over $15,000 it was $10.18. At home, Washingtonians watched *Sanford and Son, All in the Family,* and coverage of the NFL on CBS.[36]

ON SUNDAY, NOVEMBER 2, 1975, the Washington Redskins played the Dallas Cowboys at RFK Stadium on the banks of the Anacostia River. On hand to see the afternoon game between these bitter foes was Donald Rumsfeld. In his absence, Dick Cheney would handle the duties of chief of staff, as he had done so many times before under the power-sharing agreement he had worked out with his boss. But this day would be different.

Cheney reported to the White House early. It wasn't uncommon for him to work on the weekend. In fact, he did so regularly, often taking his daughters to the White House with him as he finished long-term projects or otherwise caught up on his work. Cheney came alone this Sunday. He was one of four White House officials who had been involved in the planning of a series of high-profile personnel moves that would come to be known as the "Sunday morning massacre."

The changes were sweeping. James Schlesinger was out as secretary of defense and William Colby lost his job as director of the CIA. Henry Kissinger was forced to give up his responsibilities as national security adviser, though he would remain secretary of state. And Nelson Rockefeller announced that he would not be on the ticket when Ford ran for reelection in 1976. In a sense, this was the housecleaning that Ford had put off when he first came to office.

The carefully planned announcement of the changes—Rockefeller on Monday, then Colby and Schlesinger a day or two later—was dashed when reporters for *Newsweek* learned that they were coming and told the White House it planned to break the news. The White House scrambled to carry out the changes before the story hit the newsstands.

So Ford invited both Colby and Schlesinger to the White House, Colby at eight AM and Schlesinger at eight-thirty. Colby's dismissal was expected. Cheney and a few other top White House officials had been circulating names of possible replacements for him since early July. Cheney's list was short: George H. W. Bush; Robert Bork, who was then solicitor general; Lee Iacocca, president of the Ford Motor Company; Gale McGee, a Democratic senator from Wyoming; and Byron White, associate justice of the Supreme Court.

Colby's meeting with Ford was friendly. As is customary in high-profile firings in Washington, Colby was offered another position in the administration—in his case, as ambassador to NATO—as a face-saving gesture. He declined. On the way out of the Oval Office, Colby saw Schlesinger waiting to see Ford.

"What are you doing here?" Schlesinger asked.

"I've just been talking to the president about a couple of things," Colby replied, adding, "Good luck, Jim."[37]

As the two officials exchanged greetings, Cheney visited with the president in the Oval Office. He recommended that Ford provide Schlesinger some political cover, as he did with Colby, by offering him another job in the administration.

The relationship between Ford and Schlesinger had been strained for years. Schlesinger had a reputation for being abrasive, and Ford found him patronizing in their one-on-one meetings. But it was a personal affront that ultimately led to Schlesinger's dismissal. For several months, Ford had been pushing congressional Democrats without much success for an increase in the defense budget. The chairman of the Appropriations Committee and the Defense Appropriations Subcommittee, George Mahon, a Democrat from Texas, was an old friend of the president's from his days on Capitol Hill. Ford hoped that their personal relationship would allow them to resolve the budget impasse amicably.

But Schlesinger had other ideas. At a press conference about the budget on October 20, 1975, he criticized Democrats for blocking an increase in funding. Schlesinger decried the "deep, savage, and arbitrary cuts" in the defense budget and suggested that the reductions were "driven by political considerations."[38] Mahon was livid, and Ford, who did not anger easily, was furious that his defense secretary had insulted his good friend.

Cheney, knowing all this, still urged Ford to be politic. "Mr. President, you could soften the blow and offer him an ambassadorship or something like that," said Cheney, before he brought Schlesinger in to see the president.

Ford glared at Cheney. "Dick," he thundered, "get that son-of-a-bitch in here so I can fire him." It was as angry as Cheney would ever see the president.

The new members of the team—Ford called them "my guys"—were announced quickly. General Brent Scowcroft would be national security adviser, and a former congressman from Texas serving as U.S. ambassador to China, George H. W. Bush, would run the CIA. Ford wanted his chief of staff, Donald Rumsfeld, to take over the Pentagon. Ford asked his new pick for chief of staff, Dick Cheney, to persuade his predecessor to take the job.

"I went, in a matter of hours, from being Rumsfeld's understudy and his deputy to the guy who's trying to get him to do what the president wants him to do," says Cheney.[39]

Shortly after Ford fired Colby and Schlesinger, he left for Jacksonville, Florida, for a meeting with President Anwar Sadat of Egypt. Cheney, traveling with him, telephoned Rumsfeld from Air Force One to entice him to take the job at the Pentagon. Rumsfeld was reluctant, but Cheney knew him well and made a simple argument appealing to Rumsfeld's loyalty to Ford: "The president wants you to do it. It's an important job and we signed on to do everything we can to help the man. This is what he wants."[40] Rumsfeld agreed to take the job, and the Redskins won in overtime.

After returning to Washington the following day, Ford prepared for a prime-time news conference to announce the shake-up. At a mid-afternoon prep session, Ford's top staffers peppered their boss with possible questions. On one note card was an inquiry about the president's new chief of staff: "Who the hell is Richard Cheney?"

The press conference began at seven-thirty PM. Ford fielded forty-seven questions about the changes.

The flurry of news reports that ensued covered the drama from virtually every possible angle. Is Kissinger out of favor? Will Scowcroft do Kissinger's bidding at the NSC? Was Rumsfeld behind the changes? Did Rockefeller quit voluntarily? Why now?

Not one of the questions was about Cheney. It wasn't until the next day that reporters finally got around to asking about the new chief of staff. "Is the man capable of taking over those big shoes of Rumsfeld?" a reporter asked Ron Nessen at his daily press briefing.

"You bet he is," Nessen responded, "and the president would not have given him the job if he didn't think he was going to do it." He added: "He has been, I think you know, the deputy of Don Rumsfeld almost from the beginning, and under the deputy system at the White House he has been completely interchangeable with Don Rumsfeld. They do the same work. They have equal access to the president. They speak with equal authority both in terms of bringing information to the president and in terms of bringing information out from the president and parceling it out."[41]

When Cheney received a copy of the transcript, he bracketed the section that concerned his role and sent it back to Nessen with a note that, for Cheney, was positively florid. "Thanks Ron, Dick."[42]

Six weeks later, Cheney would repay the favor by publicly praising the White House press secretary as Washington debated whether Nessen had become a liability to Ford.

The coverage of Cheney's ascent to the top staff job was largely positive. "Richard B. Cheney, the man newly designated to be chief of the White House staff, may not have deliberately cultivated a passion for anonymity but few men have risen so high with so much anonymity," wrote Charles Mohr in the *New York Times*, on November 5, 1975. "In the network of other young administrators, 'issue men' and staff assistants in the capital, Mr. Cheney enjoys an unusually high reputation for competence and skill at the subtle art of getting things done for an important boss. The praise has also come from intellectuals who are outside official circles but still familiar with the workings of the White House." A colleague in the White House told the *Times* that Cheney was conservative, but added, "He doesn't stand up and salute on conservative issues."[43]

An article by Lou Cannon in the *Washington Post,* published the following day, noted, "Cheney is not considered politically ambitious. He is both an academic and a businessman, but his chief claim is thorough and unremitting staff work which has earned him a reputation for competence within the White House and on Capitol Hill." He is regarded as a pragmatist, wrote Cannon, "rather than as an ideologue." Cheney summarized his nascent political philosophy for the *Post:* "Basically, I am skeptical about the ability of government to solve problems, and I have a healthy respect for the ability of people to solve problems on their own."[44]

If his colleagues saw the new job as a natural step, Cheney's friends from "his beer and a western" days were stunned by the news. "All of a sudden, I found out he was the White House chief of staff," says Joe Meyer, who had shared Cheney's squalid basement apartment at the University of Wyoming.[45]

Jim Little, who lived with Cheney in Berkeley College at Yale, congratulated his old friend in a letter. "After hearing nothing about you since the summer of 1962 when I stopped in Casper, I nearly fell off my chair while reading *The New York Times* last week when you became front page news as White House chief of staff."[46] Little had several conversations about Cheney with the gang from Yale. "We all reminisced about the things we did during our days in New Haven (none of which were discussed in the *Times* profile and none of which will be divulged to jeopardize your political career)."[47]

In order to persuade Rumsfeld to serve as his chief of staff at the beginning of his administration, Ford had made the job a cabinet-level position. In one of his first official acts, Cheney asked that the position be downgraded, literally giving up a seat at the table in favor of a spot along the back wall. It was a move that stunned his White House colleagues and political Washington. "A clue to the character of Richard B. Cheney, President Ford's new chief of staff," wrote Dom Bonafede, of the *National Journal,* "is evident in his refusal to accept Cabinet status as one of the perquisites of his position." Bonafede asked Cheney why he refused the higher rank. "It's just that I believe there should be a distinction between the White House staff and the Cabinet," he said. "I feel strongly about it and therefore I declined the President's offer."

Cheney's colleagues immediately sensed that under the new chief of staff the White House would be different. Rumsfeld had been a hard-driving, combative chief of staff. His fights with others in the West Wing—particularly with Ford's longtime adviser Bob Hartmann and Vice President Nelson Rockefeller—were well known. And Rumsfeld was ready with a sharp rebuke for those who did not meet his high expectations. "He was the toughest boss I ever had," says Cheney. "He could be very tough in his basic mode of operation. If you did a good job on something, you got more work. He'd just keep pushing it at you, pushing it at you."[48]

His criticism, as in his "snowflakes" during the Nixon administration, was often pointed and personal. Many of Rumsfeld's White House colleagues believed he had his sights set on bigger jobs. He seemed to enjoy the public aspects of being chief of staff. Rumsfeld, after all, had been an elected official in his own right, and many thought he wanted to run for president himself one day.

White House officials quickly noticed the differences between the Rumsfeld White House and the Cheney White House. "When Rumsfeld was chief of staff, there was something about him that raised people's hackles, something that made people feel competitive," Nessen recalls. "Under Cheney, people didn't feel competitive or threatened."[49]

Cheney, though equally demanding, was in all other respects Rumsfeld's opposite. His manner was less rigid. Few of his subordinates can remember him ever raising his voice. More than anything else, he sought to avoid the spotlight. Shortly after getting the job, Cheney told the *Washington Post,* "I really do think that a staff man should be anonymous."[50]

Cheney later noted with evident pride that he had managed to scuttle a newspaper profile of him. "At one point the 'Style' section of the *Washington Post* wanted to send a team out to take pictures of my family and my wife and myself. It sounded awful, so I said no. *The Post* did a profile anyway, but it was about the way the President worked with his staff and what he was trying to do, not about Dick Cheney the man. I had made a very determined decision to keep my head down, and I stuck to it. I thought I could get a hell of a lot more done if I was not a public figure."[51]

Because Cheney had been working interchangeably with Rumsfeld for more than a year, his role remained largely the same. The only real difference was that Cheney did not hire a strong deputy, so all the responsibilities of the job fell to him.

His most important function came as manager of the policy-making process. As full chief of staff, Cheney would determine who got to see the president, when these meetings or phone conversations would take place, and how long they would last. His role as gatekeeper had significant policy implications. Whoever had the most face time with the president had the most influence on his agenda.

With Rumsfeld across the Potomac at the Pentagon, Ford came to rely even more heavily on Cheney to handle issues big and small. Ford frequently dashed off notes to his top aide in his choppy cursive writing. One note read simply: "Dick, Haircut today or Friday AM." Another concerned transportation from a golf outing to a popular concert venue: "Dick Cheney, Glen Campbell will ride out to B.T. [Burning Tree Country Club] with me but I will change at B.T. to 'black tie' so Glen C. will need transportation from B.T. I will go directly from B.T. to Wolf Trap."

Still another included a clipping from the gossip column in the *Washington Star*. The column's first item was about Secretary of the Treasury William Simon, a strong conservative who had given a speech expressing concern about the growth of the bureaucracy under his watch. The second concerned an adviser of Ford's and a close friend of Cheney's. "The economist, Alan Greenspan, chairman of the President's Council of Economic Advisers, confirmed to an interviewer this week, in a rare burst of clarity, that he has been 'going out with' Barbara Walters. It has not been determined if the relationship is 'serious,' whatever that may mean. Greenspan is 50 and was once married. Walters, 45, was recently divorced. They recently attended a Palm Springs party."

Ford wrote: "Dick Cheney, Note p. 2. I don't believe it." He didn't specify which item he found so incredible.

Other notes from Ford to Cheney were more substantive. "Anne Armstrong—Supreme Ct.—talk with me." Or: "Dick Cheney, Status of nuclear program?"[52]

When he wasn't busy determining the status of the nuclear program or arranging Ford's haircuts, Cheney spent a lot of time with the White House press corps managing what he would later call "the politics of perception." Even as he personally sought to maintain a low profile, Cheney devoted hours of each day to the "show business" side of running the White House. These efforts involved more than just the press office.

Although the press office accounted for only 10 percent of the White House staff, Cheney said, senior officials in the Ford administration spent a considerable amount of their time on communications strategy. Cheney would have preferred that his staff spend its time on the substance of governing. But he recognized, however grudgingly, that communicating these policies was an important part of his job.

> *It's important to be able to communicate and convey, but so much of our activities and our efforts were just dominated not with quote, "policy decisions," but with policy decisions with everything wrapped around that and the question of what its impact would be on the public, how it would be perceived, how do we get the nets to cover it, what kind of coverage will they give it, what time of the day should we do it, what program are we going to knock off—is it going to be* Bonanza, *is it going to be* Police Woman *or a football game. . . . The fact of the matter is if you don't try to manage the news and if you don't have an awful lot of resources internally devoted to that question of what the viewers will see on the tube in the evening, there's no way you can begin to be effective in terms of the policies.*[53]

Cheney had good relations with the media and considered several White House reporters his friends. Journalists generally trusted Cheney because they knew that he had been involved in—or at least privy to—virtually all of Ford's decisions. Although he kept a low public profile, he frequently briefed on background to explain decisions and the process that led to them. "Everybody recognized how smart Cheney was," says Nessen. "But also, here's a young guy drinking with reporters on the plane and taking part in practical jokes."[54]

A rare exception to Cheney's general principle of anonymity came on Sunday, January 4, 1976, when he appeared on the CBS News show *Face the Nation.* Cheney sat in front of a bright blue backdrop and spoke into a large microphone resting on a wood-paneled desk in front of him. His light brown hair was receding but, with some creative combing, still managed to cover most of his head. He looked fit in a black pinstripe suit, white shirt, and red tie. For twenty-five minutes he answered—and in some cases parried—pointed questions from three veteran White House reporters.

The show was a policy wonk's dream. Many of the questions focused on unemployment statistics and budget projections. Cheney handled the questions with ease and projected an air of confidence well beyond his thirty-four years.

When Tom DeFrank of *Newsweek* accused Ford of playing politics with the budget, Cheney politely but firmly recast the question. "It's not just a potent political issue," he said. "I think that misses the point to some extent." Questions about the budget are fundamentally questions about the size and scope of government, Cheney explained, and Ford prefers small government.[55]

His answer was relaxed and authoritative, almost professorial, but without any hint of condescension. He spoke in measured tones, as if the information he was sharing was obvious and his job was simply to reacquaint the viewers with the wisdom of common sense.

Cheney's answers on *Face the Nation* were substantive and often insightful, if not always candid. In the weeks leading up to this appearance, Ford had been the subject of dozens of news stories suggesting he was clumsy. Television cameras were rolling when

Ford tripped down the stairs from Air Force One as he disembarked on a trip to Austria. In a second incident, photographers caught him knocking his head on the doorway as he exited a helicopter. For the media it was a field day, an opportunity to lampoon the president as a klutz. On a popular new television show, *Saturday Night Live,* the actor Chevy Chase made a name for himself by impersonating Ford, tumbling down a flight of stairs at the beginning of the show or, later, bumping into a desk. But Cheney and his staff were deeply concerned that the image of Ford as a bumbler would affect public perceptions of his job performance.

The situation had come to a head in December 1975, days before Cheney's appearance on *Face the Nation.* The timing was awful. Earlier that month, Ronald Reagan had announced that he would challenge Ford for the Republican nomination. And the New Hampshire primary was less than six weeks away.

Ford traveled to Vail for his annual Christmas skiing trip. At the end of one run, photographers from the White House press pool captured him as he took a spill. The image was carried in newspapers across the country and even made the evening news. It was a political moment of high absurdity. Ford, of course, had skied down the challenging Vail slopes dozens of times on this trip without a fall. Having been a boxing coach at Yale and a standout football player at the University of Michigan, he was without question one of the most athletic presidents in American history. And yet the White House had to call a communications strategy session to deal specifically with the perception that he was hopelessly uncoordinated, maybe even a danger to himself and others. It must have seemed funny for those who didn't have to deal with it. Cheney, who did, was angry about the coverage, and he privately shared his frustration with reporters. He told the White House press secretary, Ron Nessen, to do the same.

Now, as the White House staff scrambled behind the scenes to neutralize the damaging publicity, Cheney shrugged off the stories on *Face the Nation,* saying that were "irrelevant" and of little concern.

"Mr. Cheney, how concerned are you about all of the stories and cartoons that portray President Ford as sort of a bumbler?"

asked Phil Jones of CBS News. "News secretary Nessen says the staff is quite concerned that this is going to cost him votes. What is your feeling about this?"

"I don't think it'll cost him any votes. I guess I start with the conviction in my own mind, based on my own personal first-hand knowledge of the president, that the stories aren't accurate. Secondly, it's political season. It's time for those sorts of things to occur. I guess basically I'm not that concerned about it because I don't think the American people take it that seriously. It seems to me to be one of those irrelevant considerations that pops up from time to time and then fades fairly fast. . . . I think the American people have always had a bit of a sense of humor about their political leadership. I know the president doesn't take it that seriously and I don't either."[56]

Despite Cheney's efforts to calm the waters, the coverage continued. So, in order to demonstrate his competence to inside-the-Beltway reporters, Ford's staff devised a plan to send the president out to do a press briefing on the budget—a subject he knew cold. The Ford briefing on January 20, 1976, Cheney would later recall, "was just a masterful piece. It wasn't the country at large so much as it was directed directly to the Washington press corps. At that point it totally put to rest the whole question we then were faced with politically which is the bungler, stumbler, hit-your-head-on-the-helicopter type image. We did it very deliberately."[57]

It was part of a broader effort to make Ford look presidential. The next step was the State of the Union address. The responsibility for drafting the speech fell to Bob Hartmann, a prickly speechwriter who had come with Ford from Congress and served as chief of staff when Ford was vice president. Cheney received Hartmann's first draft on January 12, exactly one week before the address, and provided his feedback in a handwritten note on yellow legal paper the following morning. "This SOTU must be visionary and above all presidential. Tone and style are at least as important as substance."

And Cheney didn't like the substance. "As you know, it's too long. I wouldn't let the departments force us to use everything. The foreign policy section is too tough on the Congress and not tough enough on the Russians. . . . It needs a unifying theme or conceptual framework, which is currently lacking."[58]

The irascible Hartmann didn't take criticism well. Because he had been Ford's top congressional aide, he was overly conscious—some would say paranoid—about his status in the West Wing. He brimmed with pride of authorship and was known to fight even the smallest changes to his drafts.

With the speech only days away, Cheney thought there wasn't time to do battle with Hartmann. So he assigned a small team of Ford's aides—including Greenspan and the speechwriter David Gergen—to write an alternative address. In addition to making Ford look presidential, the speech by Cheney and Gergen sought to bolster Ford's credentials as a conservative, an attempt to deflect charges from Reagan that Ford was too moderate and lacked conviction.

Ford himself invited both teams of writers to sit with him in the Oval Office as he read through the drafts, cutting and pasting sections into a messy amalgam. The resulting speech was adequate, but not great. It was not the strong statement of principles that Cheney had envisioned when he commissioned Gergen's team to come up with an alternative, and it was not the sort of soaring rhetoric likely to inspire Republicans and lead to a series of quick, decisive primary victories.

On February 24, 1976, Ford scraped out a victory over Reagan in the New Hampshire primary, winning by a mere 1,587 votes out of more than 100,000 cast. The narrow margin was telling. After Ford won the New Hampshire and Florida primaries, Reagan responded with important victories in North Carolina and Texas. It quickly became clear that this would be a state-to-state fight, and that neither candidate would clinch the nomination before the convention in mid-August.

Ford was in trouble as the 1976 Republican National Convention approached. The Democrats had nominated Governor Jimmy Carter of Georgia at their convention in July, and Carter left New York City with a thirty-three-point lead over Ford in the Gallup poll.

Although Ford had more delegates heading into the convention, the nomination would be won or lost in Kansas City.

Cheney thought that another big battle at the convention could effectively end any chance Ford might have at being reelected. Cheney believed that even if his boss emerged as the nominee, the damage from a divided convention carried on national television would be fatal. One way to avoid such a fight, Cheney thought, was to put Ronald Reagan on the ticket as Ford's running mate. A potential Ford-Reagan ticket had been a subject of speculation in the press throughout the primaries. The talk wasn't serious in the Ford camp until Cheney quietly asked Bob Teeter to conduct some polling on possible vice-presidential candidates.

The polling confirmed his suspicion: Reagan beat the other potential candidates handily. Without telling anyone on the campaign staff or at the White House, the two men made a secret trip to Camp David to see the president. Their goal: to persuade Ford to choose Ronald Reagan as his running mate.

It didn't work. The primary battles had been tough and at times personal. Ford was in no mood to bring his erstwhile foe into his campaign.

"I still believe to this day that if we had been able to put that together we might well have won the '76 election," says Cheney.[59]

Throughout the primaries, Cheney had gradually become more involved in the day-to-day decision making of Ford's re-election effort. In March, Democrats in Congress alleged that the campaign chairman, Bo Callaway, the former secretary of the army, had used his position to obtain favorable real estate deals in Crested Butte, Colorado. Callaway insisted on his innocence, and a subsequent investigation would prove him right. But the charges presented a political problem for Ford, who had worked for nearly two years to demonstrate to voters that he was a man of integrity. Callaway would have to resign, and Cheney would have to tell him so. Peter Kaye, a press aide to Ford who attended their breakfast meeting, described it later: "Cheney played him like a guy landing a fish. Bo was flopping around, then lying there for awhile, amid long, awkward silences, and then would start flopping around again. Finally, Bo asked, 'What should I do?' 'Better get a lawyer, Bo,' Cheney told him."[60]

With the departure of several senior campaign officials, Ford came to rely even more heavily on his young aide.

In Cheney's case, that power did not translate into fame. Two years after Nixon resigned, echoes of the scandal that brought him down were everywhere. The post-Watergate campaign finance reforms left the Ford campaign strapped for cash after the primary season, so Cheney and his assistant, Foster Channock, had to finance their own travel to Kansas City. On a layover in Peoria, Illinois, Cheney stopped at a coffee shop for a snack. As he ate, he noticed a man across the dining area looking at him intently. Minutes passed and the man continued to stare. Finally, he walked to Cheney's table with the look of someone who had just had an epiphany. "I know you," the man said. "You're John Dean." Even if Cheney hadn't succeeded in forestalling a convention fight by convincing Ford to pick Reagan, he had managed to become one of the nation's most powerful political players while remaining nearly anonymous outside the political press.[61]

"By the time of the Republican convention," wrote Michael Medved in his study of presidential chiefs of staff, "his control over all major campaign decisions paralleled his absolute dominance of the White House staff."[62]

Through the spring, Ford and Reagan battled to win over enough delegates to win the nomination. Ford had done well in the early primaries, but as the campaign wore on, the Reagan camp began to focus its efforts on détente with the Soviet Union, the Ford administration's policy on the Panama Canal, and the man most responsible for those policies: Henry Kissinger.

"They'd started pounding away in Florida as I recall on Henry and on the canal," says Cheney. "And by the time we got to North Carolina it took—they got traction and they beat us in North Carolina. All of this ultimately culminated in a platform fight in Kansas City."[63]

At the convention, Reagan's supporters, led by Senator Jesse Helms of North Carolina, pressed for a plank in the GOP platform calling for "morality in foreign policy." The proposed plank was a subtle but unmistakable repudiation of Ford's foreign policy and its chief proponent, Kissinger.

Late on the night when the delegates would vote to approve the platform, Ford's top advisers gathered in the presidential suite at their hotel. "The debate was over whether we should fight this platform plank that Helms was peddling," says Cheney, "which didn't mention Henry by name but did everything but scalp him and draw and quarter him at dawn."[64]

"We must face the world with no illusions about the nature of tyranny," the plank stated, a clear allusion to Kissinger's willingness to deal with the Soviet Union. It concluded: "Honestly, openly, and with a firm conviction, we shall go forward as a united people to forge a lasting peace in the world based on our deep belief in the rights of man, the rule of law, and guidance by the hand of God."

Kissinger, not surprisingly, wanted to fight the plank. Nelson Rockefeller supported his old friend. Cheney, however, thought there was very little advantage to be gained by engaging with Reagan's forces. "Nobody ever reads the platform anyway," he says.[65]

There was an additional complication. Earlier in the evening, Ford had won a critical debate over rules. Rule 16c, had it passed, would have required Ford to name his running mate before the convention selected its nominee. The rule was pushed by Reagan's supporters to mitigate the damage Reagan had done by announcing that Richard Schweiker, a liberal Republican from Pennsylvania, would be his running mate if he were nominated. Reagan had wanted to balance the ticket to broaden his appeal. Instead, he angered conservatives who believed that Reagan's main selling point was his uncompromising conservatism. By forcing Ford to choose a vice presidential candidate before the final vote on the nomination, the Reagan camp hoped to create another issue to use against Ford.

When rule 16c was defeated, many of Ford's supporters, thinking that this victory had effectively won Ford the nomination, "scattered to bars all across Kansas City," recalls Cheney. Reagan's delegates, meanwhile, had stayed behind, anticipating a fight on the "morality in foreign policy" plank. No matter how hard they fought, Cheney argued, they could not win; they were outnumbered. "Principle is OK up to a certain point," he would say. "But principle doesn't do any good if you lose the nomination."[66]

Kissinger was adamant and at one point said he would resign

if Ford refused to fight the plank. It was a threat that didn't sit well with Tom Korologos. Korologos, the former aide of Nixon's who had written, "Do you believe this crap?" on a note mistakenly delivered to Cheney, had recovered from this early misstep to become one of Ford's top advisers. And he had had enough of Kissinger. "For chrissakes, Henry, if you're going to quit, do it now! We need the votes."[67]

Ford ultimately decided to let the plank pass without a fight. Kissinger didn't resign.

It was typical of the behind-the-scenes scrapes that characterized the 1976 Republican Convention. And it wasn't the only time Cheney found himself on the opposite side from Vice President Nelson Rockefeller. The strains of a nomination fight raised the long-submerged personal tensions among a number of Ford's top advisers, and this was particularly true of Cheney and Rockefeller.

From the beginning of the Ford administration, Rockefeller had often clashed with Rumsfeld and Cheney. When Ford first came to office, he had chosen Rockefeller, the former governor of New York, from a short list of possible vice presidents that included George H. W. Bush and Rumsfeld. Ford had spent nine months as vice president without any significant policy responsibility, and he vowed that he wouldn't do the same thing to Rockefeller; his new vice president, he said, would be a "full partner" in running the country. After consulting with Rockefeller, Ford agreed to give him wide latitude to fashion the administration's domestic policy agenda. That responsibility included full control over the Domestic Council, at the time the chief mechanism for domestic policy inside the White House.

It was a division of labor destined to create problems. Ford had inherited a troubled economy, and one of his first decisions as president was to implement a policy of "no new starts," a plan that prohibited any new government programs. (Energy was exempted from the ban.) Nobody loved new government programs more than Nelson Rockefeller. He was the archetypal liberal Republican, so identified with big government that his name came to describe a particularly spending-happy species of the GOP genus, the "Rockefeller Republicans."

Shortly after he came to the White House, Rockefeller and his staff set to work on a range of bold new programs and policies. Characteristically, these programs called for an activist role for the federal government, running counter to Ford's "no new starts" and, to some extent, his basic philosophy of government. Frustrating the policy proposals of the vice president became a significant part of Cheney's job in the Ford White House. Rockefeller, said Cheney:

> ... *would periodically produce these big proposals and he'd go in for his weekly meeting for the president and oftentimes give him these proposals. At the end of the day I'd go down for the wrap-up session and the president would say: "Here, what are we going to do with this?" And I'd say, "Well, we'll staff it out." So I would take it and put it into the system. It would go through OMB and it would go to the Treasury and all of the other places that had a say in his Council of Economic Advisers. Of course the answer would always come back, "This is inconsistent with our basic policy of no new starts," so it would get shot down.*[68]

He would later describe this role as putting "sand in the gears." The phrase "we'll staff it out" quickly became a euphemism for killing one of Rockefeller's projects.

Understandably, Rockefeller came to blame Cheney (and Rumsfeld before him) for his frustration. In their private discussions, President Ford would seem receptive to—if noncommittal about—his grand plans. But once Cheney got involved, his proposals would die a slow death.

Rockefeller was ever mindful of the problems created by Ford's chiefs of staff. In 1974, Congress ended the tradition of U.S. vice presidents living in their own homes by providing living quarters on the grounds of the U.S. Naval Observatory in northwest Washington, D.C. Nelson Rockefeller never moved into the residence, because, as Cheney puts it, "he had a bigger house up on Foxhall Road." But he used the official home extensively to fete the Washington establishment. "They had a whole series of parties for the press, for Congress, for the administration, for the cabinet, so forth," Cheney recalls. "And about the only people in town who were not invited were me and Lynne. Never got an invite."[69] Cheney would, of course, spend plenty of time in that house decades later.

These simmering disputes boiled over at the Republican convention in 1976. The first problem arose when the convention's choreographers began to plan the onstage celebration that would follow Ford's acceptance speech. An incumbent president is typically joined immediately after his speech by the vice president. But since Rockefeller was not running with Ford the convention planners agreed that the vice presidential nominee, Bob Dole, would be the first to join Ford onstage. Rockefeller saw the decision as an affront and blamed Cheney.

Things worsened from then on. Rockefeller gave the speech nominating his successor. The text was precisely crafted to allow Rockefeller to make a gracious departure from the national political stage. But when he delivered the remarks at the convention, the microphone cut out in the middle of his speech.

After the speech, a furious Rockefeller confronted Cheney underneath the podium and accused him of deliberately sabotaging his grand exit. "He thought that I had arranged to kill the sound system," Cheney recalls. It was the most intense of several confrontations—Cheney calls them "shouting matches"—between Ford's chief of staff and his vice president.

"You've got to watch vice presidents," says Cheney some thirty years later. "They're a sinister crowd."[70]

The Republican convention had been a mixed bag. Ford's acceptance speech was widely praised as serious and substantive, and his delivery as strong. But months of fending off the challenge from Reagan—culminating in the fight at the convention—had taken a toll. Ford left Kansas City still trailing Jimmy Carter by a wide margin.

After the convention, Ford went to Vail to recharge for the general election. A small team of advisers, headed by Cheney, went with him. The members of this small group—Cheney, Bob Teeter, Stu Spencer, Michael Raoul-Duval, and Foster Channock—were known as the "Cheney gang," and they would run the campaign for its final two months. Few of them had experience on a national campaign.

Several people on Ford's senior staff approached the president, warned him about banking on such a small, inexperienced team, and asked for a larger role in the campaign. Ford heard them out but ultimately set aside their concerns. He would win or lose with the Cheney gang.

Predictions from the professionals were not encouraging. "The Republican ticket of President Gerald R. Ford and Sen. Bob Dole comes out of this cheerless GOP National Convention (with only 75 days until Election Day) even further behind the Democratic ticket of ex-Georgia Gov. Jimmy Carter and Sen. Walter Mondale than Sen. George McGovern was behind President Richard M. Nixon at this stage in 1972," wrote Rowland Evans and Robert Novak. "Although this huge gap is bound to narrow by November 2nd, it must be pointed out that no candidate in modern memory has every closed this large a margin."[71]

Although presidential debates do not typically determine the outcome of elections, Ford's team knew that the president could not afford any weak performances. Preparation for the debate, led by Cheney, had begun the previous spring. At the first session, Cheney and three other advisers asked questions of President Ford. The chief of staff tried asking broad questions meant to encourage the president to discuss his general principles and to talk about his qualifications for the presidency.

"We might just want to start out on the basis of how your perspective on the presidency, and your job as president, has changed, from what it was, for example, when you were in the House, how it looks now almost two years after you've been in office," Cheney said.[72]

Cheney thought the first debate went well, and he was optimistic heading into the second debate on October 6, 1976. The questions would focus on foreign policy and national security, areas in which Cheney and the rest of Ford's team believed that the president would have a distinct advantage over the governor of Georgia. But a gaffe by Ford, followed by his stubborn refusal to correct his mistake, would slow his growing momentum and leave the campaign in turmoil for several days.

With his second question of the evening, Max Frankel of the *New York Times,* one of three journalists on the panel, told Ford he wanted to "explore more deeply" the subject of U.S.-Soviet relations. He put the president on the defensive by suggesting that his signing of the Helsinki Accords in 1975 in effect signaled America's acceptance of the Soviet Union's "dominance" over eastern Europe. Ford concluded a meandering response with a preposterous claim: "There is no Soviet domination of eastern Europe, and there never will be under the Ford administration."

In reality, the Soviet Union's domination of eastern Europe was beyond dispute; the Soviets had thousands of troops in Poland at the very moment Ford uttered those words. Frankel, stunned, interrupted Governor Carter to give Ford an opportunity to clarify his answer. Ford failed to take it.

Cheney and Brent Scowcroft, Ford's national security adviser, returned to their hotel, the St. Francis, to brief reporters after the debate. As Cheney walked into the ballroom, Lou Cannon of the *Washington Post* shouted after him: "Hey, Cheney! How many Soviet divisions are there in Poland?"[73] We're in trouble, Cheney thought.

After the debate, Ford was unapologetic about his answer, and his advisers hustled to downplay any controversy. But the staff traveling with the president began to field concerned calls from campaign headquarters and the White House. The callers all said the same thing: Ford must clarify his answer. Cheney had the unpleasant duty of telling the president that he had made a bad mistake, which would only get worse without further explanation. Ford rejected his advice, so Cheney called in the campaign adviser Stu Spencer to help persuade Ford to clean up the mess. "He threw both of us out," says Cheney.[74]

Ford twice attempted to end the controversy without actually admitting his mistake. The attempts didn't work, and Ford's intransigence became part of the growing story. The questions kept coming. Finally, before a rally at the city hall in Glendale, California, Cheney cornered the president and implored him to bring the affair to an end.

As he walked out, I was right behind him. And I said to him: "Now, Mr. President, do you have it firmly fixed in your mind what it is you want to say to the press?" And he spun around on his heels and he jabbed me in the chest with his finger and he said: "Poland is not dominated by the Soviet Union!" I just about died and thought: "This is not going to work." And then he laughed like hell and he went out and delivered his lines perfectly and cleaned it all up.[75]

Like his boss, Cheney maintained a sense of humor even in the campaign's tense final days. Despite his gaffe in the debate, Ford had managed an extraordinary comeback, narrowing Carter's lead to single digits just two weeks before voters would go to the polls on November 2.

As they traveled throughout the country on Air Force One, Ford's senior staffers had a pool to predict how many electoral votes their boss would win. Cheney made his own guess and then submitted another entry on behalf of Teeter, who, according to this entry, believed Ford would win in a landslide. He was alone in making such a bold prediction—Ford winning 371 electoral votes and thirty-six states—and it didn't take long for journalists traveling with the campaign to learn about the pool and to begin asking questions about the sources of Teeter's confidence. Cheney confessed to the prank on learning that some members of the traveling press corps were actually preparing stories on the confidence of Ford's pollster.

Ford's entire team was exhausted by the time Election Day arrived. By focusing the national discussion on Carter's fitness to be president—not Ford's—the team had managed in two months to erase a thirty-point deficit. The race was essentially a dead heat. Voter turnout would determine the outcome. Mindful of the political maxim that Republicans are more likely than Democrats to go to the polls in bad weather, the Ford campaign prayed for rain. Ford had lost his voice and struggled to speak at his final campaign appearances. Eleven months of intensity ended with one final stop in Grand Rapids, Michigan, Ford's hometown. Cheney remembers it as an emotional visit marked by a foreboding, cloudless sky:

We all went to breakfast with him. It's just the way he'd traditionally done election days. And then we all went out to the airport and the locals surprised him with a beautiful mural—all one wall inside the airport—of his life. It was a very emotional kind of moment. His mother and father were pictured there. They were both dead by then. He started to cry. Betty started to cry. The press is crying. Helen Thomas was there. She's crying. It was sort of the end, the culmination, if you will, of the campaign. And we'd had a great rally the night before in Grand Rapids—brought the local boy home, his presidential campaign.

Then we got on a plane and flew back across Michigan to Washington. And it was one of the most beautiful fall days. There wasn't a cloud in the sky. And all those Democratic precincts—you started to know it's going to be a problem in terms of turnout.[76]

On Election Day, as it had been for much of the race, Cheney's office was the unofficial campaign headquarters. Ford's top advisers gathered there periodically to receive updates on voter turnout, read the latest wire stories, and compare state-by-state results.

Throughout the evening Cheney and Bob Teeter passed along the latest numbers to Ford, who was watching returns in the White House residence with friends and family. For most of the night, Cheney and Teeter were the bearers of bad news.

But despite some early setbacks Ford's prospects began to brighten as November 2 turned into November 3. Not long after midnight, Oregon, which had already been counted for Carter, was reversed and placed in Ford's column. It was only a temporary boost.

Cheney reviewed the returns with Nessen and David Gergen until almost sunrise. He took a short nap and met with his senior staff at eight-thirty AM Wednesday to examine the vote counts one last time. But it was a formality. Most media organizations had called the election for Carter. The race was over. Carter had won.

Shortly after nine AM, Ford came to the Oval Office. He had

known when he went to bed that his prospects were not good. And a few minutes before he arrived at his desk, the White House photographer had offered his regrets on the loss. Moments later, Cheney, alone in the office with Ford, made it official: "Mr. President, we lost."[77]

Ford called president-elect Carter. "Governor, my voice is gone," Ford rasped, "but I want to give you my congratulations. Here's Dick Cheney. He will read you my concession statement." Cheney read the prepared remarks and hung up the phone.[78] Later, Betty Ford would read the statement prepared for the media and the nation.

It had been a stunning comeback, one of the greatest in U.S. political history. The fact that the final two months of the campaign were coordinated by a group of relative neophytes made the effort even more remarkable—and probably also made it possible. The collective lack of experience allowed the group to believe, against all evidence, that victory was possible.

Exhausted and disheartened, Cheney made a long list of all the things that might have been done differently to produce a victory for Ford. And then he tore it up and threw it out.

Cheney wasn't so spent that he couldn't engage in a bit of postelection mischief. Jim Naughton, the White House correspondent for the *New York Times,* had spent much of the last year masterminding elaborate practical jokes to be played on Ford's staff and his own colleagues in the press. He was overdue for some retaliation.

Naughton had once showed up at a presidential press conference dressed as the San Diego chicken. He convinced fellow reporters (fraudulently) that they had been chosen as panelists for a presidential debate. Worst, perhaps, was his prank on *Newsweek*'s correspondent Tom DeFrank. At an overnight campaign stop in Peoria, Naughton persuaded Ron Nessen to lure DeFrank to the hotel bar with the promise of a juicy scoop for the magazine's "Periscope" section. While DeFrank waited in the bar for a meeting that would never happen, Naughton herded sheep into his room. (The sheep were a reference to persistent rumors about the lengths to which students at DeFrank's alma mater, Texas A&M, would go for companionship.) When DeFrank finally got tired of waiting, he came back upstairs to find his room overtaken by a small but odorous flock, along with Naughton and several other reporters, all laughing hysterically. DeFrank's humiliation was compounded several hours later, when he awoke to answer a knock at the door, only to plant his foot in a feculent reminder that his room had been a temporary sheepfold.

On November 3, 1976, DeFrank and several colleagues plotted their revenge. Cheney's involvement was critical. Just after the election, Cheney telephoned Naughton to offer him the scoop of the year: Ford has decided to give one interview about what it felt like to lose the election, Cheney explained, and he's chosen to give it to Naughton. "Naughton was beside himself, he was so happy," says Cheney, who instructed Naughton to report to Camp David at eight AM sharp on Saturday for the interview.[79]

Naughton's superiors at the *Times* were so excited that they asked George Tames, the paper's Pulitzer prize–winning photographer, to fly back from Florida to take pictures of Ford as he poured his heart out to their ace reporter. To prevent any potential mishaps, Naughton and Tames drove to a town near the presidential retreat the day before their meeting, spent the night at a rundown motel in Thurmont, Maryland, and got up early to ensure that they arrived on time. Cheney, meanwhile, called the editors at the *Times* to tell them not to hold any space for Naughton's blockbuster.

When he arrived that morning, Naughton announced his business to the marine manning the front gate.

I'm here to see the president, he said.

The guard was confused, and told Naughton that the president was not on the premises.

Naughton explained that Ford's chief of staff, Dick Cheney, had arranged the meeting and that the interview was scheduled for eight AM. The marine again said that Ford was not on the premises before retreating to his guard booth to make some phone calls. The two journalists cooled their heels outside the gate. After several minutes, the guard poked his head out of the booth. "There's someone here who wants to talk to you," he said, extending the phone in Naughton's direction.

He put the receiver to his ear. It was Cheney. "What's the weather like up there?"[80]

In the background, the reporters who had conspired to avenge their humiliation were, by Cheney's account, "hooting and hollering and carrying on."[81] Naughton knew he'd been had. The normally quick-thinking reporter made a feeble attempt to save face. "I don't know what you're talking about," he said, standing next to the marine who had called Cheney. "I'm at home in bed."[82]

Cheney spent the final ten weeks of his tenure as the White House liaison to Carter's transition team. It was his second transition in less than three years. Carter's officials would later acknowledge that the transition was difficult; but the incoming chief of staff, Jack Watson, credited Cheney and his colleagues for being helpful. "All of our problems in the transition of 1976 were self-imposed problems. . . . I cannot imagine having had a more forthcoming, candid, open-handed cooperation from Dick Cheney and from Don Rumsfeld and from Jack Marsh than we had. I don't think they withheld any effort to help our transition and I've always been really grateful for that."[83]

Not long before Cheney left his West Wing office, his staff threw an early birthday party for him at one of the government-owned row houses across Pennsylvania Avenue from the White House. Jim Naughton, the *Times*'s reporter, intent on avenging his humiliation at Camp David, had conceived a plan to remind Cheney that reporters can appear at the most unexpected times. In his woodshop at his house, Naughton constructed a large wooden sheet cake that was placed on two tables at the party. It was topped with thick chocolate icing and covered with birthday candles. The reporter hid under the cake—directly beneath a well-concealed trapdoor—and read a novel as he waited for the appointed time. As Cheney went to cut the cake, Naughton popped out wearing the San Diego chicken head he had famously worn to one of Ford's press conferences. Cheney and most of the other guests jumped back in surprise. And Naughton made clear to Cheney that this prank was not the payback he had coming for Camp David.

"I'll get even with you."

Cheney would leave the White House with high marks. Two weeks before Carter's inauguration, White House press secretary Ron Nessen sent Cheney a note:

> *I know that as the months have passed by, you have taken upon yourself a heavier and heavier load of both official duties and political chores. I believe that your extremely able handling of your many duties was, to a large extent, responsible for President Ford's amazing comeback. . . . I know that your own very good relationship with the press reflected favorably on President Ford's good image among reporters and also helped my White House Press Office over some of the rougher spots.*

Not everyone who worked with Cheney as chief of staff thought he was up to the job. Looking back on the Ford administration, Bryce Harlow, who had been one of Nixon's staffers and an informal adviser to Ford, thought that Cheney was too young and inexperienced to be an effective staff leader. Bob Hartmann, who had served as Ford's chief of staff in Congress, believed not only that Cheney was too young for the job but that he was too conservative: "Whenever his private ideology was exposed, he appeared somewhat to the right of Ford, Rumsfeld or, for that matter, Genghis Khan."[84]

But Jerry Jones, a special assistant to Richard Nixon and director of scheduling and advance under Gerald Ford, disagreed, saying that Ford's two top aides were the class of an otherwise "mediocre" staff. "Thank God for Rumsfeld and Cheney, you know?"[85] And the pollster Bob Teeter said, "Cheney turned out to be, by what a lot of people think, one of the best—if not the best—chiefs of staff ever."[86]

On January 20, 1976, Cheney, along with three other White House staffers and Vice President Rockefeller, flew with Ford on Marine One, the presidential helicopter, as the former president left the inauguration at the Capitol for Andrews Air Force Base. As the helicopter banked to circle the Capitol, Ford said: "That's my real home."[87] It was a revealing statement from a man who had moments earlier been president.

It was one of many ways in which Cheney would be like his boss. Although he would serve as White House chief of staff, secretary of defense, and vice president, Cheney would always say that he regarded the House of Representatives as his political home.

Buried deep in the Cheney files at the Gerald R. Ford Presidential Library in Ann Arbor, Michigan, is a memo dated November 2, 1976, Election Day. The "Memorandum for the President" was written by James E. Connor, staff secretary to Ford, and covered a "set of material prepared for the transition to your second administration."[88] It had been given to Cheney for safekeeping.

The transition materials included three papers: "Planning the Transition," "Presidential Objectives," and "Modern Mid-Term Transitions: The Implications for President Ford." The White House staff prepared the first two papers, and faculty members at Harvard's Institute of Politics drafted the third.

At the bottom of the page was a handwritten note to no one in particular: "Never submitted to the President—for obvious reasons!"

It was signed: "R. Cheney, 1/20/77." Inauguration day.

CHAPTER FIVE

On the Ballot

On January 21, 1977, the former White House chief of staff woke with something unusual on his schedule: nothing. He didn't have to read the morning newspapers or double-check the day's agenda. There was no senior staff meeting. There was no early-morning session with President Ford. There were no cabinet fights to referee and no plans by Rockefeller to scuttle. The day after Jimmy Carter's inauguration, Dick Cheney had no job and no real plan to find one.

For more than two years, Cheney had worked at least six days a week, as many as eighteen hours a day. At the end of the campaign, Lynne Cheney joked with a newspaper reporter that now she knew what it was like to be a single parent: her husband had been home only one full day over the course of the past year. Sleep was a luxury and stress a constant. A smoker since his late teens, Cheney was burning through three packs a day by the end of his tenure in the White House. He ate whatever was at hand, guzzled coffee as he scurried from meeting to meeting, and often ended his day with a glass of whiskey or a couple of beers. It was twenty-nine months of controlled chaos, and it was over.

"You didn't know what you were going to be doing next, and you didn't have anyplace you had to be the next morning," Cheney recalls. "So, what the hell—we'll go on vacation."[1]

Cheney's parents flew from Casper to Washington to watch their granddaughters. Just days after he had cleaned out his West Wing office, Cheney and his wife had installed themselves in a beachfront house on the remote island of Eleuthera, in the Bahamas. It would be hard to imagine a place with fewer reminders of the intensity and anxiety of political life.

The Cheneys didn't leave Washington behind entirely, though. They shared the house with Donald and Joyce Rumsfeld. With the exception of their initial meeting—the painful job interview Rumsfeld had granted the young political science fellow back in the fall of 1968—Cheney and Rumsfeld had gotten along well. Since their early days together in the Nixon administration, Rumsfeld had gone from mentor to colleague, and by the end of their service in the Ford administration he had become a friend.

The Rumsfelds and Cheneys spent most of the week relaxing—swimming, reading, and playing tennis. Don Rumsfeld refused to let Lynne Cheney's lack of tennis experience get in the way of some friendly competition.

"You're going to be my partner," he said to Lynne. "Just stand at the net and hold the racket in front of you so you don't get hurt. . . . Just hold the racket in front of your face."

But Rumsfeld also used the quieter moments of the trip to do some hard thinking about his future. He had been in Washington for about fifteen years, and although the fire of his political ambition was undiminished, he was eager to return to his native Illinois. By the end of the vacation he had made up his mind. The Rumsfelds were leaving Washington.

Cheney, too, was considering his next move. There was at least one immediate option. Like other former chiefs of staff, Cheney had been courted by trade associations and lobbying firms hoping to turn his contacts and experience into access and money. In the weeks after he left the West Wing, he fielded several potentially lucrative offers from firms in Washington, D.C., and New York. The jobs in Washington held a certain appeal; the Cheneys had created a life for themselves there. Lynne was teaching part time and writing for a variety of respected magazines. Liz and Mary, on the verge of adolescence, were enrolled in local schools and had settled into their family-friendly neighborhood in Bethesda, Maryland.

But Cheney had little interest in becoming a full-time lobbyist. In his view, most lobbying firms were filled with people biding their time until they could get a job inside the government. It was the political equivalent of treading water, and although it would have been a comfortable life, Cheney found it unappealing.

Among the many possibilities he considered was a job in journalism. Neal Freeman, a young columnist with King Features syndicate, was interested in buying *The National Journal,* a weekly magazine that was required reading on Capitol Hill and in the White House. Freeman approached Cheney about running the magazine and its operations. Cheney was interested. *The National Journal* was a substantive publication thick with the details of policy. Perched comfortably on the line between politics and academe, the writing was straightforward and accessible, intellectually rigorous but never pedantic. But after several promising discussions, the owners decided not to sell.

Cheney had some time to consider his options. He knew he would be in Washington for several more months, at least until his daughters completed their school year. He picked up some part-time work as a business development consultant for his former employer, Bradley Woods and Associates, and also did some short-term consulting for Nelson, Harding, Yeutter, and Leonard, a law firm with offices in Washington and Nebraska.

As he considered his options, there was one possibility that kept presenting itself: he could run for office. In certain ways, Cheney was an unlikely candidate. It is one thing to revel in politics from behind the scenes, as Cheney always had done, and quite another to subject yourself to the enormous scrutiny that comes with running for public office. Politicians often set themselves on a path toward elected office from an early age and make careful decisions along the way with that goal in mind.

Cheney, on the other hand, had last run his own campaign as a candidate for senior class president at Natrona County High School, a social outlet. And his activities after graduation were hardly the stuff of a political résumé: he had failed out of Yale, racked up two arrests, worked as a lineman, and—when he got his act together—pointed himself firmly in the direction of a life in academe. He had never served in the military, and he had received five deferments during the Vietnam War. He had gotten involved in electoral politics by accident and had always seemed indifferent—if not allergic—to the sort of public admiration that motivates so many politicians.

He had, however, worked in campaigns and government for more than a decade. He had ample connections, and he liked politics. And his many conversations with Ford had polished his image of congressional service. "Being around him as much as I was for that two and a half years he was president," Cheney had "a lot of opportunity to talk about Congress and listen to his old war stories. This was a guy who loved and revered the institution."[2]

Cheney also drew motivation from another aspect of his experience with Ford. "Part of it, I suppose, was hooked into the fact that we'd lost the election," he says. "I didn't like losing. . . . If I was going to do it, you know, I wanted to be the guy. I wanted to have my name on the ballot. I wanted to be the principal. . . . I didn't want to lose because somebody else got beat at the polls. I wanted to control my own destiny."[3]

In the summer of 1977, Cheney returned to Wyoming to prepare for a run for office. Lynne and Liz flew back to Wyoming, while Dick and Mary drove a U-Haul and towed one of the family cars behind.

Since Wyoming was admitted to the union in 1890, the state has had two seats in the U.S. Senate and only one at-large seat in the U.S. House of Representatives. In the summer of 1977, entrenched incumbents occupied two of those seats, but the third was open: Senator Cliff Hansen, a two-term Republican and a member of the GOP leadership, announced his coming retirement in 1977.

Cheney sought the counsel of Stan Hathaway, a former governor of Wyoming who had been chairman of the Wyoming Republican Party when Cheney took his first political job in the state legislature. Paying obeisance to Hathaway was one of the prerequisites of running for office as a Republican in Wyoming, so Cheney made the three-hour drive from Casper to Cheyenne. There was more to the trip than ritual, though; Hathaway knew politics in Wyoming better than just about anyone around at the time. If Cheney could win his support, it would be a major boost for his prospective candidacy.

Although they overlapped in Cheyenne in the mid-1960s, Cheney knew Hathaway only from a distance. Cheney was a lowly intern when Hathaway was the state's top Republican operative. Ten years later, when Gerald Ford was looking for a new secretary of the interior to replace Rogers C. B. Morton, who was leaving to run Ford's reelection campaign, Cheney, then White House chief of staff, suggested Hathaway, whom Ford nominated on April 4, 1975. The former governor of Wyoming survived a brutal confirmation battle, but once he was in office, "the environmentalists went after him," says Cheney.[4] After less than four months on the job, Hathaway resigned the position and returned to Wyoming.

When Cheney visited him in Cheyenne, Hathaway was less than sanguine about his prospects. By that point, Hathaway had already spoken to another possible candidate, Al Simpson, who had served in the Wyoming state house of representatives for more than a decade. His father, Milward Simpson, was a well-known and respected politician in Wyoming, having served as both governor and U.S. senator. Al Simpson was the natural candidate for Hansen's seat.

Cheney remembers Hathaway's words when he mentioned running for it himself. "Well, you could do that," Hathaway told him, "but of course if you do, Al Simpson is going to kick your butt."[5]

The Cheneys settled into a roomy house at 902 South Beech Street in Casper. Their new home needed work, and Cheney fixed it up himself as he mulled over his political future. A Wyoming businessman and former U.S. representative, John Wold, approached him about joining a uranium venture he was starting with his sons. But as Cheney considered that opportunity, another one presented itself unexpectedly.

On September 17, 1977, at halftime during a football game between the University of Wyoming and the University of Texas–El Paso, Wyoming's sole member of the House of Representatives, the Democrat Teno Roncalio, announced over the public-address system that he would not seek another term. The announcement took Wyoming's political establishment by surprise. Roncalio was a popular politician, despite the fact that he was in the state's political minority, and he had been mentioned as a possible candidate for the Senate seat that opened after Hansen's retirement. According to a news report after his announcement, Roncalio, age sixty-one, had "grown weary of the effort required to be reelected in a congressional district that encompasses an entire state."[6]

Cheney was interested in the seat the moment he heard the news. It wouldn't be an easy run; the two men Cheney would face in the GOP primary had higher profiles than the anonymous former White House chief of staff. Ed Witzenburger, a former Air Force pilot who had worked in Washington in the 1950s as the Air Force liaison to the U.S. Senate, had been elected Wyoming's state treasurer in 1975; Jack Gage was the son of a former Wyoming governor of the same name.

Cheney had another liability, too. Republicans in Wyoming had been deeply divided by the fight between Ronald Reagan and Gerald Ford in the presidential campaign of 1976. Those fissures remained, and Cheney was, naturally, identified as a Ford Republican. At the Republican National Convention in 1976, Cheney battled directly with some of the very people whose support he would need to win the congressional primary in 1978.

In 1976, Ronald Reagan ran as a leave-us-alone conservative. And as the former governor of a western state who favored cowboy boots over wingtips, he was popular in Wyoming. After several strong performances by Reagan in Republican primaries in the spring of 1976, the Ford campaign began to fashion a defensive strategy designed to prevent Reagan from winning the GOP nomination at the convention in Kansas City. Even small states like Wyoming were considered potentially decisive.

On July 7, 1976, six weeks before the Republican National Convention, President Gerald Ford hosted a white-tie state dinner to honor Queen Elizabeth II and Prince Philip. Held in connection with the recent national bicentennial celebration, the gala was planned to be one of the grandest parties ever held at the White House. The largest party rooms at the White House could not accommodate all the guests, so the first lady, Betty Ford, ordered the staff to construct an extravagant outdoor party room in the Rose Garden, complete with a tented roof and a wood floor. Comedian Bob Hope entertained the many foreign dignitaries and VIPs from across the country. A disproportionate number of those VIPs came from a small state in the frontier West. "If you look at the guest list," recalls Cheney with a smile, "you'll find there are a few Wyoming cowboys on there, because we needed their votes, and it was shortly before the 1976 convention. We worked it so hard, that a guy named Dick Jones, from Cody, he was the Reagan chairman, and Dick ended up—he got so angry, he finally stormed out of the convention, and we split the delegation. And so there was bad blood in Wyoming over the Ford-Reagan fight."[7]

Cheney's close association with Ford would present a host of additional difficulties. Chief among them was how to emphasize his considerable experience in national politics without alienating voters in Wyoming who were intensely skeptical of Washington. It would be a delicate balance.

Cheney had been away from Wyoming for more than a decade. He had cast his most recent vote—in the election of 1976, for Gerald Ford—in suburban Maryland and was concerned that he would have to face questions about whether he was a true Wyomingite.

Cheney formally entered the race on December 14, 1977, in Casper. At a press conference after his announcement, he fielded several questions from local reporters about his ties to Wyoming. "My wife and I were raised here, my parents live here, and I have no hesitation in calling myself a Wyomingite," he insisted. "Wyoming is my home."

Cheney campaigned on his political experience, but rarely spoke of his time in Washington without reminding voters that his roots were in Wyoming. "You had to do it in Wyoming style. The last thing I needed to do was to come in here and, sort of, big-foot it around—'I had been White House Chief of Staff, and you guys are obviously going to want me for this job.' Worst possible thing I could do."[8]

Cheney's first television ad, a biographical spot designed to introduce Cheney to the voters, opened with a still photograph of Frederic Remington's famous bronze sculpture *The Bronco Buster*, a cowboy on a bucking horse. As the shot widened, viewers saw that the picture was taken in the Oval Office and featured Ford and Cheney talking at Ford's desk, with the work of art in the background.

The narrator intoned: "In 1975, a young Wyomingite was named to one of the most important posts in the country, chief of staff to the president of the United States. His name? Dick Cheney, the youngest person ever to hold that responsible position. But to the people who watched Dick Cheney grow up in Wyoming, such leadership came as no surprise." The ad cut to a photograph of Cheney as a boy, in his baseball uniform with "Casper" emblazoned across the chest. Several Wyoming voters, including one of Cheney's high school teachers, testified to his roots in the state.

The ninety-second ad twice noted that Cheney had graduated from the University of Wyoming. As the narrator informed viewers that Cheney had received two political science degrees from the state school, a shot of his diploma flashed across the screen; it was followed by a clipping from the University of Wyoming newsletter that boasted, "Alumnus Chief of Staff." There was no mention of Yale.

The Wyoming political landscape was littered with the failed candidacies of Ivy League graduates, including Cheney's benefactor, Tom Stroock, a Yalie; and one of his most generous supporters, a failed gubernatorial candidate named Warren Morton, who had graduated from Harvard.

"If you were in politics in Wyoming, the biggest political plus is if you went to the University of Wyoming," says Morton's wife, Kathy.[9]

Although there were few occasions for Cheney to be grateful for the troubles he had at Yale, his congressional campaign was one. "I was worried about anything that could be used against me by way of a sense somehow of the entitlement to this post, or that I was some guy that they had brought up in Wyoming, but I had become this slick, smooth operator. It was a good thing that I hadn't graduated from Yale."[10]

To avoid any sense of East Coast sophistication, Cheney's ads were produced in Sheridan, Wyoming, despite having been created by Bob Gardner, an advertising guru from San Francisco whom Cheney had come to know in the Ford White House. "They looked like they were produced in Sheridan instead of San Francisco, and that was deliberate. That's exactly the way we wanted it," says Cheney.[11]

At the press availability following the formal declaration of his candidacy, Cheney was asked about the timing of his announcement, an opening many politicians would use to reiterate talking points on the rationale for their candidacy. His answer was less complicated. "I have made up my mind and would not have felt comfortable posturing as if I hadn't."[12]

This down-home style would characterize Cheney's first campaign. Shortly after his announcement, he sent a six-page, oversize mailing to all voters in Wyoming. The brochure was full of the same short, straightforward language.

Above a large picture of the smiling candidate in a wool-lined jacket was a banner headline posing the question that would be answered on the following pages. "Who Is DICK CHENEY and Why Is He Running for Congress?"[13]

In some respects, the mailing read like a typical campaign pamphlet for a conservative Republican congressional candidate in a midterm election. For Cheney, though, the booklet was the first public statement of his political philosophy, and it included a detailed enumeration of his positions on a variety of issues, views he had largely kept to himself as a staffer for the previous decade.

"We need a Congressman who respects the idea of limited government," he wrote. "That is a concept which our Founding Fathers never forgot and an idea that we in the West have always valued. Dick Cheney is running for Congress because he wants to make the American system work the way we here in Wyoming know it should work."

The pamphlet contained short policy statements on subjects from taxes and government waste to agriculture and water, the last of these eliciting a declaration of proud parochialism. "The people of this state have more than a century of experience in managing water as a scarce and valuable resource, and instead of letting Washington tell us what to do with our water, we should be educating the rest of the country on how to take care of theirs."

Drawing on his experience in the Nixon administration, Cheney maintained that solutions offered by the government are frequently as harmful as the problems they seek to solve. "I want Congress to realize that the quality of life of the people in Wyoming is being as adversely affected by unjustified and unwarranted accumulations of governmental power as by pollution, discrimination, or unsafe working conditions."

On national security, Cheney offered little more than Republican boilerplate. "I will make it abundantly clear to the administration that the people of Wyoming expect U.S. foreign policy to be conducted from a position of strength, not weakness."

Most interesting were the comments on taxes, in which Cheney proclaims his belief in supply-side economics two years before Reaganomics popularized the theory. "Major tax cuts would increase government revenues. As people keep more of what they earn, they'll be encouraged to work harder, to save, and to invest. The economy will expand and grow and government revenues will increase."

Because both of his opponents were from Cheyenne, Cheney's strategy for the Republican primary in August was simple: split the Cheyenne vote three ways and win Casper outright.

To do this, he would have to mitigate the residual anger from the battles between Reagan and Ford. Wyoming is a state small enough that endorsements and campaign personnel matter. In Cheyenne and Casper, with populations of less than 50,000, recruiting a few top Republican Party activists could mean the difference between winning and losing.

So it meant something when Peggy Mallick, a resident of Casper who had been heavily involved in Reagan's campaign, joined Cheney as a volunteer coordinator. Bill Thomson and his wife, Toni, supporters of Reagan from Cheyenne, also lent their political heft. Bill was the son of Thyra Thomson, who served six terms as secretary of state, and former Congressman Keith Thomson, the closest thing Wyoming had to a power couple.

Toni's family had for decades owned and operated one of Cheyenne's most successful ranches.

Cheney's campaign virtually defined the maxim that all politics is local. In Park County in northern Wyoming, he won the
backing of Mildred Covall, a powerful former superintendent of
schools in Cody. (Covall checked with Al Simpson before making a commitment to Cheney. Having by that time consulted
with Stan Hathaway about Cheney and attended a reception for
him in Casper, Simpson gave her his approval.)

"If Mildred said, 'This is my guy,' I would have 1,000 votes—
just because she was so respected in the community," Cheney said.
"It was those kinds of connections that put it all together."[14]

Meanwhile, down in Torrington, a town of 4,000 near the Ne-
braska border, John Vandel took a call from his brother-in-law,
Greg Goddard, in Buffalo, Wyoming, at the base of the Bighorn
Mountains. Goddard was excited about Cheney's run and asked
Vandel to consider supporting him. Vandel, a local pharmacist,
made arrangements to meet the candidate.

On his next trip to Torrington, Cheney picked up Vandel in a
rented Chevrolet Vega. "I wasn't in his car for three blocks, and
I realized this is an impressive guy," says Vandel, who is now the
dean of the school of pharmacy at the University of Wyoming.
Vandel became Cheney's chairman in Goshen County. "Wherever he went, in the parties I put on for him, people knew immediately that he was a man of substance."[15]

In other parts of the state, Cheney relied on old friends to boost
his efforts. In Teton County, it was a recent transplant from Washington, D.C., who had worked for Jim Baker's delegate-counting
operation at the 1976 convention; and in Laramie, a teacher friend
of Lynne's from her days at the University of Wyoming.

The campaign was operating on a shoestring and almost everyone working on it was a volunteer. One of the few hired hands
was Gardner, the adman; in keeping with the do-it-yourself ethic,
he slept on the sofa at the Cheneys' house when he was in town
from California.

Gardner remembers the operation as efficient but a bit unusual. "It didn't have a typical campaign structure," he recalls.
Lynne Cheney served as the unofficial campaign manager, run-

ning many of the day-to-day operations and supervising everything from the essential to the picayune. "She corrected my
grammar," says Gardner.[16]

By early summer, the campaign had hit its stride. Cheney
drove himself around the state in a rented Ford Mustang, sometimes taking one or both of his daughters and occasionally hauling
the entire family from event to event. Although they complained
about their father's driving music—an eight-track tape of the
Carpenters—the girls liked to go along.

Only Mary was with her parents on a campaign swing to
Cheyenne in mid-June. (Her sister was in Laramie with family friends.) The three Cheneys were staying with Joe Meyer—
Dick's former Wyoming roommate and his high school football
teammate—and Joe's wife, Mary.

At four AM on June 18, Father's Day, Cheney was awakened
by a tingling sensation in the ring and pinky fingers on his left
hand. It felt as though he had bumped his elbow, but he was
in bed and until moments earlier had been fast asleep. He was
only thirty-seven years old; heart trouble seemed like a remote
possibility. And he wasn't suffering any of the other symptoms
of heart trouble—chest pain, shortness of breath. But as he lay
awake thinking of possible explanations for the strange feeling,
he recalled that his first cousin, Jamie, had suffered a serious heart
attack just two weeks earlier. Jamie was ten years older, but still
young for a devastating heart attack. Concerned, Cheney woke
his wife, who hurriedly fetched Joe Meyer to drive them to the
hospital.

"I walked into the emergency room, sat down on the examin-
er's table, and passed out."[17]

Cheney remained hospitalized for eleven days. It was a mild
episode, but any heart attack at such a relatively young age suggests a serious cardiovascular problem, and a battery of tests
confirmed the diagnosis. It was hard to believe. "When I heard
'Dick Cheney's had a heart attack,'" says Cheney's brother-in-
law, Mark Vincent, "I thought it was his father."[18]

For Cheney, it was time to reevaluate his priorities. "The
first question you ask is, My God, can I keep going with the
campaign and everything, having had a heart attack? The policy

fallout from it—are people going to say, I can't work for him, he's not up to the job? Basic fundamental question of whether or not this means you're going to have to lead some kind of sedentary lifestyle now and give up your political aspirations. The smart thing that some prudent person would do is quit this crazy life you're leading. You had a heart attack. Just bag it, go home, take it easy. You can't live your life like this and expect to survive with family obligations."[19]

Cheney had serious decisions to make. Would he continue his campaign for Congress? If so, how would he deal with the heart attack, a potentially decisive setback in a competitive three-way primary? And how would his condition affect his family? "I had to sort of pause and reflect and think about what I was going to do with the rest of my life."

As he lay mulling over his future, Lynne walked into his hospital room one afternoon with a telegram from Jim Naughton, the reporter at the *New York Times* whom Cheney had dispatched to Camp David for the bogus interview with President Ford. She was laughing out loud.

"I was in a hospital in Cheyenne . . . tubes running from every part of my body, life's passing before my eyes. I'm thirty-seven years old and had a heart attack, and I'd stopped the campaign and so forth. . . . I thought, 'Now what the hell's funny? There's nothing funny about this.' She handed over the telegram, which was signed only 'Naughton.' It read: 'Dear Dick, I didn't do it.'"[20]

As he began to mend, Cheney consulted with his doctor, Rick Davis. "He said, 'Look, hard work never killed anybody.' He said, 'What is bad for you, what causes stress is doing something you don't enjoy, having to spend your life living in a way you don't want to live it.'"[21]

It was just what Cheney wanted to hear. Though already inclined to stay in the race, he worried about the potential consequences, for his personal health and consequently for his family's well-being, of the inevitable stress that comes with public office. The heart attack would mean other dramatic changes. Cheney quit his three-pack-a-day habit cold turkey. He also cut back on coffee and began to watch what he ate.

On Cheney's release from Cheyenne Memorial Hospital on

Thursday, June 29, Dr. Davis told reporters that the prognosis was "excellent for Dick's full and complete recovery. After a period of rest and recuperation at home he can expect to be able to resume a full and active schedule."[22]

Having decided to continue his campaign, Cheney and his small group of advisers considered how they would handle the publicity generated by the heart attack. One option was a frank public discussion of the episode and its possible consequences. So the Cheney campaign filmed a thirty-second television ad known internally as "Heart Attack." The spot featured Cheney, wearing a blue shirt, brown pants, and cowboy boots, sitting on a patch of grass speaking extemporaneously to five supporters in a circle around him. In a voice approaching a murmur, Cheney matter-of-factly tells the group that the heart attack made him think twice about running for Congress but that he is determined to forge ahead because he enjoys being part of something "beyond personal self-interest." The ad closes with an odd history lesson from Cheney. "There's ample precedent certainly for people who've had heart attacks—Dwight Eisenhower, LBJ—for pursuing active political careers, and I expect to be one of those."

After reviewing the ad for several days, Cheney and his advisers decided the grim subject matter would be jarring for voters. The spot was never aired.

In the end, Cheney decided to write a two-page postal patron letter to all potential voters in Wyoming. The letter explained the incident, provided a brief tutorial on managing cardiovascular disease, and described the various lifestyle changes the heart attack had already produced.

Cheney's doctor recommended taking a month off from active campaigning in order to recuperate. So Cheney hunkered down at his house in Casper, balancing plans for his return to the campaign trail with time spent sitting under an old spruce tree in his backyard, reading Richard Nixon's newly released memoirs. He began a light exercise regimen, walking the five blocks from his house to the campaign headquarters and back. But with the primary less than three months away, Cheney could hardly afford to be out of circulation completely. Thus, while he was convalescing, his wife Lynne became something of an unofficial stand-in candidate, making appearances and short speeches on her husband's behalf. By all accounts, she filled in well.

The entire ordeal had an unanticipated benefit. "In a funny sort of way, the heart attack helped the campaign. . . . It got a lot of press, a lot of free media we didn't have to pay for. All of a sudden people know there's some guy named Cheney out there running for Congress. But I think it also let people see me react to a significant event in terms of how you handle it, respond and so forth. So I think on balance it was a net plus."[23]

In the short term, Cheney's medical setback did not impede his fund-raising efforts. Cheney's former aide Foster Channock told his friends and colleagues to send checks rather than flowers. Bill Steiger, Cheney's former congressional mentor, also helped. Without a serious challenger in his bid for reelection to the House, Steiger asked his top donors to redirect their contributions to his friend in Wyoming.

Despite this additional money, it didn't take long for Cheney to empty his campaign account. As the primary approached, the Cheneys opted to take what little remained of their personal savings and invest it in the campaign. They had two urgent needs: a poll and additional television advertising. They wouldn't be able to afford both.

"You'll know exactly where you stood if you go do a poll, answer all those questions and find out," Cheney says. But conducting a poll is costly. "If you did that, then you didn't have any money to do anything about it, and so we said, the hell with it, we'll buy ads."[24]

The campaign bought time for a spot they called "Rope Trick," another attempt to link Cheney's Washington experience and his Wyoming background. The ad opened with a close-up shot of a man's hands knotting a thick lasso rope.

"If you want to get the job done as Wyoming's congressman, you've got to know the ropes," said a narrator. "The problems Washington is making for us here in Wyoming are bigger than ever. We need a congressman who'll stand up for Wyoming, a congressman who has had experience, not an amateur getting tied up in on-the-job training. Wyoming's Dick Cheney knows the ropes. He's had experience with Congress, the executive branch, and as president Ford's chief of staff." The shot widens to show that the hands belong to Cheney, who has worked the rope into what looks like a knotted mess. But he yanks on both ends, the rope unravels, and the narrator says: "Dick Cheney. With your help, he'll make a great congressman."

The ads worked. Cheney outpolled his two opponents, winning 42 percent of the vote, with 31 percent going to Witzenburger and 27 percent to Gage. He would face Bill Bagley—a Democrat and a former aide to Roncalio—in the general election. Bagley was a Mormon from Star Valley, and because Mormons typically vote Republican there was a possibility that Bagley would steal some Republican votes from Cheney.

The general election focused on local issues, and with few exceptions the two candidates found themselves in agreement. The campaigning was largely substantive, and relations between the candidates were cordial. "I remember we'd be walking along in a parade or something and he'd be on one side of the street and I'd be on the other," says Bagley. "And he'd walk over and we'd visit a bit. It wasn't an unfriendly thing."[25]

But Bagley, like Cheney's opponents in the primary, pointed out repeatedly that Cheney had been out of the state for several years. Bagley's four-page campaign brochure promised that he would follow the tradition established by his boss: "A Congressman *for* Wyoming and *of* Wyoming, not a Congressman *from* Wyoming." Bagley sounded the same themes in his announcement, adding that his party affiliation would give him a voice that a Republican would not have. "Wyoming needs a voice in the majority part that will be heard in Washington by the Democratic majority in Congress, and by the Democratic administration in the White House." Bob Reese, Bagley's campaign manager in Cheyenne, told the University of Wyoming's student newspaper, *The Branding Iron*, that Cheney was guilty of "carpetbagging, returning to Wyoming so he can return to Washington."[26]

Those charges were the nastiest in what was generally regarded as a clean campaign. "I figured I wouldn't want to do anything in the campaign that I wouldn't want to live with afterward," says Bagley.[27]

Cheney's victory in the primary had brought in a fresh round of backing, and as the general election drew closer, Bagley was at a significant financial disadvantage. So he went to Cheney with a proposal.

"I realized one time that I didn't have enough money to get into the billboard business," Bagley recalls. "So I called him up and proposed that we would agree not to advertise on billboards. He agreed to that. I'm not sure that I disclosed the fact that I couldn't afford the damn things anyway."[28]

Cheney's decision was based less on magnanimity than on confidence. He'd entered the general election believing that if he won the primary, he'd win the seat. "In the fall campaign," he says, "there was never any question in my mind who was going to win."[29] In the end, Cheney received 75,855 votes, 59 percent of the total votes cast. His bid to "control his own destiny" had succeeded.

Within two days of the election, Cheney made his first move as congressman-elect. He placed a call to the House minority leader, John Rhodes. Cheney, as Ford's chief of staff, had gotten to know Rhodes well when they worked to advance the president's agenda in Congress. Rhodes, he knew, would play an influential role in determining committee assignments for incoming freshmen. But unlike most House members, who jockey for seats on Ways and Means or Appropriations, the power committees, Cheney told Rhodes he wanted Interior, a committee with oversight of the nation's natural resources and public lands. On its face, the request was unusual. On closer inspection, it was an obvious choice. The federal government owns some 50 percent of Wyoming, and a seat on Interior would give Cheney a voice in how that land was used.

Rhodes approved, on one condition. He wanted Cheney to serve on the Ethics Committee, a thankless, low-profile assignment that House members avoid because it requires them to sit in judgment on their colleagues. "That was the trade-off," says Cheney. "He had been thinking about it."[30] Cheney agreed to the deal.

He quickly set about making arrangements for his staff. He plucked Patty Howe, a highly regarded senior aide to the retiring senator, Cliff Hansen, before Al Simpson, Hansen's replacement, could hire her. Cheney asked his longtime friend Joe Meyer to move to Washington as his chief of staff. Meyer wasn't interested, so he turned to David Gribbin, a friend from his days as a graduate student at the University of Wyoming. Gribbin had taken a political philosophy class with Cheney and shared his low-key style.

Gribbin was not expecting a life in politics. He had earned a master of divinity degree from Wesley Theological Seminary in Washington, D.C., and in 1978 he was seeking to join the ministry full-time. With his wife in a good job, and his young children enrolled in local schools, Gribbin went to the Baltimore Methodist Conference to request a placement in the greater Baltimore-Washington area so that his family wouldn't have to move. "Laurie and I were praying for the right phone call," he recalls. Unexpectedly, that call came from Dick Cheney.[31]

Gribbin, who was, like Cheney, a graduate of Natrona County High School, agreed to take the job, despite his concern that he didn't have much to offer. "What am I going to tell a guy who was just chief of staff at the White House?" It was a concern that would be shared by Cheney's staff thirty years later.[32]

Cheney took time off from his preparations for a vacation in Hawaii. His family spent a week at a friend's house on the Big Island. While they were there, Cheney had a favor to repay. Al Abrams, an acquaintance from Cheney's days at the OEO who had written Cheney his first PAC check several months earlier, was running the National Association of Realtors. He asked Cheney to give a speech at its annual convention in Honolulu. It wasn't difficult duty. And as a bonus, Abrams had also invited Cheney's friend Bill Steiger to address the delegates. The Cheneys and the Steigers spent hours around the hotel swimming pool just "de-stressing from the campaign," says Cheney. For reasons no one could have anticipated, it would be a memorable vacation.

Cheney returned to Wyoming and prepared for another cross-country move. The family, accompanied by Cheney's father, hitched the car to the back of a U-Haul and headed east. The trip was interrupted twice: once by design, with a visit to the Rumsfelds in suburban Chicago; and again by car trouble in the middle of a massive blizzard that buried the upper Midwest in several feet of snow. The truck that the Cheneys had rented broke down in South Bend, Indiana, and the delay meant that Congressman-elect Cheney might miss the swearing-in ceremony for his freshman class. Leaving his wife and children at a Holiday Inn, Cheney pressed on in the car with his father. A break in the weather allowed him to catch a plane from Cleveland to Washington, where he took the oath of office with his newly elected colleagues. His wife and daughters arrived in Washington several days later, having missed the big event.

Cheney was in demand right away. On December 4, 1978, after the first caucus of new members at the Cannon House Office Building, Cheney and several classmates met with a photographer from one of the major newsmagazines to have their picture taken for an upcoming article. As they stood on the steps outside the Cannon building, Cheney noticed that the flags across the street at the Capitol were being lowered to half-mast. He asked the photographer who had died. The response stopped him cold.

"Bill Steiger."

Steiger was forty. He had just had a full physical and received a clean bill of health. They had been in Hawaii together less than a month before, and there was no sign of trouble then.

For Cheney, the loss of his good friend and mentor served as a stark reminder of his own mortality. Steiger had died in his sleep after a massive heart attack.

The tributes came quickly. "It was not chance that advanced Steiger to the front rank of his party and his profession," wrote the columnist George Will. "But chance has done what only chance could do: It has prevented him from what would have been one of the most distinguished careers Congress has known."[33]

James Reston of the *New York Times* noted that even by Washington standards, Steiger's death evoked "an exceptional outpouring of respect and affection for one of the most promising young men in Congress," responding to "the magic of his personality, the gifts of his energy and intelligence and the shock of his premature death."[34] And David Broder of the *Washington Post* regretted not having told Steiger's story before he died: "The most uncomfortably remembered stories are those where you might have said—but did not—that somebody is doing a helluva a job in public office. Bill Steiger had done that kind of a job."[35]

Cheney had counted on Steiger to guide him through his first days in Congress, and without that counsel Cheney made a point of attending the formal orientation sessions for new members. He had not worked with national party leaders during his campaign, declining the advice and support offered by the National Republican Congressional Committee and the Republican National Committee. ("I knew how I wanted to run my campaign, and I didn't want to get a lot of advice out of Washington I didn't think I needed," he says.)[36]

The party that does not control the White House typically does well in off-year elections, and 1978 was no exception. Thirty-five new Republicans across the country swept into office on a strong current of anti-Carter and anti-Washington sentiment. "There were a lot of eager beavers in the crowd, people that were absolutely convinced now that they've been elected, they're going to get the place straightened out," recalls Cheney.[37]

A freshman orientation at the Dulles Airport Marriott Hotel provided an outlet for their enthusiasm. During their breakout meetings, the new members talked about their plans, often at length, in an effort to impress their colleagues. Representative Bill Thomas, a newly elected member from an agricultural district in central California, remembers the scene. "There was this one kind of huffy guy with crazy hair, who's running around talking to everybody like he knows everybody," he says. "And that was Newt Gingrich. Here's this guy from Texas telling us all of this stuff, he knew all of this stuff about Congress and the White House. That's Tommy Loeffler. He was legislative liaison for Gerald Ford."[38]

Cheney said almost nothing, recalls Thomas, but when he did contribute he was direct and to the point. He never boasted about his experience as White House chief of staff. As the group members got to know one another, Cheney was cordial but kept largely to himself. "He was not a glad-hander," says Thomas.[39]

Among the new members were several legislators who would

still be serving in Congress almost thirty years later: Jerry Lewis from California, Olympia Snowe from Maine, and James Sensenbrenner from Wisconsin. Others in the class included Dan Lungren from California, who returned to Congress after a stint as his state's attorney general; and Carroll Campbell, a future governor of South Carolina.

It was an ambitious group. "We have the seeds of a new majority," Lewis told the *Washington Post* after one month in office.[40]

The beginning of the session made it official: in two years, Cheney had gone from being one of Washington's most powerful men to being one of 435 individuals pursuing the parochial interests of their districts. As chief of staff at the White House, Cheney had spent his days in the center of power in America, even the world, involved in decisions that would shape his time. As a freshman member of the House of Representatives, he would spend his days thinking about water rights for Wyoming's farmers or lobbying to change the Social Security number of a constituent who was upset that her government identification included "666."

So Cheney kept his head down as the new session began, wanting to avoid giving his colleagues the impression that his experience made him anything more than just another freshman representative. He focused on Wyoming issues and the day-to-day requirements of setting up a new office, with some success in the former arena and less in the latter. Gribbin says the first six months of Cheney's first term were "chaos, the kind of environment a good management book would never allow."[41]

The first order of business was formulating official positions on the many issues that might require his attention. "A member of Congress has to have policy views on everything from Nicaragua to the fix for Social Security," says Gribbin.[42] Cheney held several staff meetings designed to determine his legislative priorities and to develop position statements to send to constituents.

On subjects like grazing on public land and public water rights, he drew on the experience of his staff. On national issues,

Cheney simply dictated his thoughts, developed over ten years in the executive branch.

Over the course of his first term, Cheney's aides picked up on his preferences and his pet peeves. He liked to receive advice about policy in written memoranda that included background on the issue, recent developments, and a formal staff recommendation. "That was his way of doing business," recalls one veteran of Cheney's congressional staff. "Don't talk to me. Let me read it."

Cheney encouraged his staff to take strong views, even when to do so would mean challenging their boss. "We learned quickly that we couldn't send him mush," says Gribbin. "He appreciated honest disagreement."[43] The staff members also understood that Cheney would often know more about a particular issue than they did, even if it was an issue that they had been working on for years.

In keeping with his low-key presentation, Cheney never wanted an entourage. He preferred to walk the short distance from his office, Room 427 of the Cannon House Office Building, to the Capitol without a passel of aides scurrying along, and he usually traveled alone. A columnist from the *Lander* (Wyoming) *Journal* took note during a chance encounter at the Denver Airport. The fact that the young representative was alone, "just himself, lugging numerous suitcases and garment bags," was a marker of his "lack of pretention."[44]

Cheney made trips back to Wyoming nearly every weekend and traveled throughout the state in an RV—his "mobile office"—during congressional recesses. In a typical week, Cheney would set up for meetings with constituents in the parking lot of the Jeffrey City Grocery or the Big Horn IGA, or park outside the Irmy Hotel in Cody. He would sit for interviews with local radio stations and weekly newspapers like the *Thermopolis Independent Record*, the *Basin Republican Rustler*, or the *Greybull Standard Tribune*.

"The philosophy I started with in terms of Congress is I wanted to take that first two years, that first term, and really nail my base," says Cheney. I wanted to "make absolutely certain that I'd done everything I could here at home to lay the foundation so that I would be free in the future to pursue the issues I wanted to pursue in Congress."[45]

Not everyone was sold. Despite his regular cross-country trips, Cheney could not entirely shake criticism that he was a carpetbagger. Bernard Horton, a columnist for the *Cheyenne Eagle*, wrote that Cheney was the congressman "just barely of Wyoming." Cheney should not seek reelection in Wyoming, Horton wrote: "We suggest it would be more appropriate for him to seek some office in the District of Columbia. He has spent a lot more time in Washington, at least in recent years, than he has in Wyoming."[46]

It would prove to be a delicate balance. "I didn't want to be a slave to the district," says Cheney. "I had to represent the district, and that was important, but I thought I could be a lot more effective if I could say what I believed and speak my piece. . . . I wanted to be free to operate and then to pursue my interests."[47]

Among those interests was the study of Congress. Although Cheney had long ago abandoned the life of an academic for politics, he sought out opportunities that would allow him to keep one foot in the world of ideas. He frequently spoke to outside groups about the role of the legislative branch in shaping the national agenda, and he sometimes asked his staff to send articles he had written to political scientists around the country. Cheney moved easily in the world of Washington think tanks, a sort of middle ground between academe and practical politics.

Shortly after he arrived in Washington, Cheney was invited to participate in a series of dinner discussions among a select group of newly elected House members, chosen for their understanding of the theoretical underpinnings of deliberative democracy. Among the other participants were the Republican Newt Gingrich, who had received a PhD in history from Tulane University; Geraldine Ferraro, a public school teacher from New York who had gotten her law degree at night; and Democrat Martin Sabo, who spent a decade as a leader of the Minnesota state house of representatives before his election to Congress. The occasional dinners were a setting for free-flowing discussions about their experiences in the House, focusing on process and institutional integrity. There was little discussion of politics, and no partisanship. The conversations were recorded, edited, and eventually published as a monograph called *Congress Off the Record: Candid Analyses of Seven Members.* The elected officials were not identified by name, and the publication of the brief book resulted in a short-lived name-that-member guessing game on Capitol Hill.

Cheney also sought to carve out a role for himself in the conduct of national security policy. Cheney's committee assignments required that he devote most of his time to "Wyoming issues" and congressional governance. But in his spare time, he focused on foreign policy and defense. He sought out experts in diplomacy, military affairs, and geopolitics to supplement his experience on the job in the Ford White House. Even members with little interest in international affairs would have had trouble ignoring two significant developments during the fall and winter of 1979.

On November 4, student radicals in Iran stormed the U.S. embassy in Tehran. Officials in the State Department at first declared that the takeover was "a peaceful demonstration" and denied that the captive U.S. personnel were "hostages."[48] Within days, though, newspapers began carrying photographs of the captive Americans, blindfolded and with their hands bound behind their backs. The radicals, followers of Ayatollah Ruhollah Khomeini, declared that they would continue to hold the diplomats until the deposed shah, Mohammad Reza Pahlavi, then in the United States to receive treatment for cancer, was returned to Iran so that he could be tried for crimes against the Iranian people. In Carter's next address to the nation, the perpetrators were no longer described as student activists; they were now called "militant terrorists" acting with "the support and approval of Iranian officials."[49]

The foreign policy situation would quickly become even more perilous. On December 25, 1979, the Soviet Union invaded Afghanistan. Despite the frenzied activity of the Soviet military in the months leading up to its move into Afghanistan, the U.S. intelligence community, with few exceptions, had downplayed the possibility of such an invasion.[50] In response, the CIA moved to support rebel groups of anti-Soviet mujahideen forces.

Cheney made a mental note of the intelligence failure and backed the decision of the Carter administration to arm the rebels, whom he regarded as "a tough and able people who want to fight for their country."[51] Arming the mujahideen, Cheney said, is "the best way to make Russia pay, and it will also keep them tied down so that they cannot move into Iran."[52]

Through the winter the media gave constant attention to the hostage crisis in Iran. After several harrowing months of captivity, the Americans were still being held, with little immediate hope of release. By spring, the United States was threatening to boycott the 1980 summer Olympics, held that year in Moscow. Although Cheney agreed, "reluctantly," with Carter's call for the boycott, his disagreement with the president's handling of the hostage crisis was plain: "My major criticism is his repeated public statements that he will not use force to resolve the situation," Cheney said in an interview with the *Pinedale Roundup.* "Although he might decide not to use force, he should not make his decision known. Once you remove the threat of force you remove any incentive for the Iranians to free the hostages. Every single President for the last half-century at some time has had to use force to safeguard American lives. This is Carter's crisis. It's his test."[53]

Despite his strong opinions, Cheney had no more role in the conduct of U.S. foreign policy than any other freshman member of the House of Representatives. And while he followed closely the developments in Afghanistan and Iran, he spent much of his first term on matters relating to the Ethics Committee, fulfilling duties he had accepted as part of the deal struck with Rhodes in November 1978. Several of Cheney's colleagues in the House had pimped themselves for money and gifts, some of them through complicated kickback schemes and others for quick cash handouts. These were dark days for the House of Representatives and active days for its in-house ethics police.

Representative Charlie Diggs, a Democrat from Michigan, was referred to the committee for misusing funds. Diggs had been convicted in federal court on charges that he had lined his pockets by getting kickbacks from his staff. He was appealing the conviction and refused to resign his seat. Newly elected Republicans, led by the firebrand Newt Gingrich, were pushing hard for his expulsion from Congress.

Cheney found Diggs's behavior reprehensible but thought that Gingrich and others advocating expulsion were seeking to "demagogue" the issue. For Cheney, expelling a member of Congress presented the committee with profound questions about the nature of a representative democracy.

"There's an inherent conflict built into the two provisions in the Constitution," he says. "One is that voters have a right to pick whoever they want to represent them, and they do it through the direct election for the House of Representatives. The only way you can get to be a member is by election of the constituents. But whoever the constituents elect, as long as they're twenty-five years old and a citizen, and so forth, gets to serve. And it is the other provision that said that with a two-thirds vote you can expel a member. And those two things come into conflict. . . . You've got a lot of people in the Congress that you and I might think, what the hell is he doing here? He got elected, the same as anybody else. And it doesn't matter if he's stupid, or what his philosophy is, or if he doesn't show up; he's got every right to be there. So when you get to the point where you say two-thirds of the members can come together and deny to the voters the right to be represented by this guy they picked, that's a big deal."[54]

On the day that the Ethics Committee was to begin hearings on the matter, Diggs admitted that he had misused funds and agreed to a formal censure from the House. Cheney voted with the overwhelming majority of his colleagues to censure Diggs.

In September 1980, Cheney and others on the committee again considered expelling a member of the House. The issue this time was Abscam, the result of an investigation by the FBI into payoffs to public officials. The FBI had conducted a successful sting operation against local politicians in Philadelphia and, acting on a tip, broadened the investigation to include members of Congress.

The result was a Hollywood-style bust of several top lawmakers, revealing naked bribery that would have been thought unbelievable had it not been caught on videotape. Representatives from New Jersey, Pennsylvania, and Florida were filmed accepting briefcases full of cash in exchange for their promises to support immigration legislation pushed by a wealthy—but entirely fictitious—Arab sheikh. The damning evidence of their corruption led most of the conspirators to resign from office in disgrace. But a former longshoreman from Philadelphia in his first term, Representative Michael "Ozzie" Myers, stubbornly refused to give up his seat. The Ethics Committee met to decide whether he should be forced out. As in the case of Charlie Diggs, the panel had considered such punishment before, but the Congress had not expelled one of its own since the Civil War.

The committee met in closed session, and after watching a videotape of Myers telling undercover FBI agents that "money talks in this business and bullshit walks," voted ten to two (with Cheney in the majority) to recommend expulsion. Cheney cast the same vote on the House floor on October 2, 1980, where Myers was ousted after a vote of 376 to 30.

The cumulative effect of the endless scandals, together with a sagging economy at home and troubles abroad, contributed to an anti-incumbent cloud that hung over the country for most of 1980. Ronald Reagan, barely defeated in the 1976 Republican primary, faced another stiff fight; this year's roster was packed with seasoned Republicans vying for their party's nomination for the presidency, including George H. W. Bush, Howard Baker, John Connally, and John Anderson.

In his home state, Cheney ran unopposed in the Republican congressional primary and would face the winner of a four-way Democratic contest. Like much of the competition he would face over the course of his congressional career, it was a field that could be assembled only in Wyoming. There was Al Hamburg, a sign painter from Torrington, who would later gain local notoriety when he unsuccessfully sued a woman who bought his car for $100 and fifty sexual favors and stopped payment after just thirty-three assignations.[55] His main selling points, it seems, were flexibility and persistence. "I ran twice as a Republican," he said. "This is my third try as a Democrat."[56] Hamburg finished second, a result that ensured this campaign would not be his last.

Although Hamburg didn't win, he outpolled Ted Hommel, who declared before the race: "I believe I stand a chance of making it through the primary, but right now my hopes aren't too good for the general." Hommel finished third, but two years later won the right to face Cheney.[57]

Sidney Kornegay likewise predicted that even if he were to win the primary, he wouldn't "have a snowball's chance against Cheney in the general." Voters might have sensed his lack of motivation. "I waited for a viable candidate to come forward, and when none did, I filed so there would be some names on the ticket,"[58] he told a reporter. He finished third.

The Democratic winner in Wyoming that year was Jim Rogers, a first-time candidate from the town of Lyman, who had to drive fifty miles to Green River to watch the primary returns on television.[59] Rogers owned the Branding Iron Motel and Lounge, where he also worked as the town bartender. "I've sat in this bar for two years and listened to the people talk, so I know what they want," he said before the primary.[60]

By the time he won, he had apparently forgotten. Asked what Wyoming voters wanted, he said: "I'm going to go out and find out. After I find out, I'll let everybody know."[61]

Cheney said that he was open to public debates. Rogers, though, "seemed taken aback at the suggestion of confronting Cheney in a debate before the voters," according to a report in the *Casper Star-Tribune.*

"I haven't any business doing anything like that," he said. "What for?" Pressed about his reluctance, he added: "I don't want to get on TV tomorrow and discuss things I have no idea what I'm talking about."[62]

The time Cheney did not have to spend on his own race he devoted to national politics, where for a junior member of the House from a state with three electoral votes, he played an outsize role. For six months, from the middle of the Republican primary season through the GOP convention of 1980, Cheney counseled his former boss, Gerald Ford, as Ford became a potential challenger of Reagan, then a possible running mate, and eventually a would-be "copresident."

George H. W. Bush won the Iowa caucuses, delivering what the *Washington Post* called a "major setback"[63] to Reagan, but six weeks later Reagan regained his footing with his unexpected twenty-seven-point victory over Bush in the New Hampshire primary. Although he was gaining among GOP voters in the primaries, polls still showed Reagan losing a general election contest to Jimmy Carter. Even more troublesome, those same polls showed Ford, a suntanned retiree living and golfing in southern California, beating Carter handily in a head-to-head race.

Ford was well aware of the polls and told friends that he was concerned that Reagan, without a stronger moderate Republican challenger, would cruise to the GOP's nomination but lose the general election to President Carter. His private musings became public on March 2, when he told the *New York Times* that Reagan was unelectable. "A very conservative Republican can't win in a national election," said Ford, before making the obvious point that Ronald Reagan was a very conservative Republican.[64]

At least initially, Cheney did not volunteer his thoughts on the possibility of Ford's running. But at a Lincoln Day dinner in Hudson, Wyoming, on February 15, Cheney declined to endorse any of the Republicans in the race, saying that it would be disloyal for him to back anyone other than Ford unless his former boss officially removed himself from consideration. "There are enough people pounding on him about that that he doesn't need to hear from me, too," he told the *Rocky Mountain News* in early March.[65]

Ford began to publicly weigh a potential run, meeting with key advisers, encouraging public officials to announce their support, and meeting with the press. In an interview in early March with ABC's Barbara Walters, he put the chance of his candidacy at "around fifty-fifty."

Reagan did not appreciate Ford's vacillation, at one point saying he hoped Ford would "hang up his golf clubs" to take him on in the race. He was not happy that his former rival had called him unelectable. "I don't believe that any Republican or any Republican candidate should say that about any other candidate," Reagan said.[66]

Throughout March, the Associated Press filed dispatches almost every day from Ford's hometown, reporting that nothing had changed. "Ford Still Hasn't Decided," was a typical headline. In mid-March, Ford met in Washington with a group of six close advisers to assess his prospects. The views were mixed, but there was agreement among the participants that if Ford were to run, his late entry would make the race exceedingly difficult.

Cheney was skeptical. Although he remained fond of his former boss and promised to support him if he decided to run, Cheney thought it a bad idea. With the primaries well under way, Ford's prospect of winning the nomination was not good, thought Cheney, and his candidacy might fatally split the Republican Party, just as Reagan's had in 1976. And unlike many of Ford's other advisers, Cheney liked Reagan.

"I finished the '76 campaign with a fair degree of respect for Reagan," says Cheney. "I never bore him any ill will or animosity for having challenged Ford, like some of my colleagues did."[67]

Most important, Cheney disagreed with Ford about Reagan's electability. "I thought he could win."[68]

Ford had promised a decision by late March. He asked Cheney, Bob Teeter, and Stu Spencer to fly to California for a strategy session. "It was a fairly short meeting,"[69] says Cheney. The three advisers gave Ford a realistic assessment of his chances; they were bleak. Cheney remembers Ford's response.

"He said something to the effect of: 'Look, I can sit here with you guys today and decide to do it, but,' he said, 'I know when you come in here tomorrow to tell me I've got to get on the airplane and where I've got to go and what I've got to do and who all I've got to speak to, and so forth, I'm going to be mad as hell at you.'"

"So that was it," says Cheney. Ford would stay retired. "And it was the right decision for him."[70]

But it was a decision Ford would revisit just five months later, this time not to stall Reagan's drive toward the presidency to but to help it. Cheney attended the 1980 Republican National Convention in Detroit, as a delegate from Wyoming. His only official responsibility was to cast a vote for Reagan, who would be the party's nominee. It was quite a contrast from Cheney's role in 1976, when he had a hand in virtually everything that took place at the convention in Kansas City. One sign of his relative insignificance was the location of his hotel. The VIPs and delegations from big states are assigned to hotels close to the convention center; states with less influence, like Wyoming, are relegated to the leftover hotels. "We were clear out in the boonies outside Detroit," says Cheney, "halfway to Ann Arbor or something."[71]

On the morning of July 16, a Wednesday, Cheney's drive to the city took on greater urgency. He had received a call beckoning him to the Plaza Hotel, where Ford was staying. For weeks there had been whispers in Republican circles about a potential "dream ticket" of Reagan and Ford. Cheney didn't take the rumors seriously. Ford's hard feelings toward Reagan had diminished, but he knew that they had not disappeared altogether. Cheney also remembered Ford complaining that his nine months as vice president under Richard Nixon were "the worst nine months" of his life. And yet, despite all this, Cheney was told that Ford was seriously considering joining Reagan on the ticket. The drama that ensued, noted a front-page story in the *Washington Post*, "transformed this convention from one of the most predictable into one of the most memorable."[72]

Before the convention, Reagan quietly asked his campaign lawyers to prepare an analysis of the constitutional issues that could arise if he invited Ford to join him on the ticket and how best to avoid them. Since Ford had relocated to California from Michigan, both candidates came from the same state, a problem specifically addressed by the Twelfth Amendment to the Constitution. Reagan gave the memo to Ford.

"I don't want to quantify the chances, but please," Ford said to Reagan, "I think I can help you more on the outside."

"Well, if you'd just take the memo and think it over," Reagan replied, "and don't answer right now."[73]

Ford agreed. As he contemplated his future, representatives from Reagan's team made clear to Ford's top advisers that Reagan wanted Ford as his running mate. Other senior Republicans—including the party chairman, Bill Brock, and several of Ford's former colleagues from Congress—encouraged Ford to take the job. Throughout the first two days of the convention, Ford consulted with his advisers, who were divided on the issue. Jack Marsh and Henry Kissinger wanted him to join the ticket.

"Henry saw this—would have seen this as an opportunity for him to reengage in a major way politically," says Cheney.[74]

The invitation to join the deliberations of Ford's brain trust came not from Ford, but from Senator Howard Baker and Representative John Rhodes, the Republican leaders in Congress. They asked Cheney and Teeter to join the discussion "because they knew of our past relationship, because we weren't part of the cabal, if you will, that was trying to engineer this thing." Cheney and Teeter, as they had been when Ford considered challenging Reagan in March, were skeptical about a Reagan-Ford ticket now. Cheney believes that Baker and Rhodes invited them to add to the ranks of those opposed to Ford joining Reagan. "Howard and John had reservations themselves."[75]

The meetings on Wednesday lasted throughout the day and came at a time when momentum seemed to be building for a Reagan-Ford ticket. By refusing to rule out serving as vice president, Ford seemed to be inviting further overtures from Reagan. He had spent hours earlier in the week weighing the pros and cons with his advisers. His conversation on Monday with Henry Kissinger lasted several hours, continuing until three AM Tuesday.[76] Ford sent word to the Reagan camp that he was prepared to consider accepting a spot on the ticket if he was guaranteed a substantive role as vice president.

Ford "put down a list of demands which he thought the Reagan people would never meet, but were so sweeping that if in fact they did meet them, he'd have a hell of a job," says Cheney. "I mean, it was everything. There was a major hand in personnel; he wanted to have a major say over the National Security Council and budgets. I mean, he just sort of threw everything out there."[77]

Cheney arrived at Ford's suite in the Plaza Hotel to find several of his party's leading lights sipping coffee and chatting among themselves. Over the course of the day, Cheney, Teeter, Baker, and Rhodes would be joined by Governor Jim Thompson of Illinois; Senator Bob Dole of Kansas, Ford's running mate in 1976; Brock from the RNC; and Representative Bob Michel, a close associate of Ford's who was serving as Rhodes's deputy in the leadership. Ford joined the group, followed a short time later by William Casey, Reagan's longtime adviser, who was there on behalf of the nominee-in-waiting. The discussion turned serious.

Casey came with specifics. The Reagan camp was willing to give Ford nearly everything he wanted. "Casey comes in at one point with a piece of paper and a list of the stuff they were willing to do," says Cheney, "which was mind-boggling. They went quite a ways in terms of what they offered Ford. . . . I can just remember being struck by how much they were willing to give."[78]

The more power Ford was offered, the more doubtful Cheney and Teeter were about the arrangement. "I remember Bob and I talking about it with each other, saying, 'This is never going to work. You can't delegate the presidency to somebody else. . . . It became a problem primarily because you can't share the presidency in that sense. It's hard to formalize that kind of relationship between the president and the vice president, sort of a treaty of who's going to run what."[79]

In the end, the negotiations collapsed and Ford withdrew from consideration. "It was doomed to fail," says Cheney, who believes that Ford never wanted the job. "I always felt Ford was asking for a lot with the expectation that he would not get it and he'd have a legitimate reason not to accept."[80]

Cheney's participation in these negotiations gave him occasion to contemplate the proper role of a vice president and the difficulties that would arise in trying to split what is essentially an indivisible job. It was an experience that would shape his thinking on the vice presidency as an institution and the kind of leaders who can serve effectively in the position.

Reagan turned to George H. W. Bush, whom he had defeated for the nomination. The ticket went on to one of the most lopsided victories in recent U.S. history. Reagan and Bush won 489 electoral votes to just 49 for Jimmy Carter and Walter Mondale. Three of the 489 electoral votes came from Wyoming, where Reagan trounced his opponent, 63 percent to 28 percent, with the remaining votes going to the Independent, John Anderson, and to minor parties.

In the congressional race, Cheney outpolled even Reagan,

winning 69 percent of the vote. He had outspent his Democratic opponent, Jim Rogers, $97,959 to $0.[81]

One of the participants in the discussions about Reagan and Ford, the House minority leader John Rhodes, had announced that he would relinquish his leadership post after the Ninety-sixth Congress. As a result, one of the first postelection orders of business for Cheney and his Republican colleagues was to choose new leadership. By early December, the race had come down to two men: Guy Vander Jagt, from Michigan; and Bob Michel, from Illinois.

Cheney's freshman classmates quickly rallied behind Vander Jagt, whom they had come to know when he ran the National Republican Congressional Committee during the election of 1978. Cheney and Representative Tom Loeffler of Texas supported Michel. Both Loeffler and Cheney had gotten to know Michel during the Ford administration when Michel was the House Republican whip. Loeffler, as a legislative liaison for the White House, worked closely with Michel to help move Ford's agenda through a Democratic Congress. Cheney watched this process from a greater distance, but he was impressed with Michel's considerable legislative skills and more impressed that Ford thought so highly of Michel.

Not long after Cheney threw his support behind Michel, a top aide to the Illinois Republican named Walt Kennedy suggested that Cheney consider his own bid for a leadership position. Kennedy told Cheney that he could win the chairmanship of the House Policy Committee, a position that would make him the fourth-ranking member of the Republican leadership, and encouraged him to run. It was something Cheney had not considered. He had assumed that perhaps one day he would seek a leadership job, but he was only a second-term congressman. Cheney worried that his colleagues might see such an effort as overly ambitious or "presumptuous." He nonetheless agreed to consider it.

As Cheney thought about his prospects, he became more encouraged. Two other young Republicans—Jack Kemp from New York and Trent Lott from Mississippi—had already declared their intention to seek other leadership posts and were favorites to win. Although they had been around longer than Cheney, their campaigns signaled something of a youth movement among Republicans eager to support Ronald Reagan, soon to take up residence at the other end of Pennsylvania Avenue. The two candidates who had already declared their intention to run—Marjorie Holt from Maryland and Elden Rudd from Arizona—were considerably older. Rudd "wasn't very impressive," says Cheney, and although Holt was a respected conservative, she was beatable.[82]

More important, Cheney understood that the overture from Kennedy meant one thing: Bob Michel wanted him to run. Kennedy didn't freelance. The only way Kennedy would encourage Cheney to make a bid for leadership was if Michel instructed him to do so. And Michel had done just that.

"Frankly, I didn't want Marjorie Holt to know that I was bucking her," says Michel, recalling that he dispatched Kennedy to talk up Cheney to others in the Republican caucus, too. "Walt was one of those who could, just in casual conversation, he could tell them: 'Bob thinks Dick would be a great guy for the job,' without my having to go ruffle people's feathers."[83]

Bob Michel supported Cheney because he thought the young Wyomingite would bring much-needed ideological balance to the new leadership team. Most of the party's other candidates were well-known conservatives. "I just thought we were collectively—as a group—we were really so conservative that we needed a moderate," Michel recalls with a laugh. "I think my own thinking was colored to a degree by the association with Jerry Ford."[84]

Cheney defeated Holt, ninety-nine to sixty-eight, and in just his second term became the youngest member of party leadership in a century. Michel, Lott, and Kemp won their races and together with Cheney formed the core of the new Republican leadership in the House.

Michel's sense of Cheney's role on the new team—"giving a voice of moderation to what was really a right-wing conservative leadership group"[85]—was echoed in the media. *The National Journal* reported that both Democrats and Republicans had "chosen a roster of leaders for the 97th Congress who are skillful at bipartisan compromise,"[86] and *Newsweek* described the new Republican leaders as "popular pragmatists."[87]

"Right-wing Republicans may be on the resurgence just about everywhere else on the national political scene," noted the *Christian Science Monitor,* "but they fared poorly in this week's GOP leadership elections in the US House of Representatives," as evidenced by the victory of "the more moderate Richard Cheney."[88]

Although Cheney had supported Ronald Reagan in 1980, his reputation as a moderate followed him from his days in the Ford administration. His voting record during his first term fit this image. According to voting records compiled by the *Congressional Quarterly,* Cheney supported policies advocated by Jimmy Carter 30 percent of the time in 1979 and 38 percent of the time in 1980.

A few hours after Reagan was inaugurated, following four months of exhausting negotiations led by the Carter White House, Iran finally freed the American hostages, who had been in captivity for 444 days. Ultimately, the United States agreed to unfreeze nearly $8 billion in Iranian assets in exchange for the safe delivery of the fifty-two hostages.

The returning hostages told horrific stories of their mistreatment at the hands of the government-backed terrorists in Iran. One of the Carter administration's negotiators, Ambassador Ulrich Haynes, warned that the deal could set "a dangerous precedent" by encouraging future acts of terrorism.[89]

But Cheney, in an interview with the Associated Press shortly after the hostages were released, discouraged any retaliation by the United States government against Ayatollah Khomeini's regime.

"I guess I feel the anger everyone else does, but that's been there for 444 days—that's nothing new," he said in an interview with the Associated Press. "We've got some long-term strategic interests in the Persian Gulf, so I think it's important that we put it [the hostage issue] behind us as quickly as possible. . . . We should not let the emotion of the moment dictate our responses."[90]

Cheney was not alone in this view. Within days, President

Reagan warned of "swift and effective retribution" for future acts of terrorism, but rejected any further response toward Iran. "I'm not thinking of revenge," he said at his first presidential news conference on January 29, 1981. "What good would revenge do? And what form would it take? I don't think revenge is worthy of us."[91]

It was the first of many times Cheney would find himself agreeing with Reagan.

When Dick Cheney began his first campaign for the House of Representatives in 1978, he was a Ford man. He was asked to run for the House leadership because Bob Michel thought of him as an extension of Ford. But over the course of the next eight years, and for much of his subsequent public career, Cheney would be closely identified with the policies and the political philosophy of Ronald Reagan. As he would proclaim more than five years into his vice presidency:

"We are all Reaganites now."

CHAPTER SIX

Leadership

When Cheney was elected to congressional leadership on December 8, 1980, his was no longer just one voice among hundreds. As chairman of the Republican Policy Committee, Cheney became the fourth-ranking member of his party in the House, with responsibility for shaping the Republican agenda that would emerge from it. In a town that glorifies power above all else, Cheney was once again a player. Even Cheney, never known for magniloquence, would later acknowledge the significance of his new position:

"It was a big deal."[1]

So, too, was his first move as a member of the House leadership: a clarion call for the weakening of the institution he now served as a leader. "Congress has been a big part of the problem," Cheney explained to Representative Richard Gephardt, a Democrat from Missouri. The two men, joined by Democratic Senator Wyche Fowler and Newt Gingrich, both from Georgia, had come to the American Enterprise Institute as participants in a panel discussion, "Revitalizing America." What started as a rather ordinary conversation about national priorities quickly evolved into an intense debate about the proper role of the executive branch in the American constitutional order. Cheney stunned his colleagues with his aggressive attack on Congress.

"A fundamental problem has been the extent to which we have restrained presidential authority over the last several years. Consumed with the trauma of Watergate and Vietnam, we have tampered with the relationship between the executive branch and the Congress in ways designed primarily to avoid future abuses of power, similar to those that are alleged to have happened in the past. We have been concerned with the so-called myth of the imperial presidency. . . . We must restore some balance between the Congress, on the one hand, and the executive branch, on the other."

"What we need is a stronger Congress, not a weaker Congress," Gingrich countered. "The greatest danger of the Reagan administration is that conservatives will decide they can trust imperial presidents as long as they are right-wing when they are imperial."

It was an extraordinary exchange between two men with little in common other than the date of their arrival in Congress and their shared belief in limited government. Cheney was reticent, almost quiet; Gingrich was bombastic. Cheney was a strong believer in the legislative process, in bipartisan compromise, in working within the system; Gingrich preferred confrontation and wanted to bring fundamental change to Congress by dismantling the existing power structure.

"Dick was not subject to knee-jerk arguments," says Leon Panetta, a Democrat from California who served with Cheney in Congress and went on to become chief of staff in the Clinton White House. "He hated people who approached things with an ideological bent. He always recognized the importance of the process, of getting something done, in contrast to Newt Gingrich, who stood off to the side and just threw bombs."[2]

The dialogue between Cheney and Gingrich on the panel provided not only a clear picture of two dramatically different interpretations of republican government, but also a preview of how two of the most consequential politicians in recent history would shape the course of politics as America entered its fourth century. In Cheney's comments were hints of the coming expansion of presidential powers in wartime, and in Gingrich's were the seeds of the Republican takeover of Congress in 1994.

"Unlike Dick Cheney, I have no faith in electing presidents who are better than congressmen," said Gingrich. "I do not think they are better than congressmen, frankly. I think it is dangerous to think of presidents as being more courageous than congressmen. They just have better speechwriters and better makeup. . . . It is good for the country to have two strong central parties that are capable of drawing lines, recruiting candidates, training candidates, and thinking in terms of national policy."

"You may be concerned about danger," Cheney shot back, "but in times of great national peril in the past—during the Civil War, or during the Depression, or during World War II—we have basically responded, as a society, to those crises by having the president, whoever he was, assume extraordinary authority. If you are as concerned as I am about the 1980s, we are talking about a situation again, when, at a minimum, we must reduce the trend of the last few years, or we will have undermined presidential authority rather than granted additional authority. . . . We in the United States are likely to find that sometime during the decade of the 1980s we will have to resort to force someplace in the world. If we do not, it will be a decade unlike any we have seen in the last half century. While certainly we want to advocate those principles by which we, ourselves, live—democracy, freedom, individual liberty, and human rights—it is also important not to let that create a smoke screen in front of our own eyes when it comes time for us to assess the national self-interest and the interests of our allies overseas."[3]

Cheney was saying the same things in private that he was saying in public, and to a far more receptive audience. Shortly before Ronald Reagan was inaugurated on January 20, 1980, Cheney spoke with an old friend who was about to become very powerful.

Jim Baker had been an eager young political appointee at the Department of Commerce when he was chosen to work as a delegate counter for Gerald Ford at the Republican convention in 1976. Cheney was impressed with his performance and invited him into the campaign's inner circle for the two months before the general election.

Baker remembers Cheney as an effective and unassuming

chief of staff, "a far cry from . . . some hard-eyed chief of staff regimes in previous administrations."[4] He came to this view from personal experience. During the 1976 campaign, Baker told a group of Ford's supporters in Oklahoma that Henry Kissinger would not be likely to have a role in a second Ford administration. His off-the-record comments ended up in a story published in the *Daily Oklahoman* and were later distributed nationally in wire reports. Kissinger was not happy, and Baker was called in to see Cheney. "Don't worry about it," Cheney told him. "Just make it right with Henry."

"Another chief of staff might have reamed me out on the spot—and would have been justified in doing so—but I remember thinking that Dick had taught me a lesson about being cautious in a very graceful fashion," Baker said.[5]

Four years later, as Baker was preparing to take over Cheney's old West Wing office as Ronald Reagan's chief of staff, he reached out to his former boss for advice. Baker was coming in as one-third of a powerful three-man team that would run the White House. The other two—Edwin Meese III and Michael Deaver—had been with Reagan longer and knew the president-elect better. If Baker were to avoid being shouldered to the side, he would have to control the system, and Cheney knew how to do that.

Baker took four pages of notes during their long discussion. Some of the advice was technical: "Orderly schedules and orderly paper flow [are] the way you protect the president," Cheney had said. "Don't use the process to impose your policy views on other people."[6]

Cheney advised Baker to "keep a low profile" and "talk to the press always on background." He warned: "If you become a major public figure you lose credibility, feathering your own nest rather than serving the president."[7]

Of the many things they discussed that day, nothing was more important to Cheney than the balance of power in the federal government. "Restore power and authority to the executive branch," Cheney told Baker. It was the first item on his long list.

It is altogether fitting that Cheney would begin his service in the leadership of the legislative branch with an admonition about strengthening executive power, as he would end his congressional career with an almost single-minded devotion to preserving presidential prerogatives, first through his role on the committee investigating the Iran-contra scandal and later with a parting blast at "congressional overreaching."

The years in between were momentous. Ronald Reagan brought to Washington, D.C., a worldview that was a stark contrast to the prevailing ideologies of the twentieth century. After fifty years of activist government, with its New Deals and Great Societies, Reagan vowed to reverse course. Not only is government not the solution, he would declare; it is often the problem. Reagan's tax cuts would make the federal government leaner and spending cuts, his critics complained, would make it meaner.

And after a decade of détente, Reagan would choose to confront communist expansionism and to reframe the debate in distinctly moral terms. Communism was not just different. It was wrong. The Soviet Union was not just a geopolitical rival. It was evil. Totalitarianism was not to be tolerated. It was to be challenged.

It was a bold new conservatism in both domestic and foreign policy, and one that Cheney embraced.

Cheney's conversation with Jim Baker marked a new phase in his congressional career. The first lasted from December 1977 through December 1980. It began with a competitive and successful bid for his House seat and ended with a competitive and successful bid for House leadership. In between, Cheney focused his efforts on matters of particular interest to Wyoming and issues related to his committee assignments. In that sense, he was like many Wyoming politicians who had preceded him.

The second phase began with his election as chairman of the Republican Policy Committee and ended with his resignation from the House in March 1989. In between, he would steadily accumulate power and take on more visible roles in conducting the business of the House. He rose from the fourth-ranking member of the leadership to become Bob Michel's number two, won a seat on the intelligence committee, and served as the highest-ranking Republican House member on the committee established to investigate Iran-contra. Gone were the days when Cheney was stopped in an airport in Peoria and mistaken for John Dean. By the end of his time in Congress, Cheney was becoming a well-known national political figure. When he was on a trip to Italy in the late 1980s, a family of Americans stopped him in the train station in Rome because they recognized him as a chairman of the Iran-contra hearings.

In the 1980s, Cheney was his own man. It was Dick Cheney in his purest form. During that period, Cheney never had a serious challenge for his congressional seat or a leadership job. Those leadership years multiplied Cheney's day-to-day responsibilities and required that he involve himself in a much broader range of issues.

"It made the House a lot more interesting for me," he says.[8]

On January 30, 1981, Cheney's fortieth birthday, he had an appointment at three PM to interview a prospective legislative aide. He had not had many such openings during his short congressional tenure. He had tried to hire Fritz Steiger, a nephew of his late friend, who didn't want to relocate to Washington, D.C.

Shortly before the appointed time, Jim Steen reported to Cheney's office, Room 225 of the Cannon House Office Building. Steen had worked alongside Cheney in Steiger's office in the late 1960s, and Cheney remembered him fondly. During the intervening years, Steen had gained experience working on the House Interior Committee, where Cheney continued to serve.

Cheney was uncharacteristically late for the interview, showing up at about three-twenty. He invited Steen in and they chatted about the job and Steen's background. The interview was going well when Cheney stopped it.

I've got to give a speech at three-thirty, he said. It's upstairs and it won't take long. Would you like to come along? Steen agreed, and Cheney promised that they'd finish on the other side of the speech. Already, Steen thought it was a job interview unlike any he'd ever had before.

In a large room one floor above Cheney's office, the congressman met with a group of women government executives. "I sat there as Dick talked to this rather hard-bitten group of women government employees, who were very upset at something that Reagan had proposed that would impact them negatively," Steen recalls. "Cheney listened carefully, but then in effect said, 'Look, we're going to have to agree to disagree on this one.'"[9]

It wasn't what they had wanted to hear, says Steen, but "they grudgingly made it clear that they admired his candor—and not that he completely won them over, but they were clearly taken aback by plain speaking." It was not what Steen had become accustomed to seeing on Capitol Hill, and he decided at that moment that he would take the job if Cheney were to make him an offer.

They returned to Cheney's office to resume the interview. Lynne was waiting there with one of the Cheney girls and the family dog to help celebrate his birthday. Cheney asked them to wait outside so that he could finish talking to Steen. They obliged, but left behind Cyrano, a friendly basset hound that made himself comfortable at Steen's feet as the interview proceeded.

In the tense environment of a job interview, where saying or doing the wrong thing can kill the deal, Steen now had something else to worry about. Do I acknowledge the dog? Maybe pet him? Or do I just ignore him? Steen chose the final option and Cyrano remained quietly at his feet, looking at him plaintively, spurned.

When the interview was over, Cheney invited Steen to stick around for some cake with his family and the staff. He did, and he got the job.

Steen and his legislative colleagues would spend a good part of 1981 working to pass what would become the signature policy of Reagan's first presidential term: tax cuts. Marginal rates had climbed throughout the 1970s, and the top rate when Jimmy Carter left office was 70 percent. Reagan campaigned promising massive tax cuts—in 1976 and again in 1980—and his landslide victory gave him momentum going into the debate. Republicans had gained control of the Senate on Reagan's coattails; and although they remained a minority in the House by more than sixty seats, a good number of southern "boll weevil" Democrats were conservative and sympathetic to the White House's economic policy.

Democratic leaders on Capitol Hill, led by the Speaker of the House, Tip O'Neill, initially fought Reagan's tax proposal as irresponsible at a time when the country was struggling with inflation. But the tax cuts, like the president pushing them, proved popular. Democrats eventually offered an alternative that would have lowered rates, but not nearly as dramatically as the White House plan. The fight intensified over the summer, as members of Congress prepared for the recess in August that would take most of them back to their districts, where they would hear from the voters.

The Republican National Committee, together with Cheney's Republican Policy Committee and the House Republicans' campaign committee, launched a $600,000 advertising campaign—labeled an "extravaganza" by one newspaper—designed to sell the tax cuts.[10] Cheney held a press conference to announce the effort, which included TV spots to run in the Washington, D.C., media market and radio ads to be broadcast in congressional districts of selected Democrats. The ads depicted O'Neill as a deceptive Santa Claus, hawking phony tax cuts. Cheney explained the strategy to reporters.

"It's obviously useful to us, to get our point across, to focus on the speaker. It's not an attack on him personally," Cheney insisted. "I think he's a fine gentleman."[11]

Cheney's work on the tax cuts earned him praise from the president, who thanked him by name at a ceremony on Capitol Hill to launch the final push to enact what Reagan called "the most crucial item left on our agenda for prosperity."[12] In the end, forty-eight Democrats joined their Republican colleagues in passing a 25 percent tax cut. A vote one week later on the compromise bill that emerged from the Senate passed with an even greater margin.

If Cheney supported Reagan's domestic agenda, his initial impression of Reagan's foreign policy operation was less favorable. At a meeting of public policy analysts, Cheney was disappointed that Reagan was not interested in international affairs. "Most presidents are absolutely enthralled with foreign policy, but that hasn't happened with Reagan. He doesn't seem to have the hunger to get into that area as Nixon and Ford did. He does what he has to do and that's all." Cheney was equally critical of the White House foreign policy apparatus. "I don't see any creative or aggressive effort by the Reagan security council staff. They have downgraded the NSC too much."[13]

Although Cheney did not have committee assignments that

required him to focus on national security, he continued his education—and burnished his credentials—by participating in several groups examining those issues, both inside and outside Congress. Cheney served as cochairman of the "Georgetown Study Group on Strategy," a collection of academics and policy makers—Republicans and Democrats—examining the future of American foreign policy. Cheney believed that the major obstacle to producing such a national strategy was a lack of consensus on two basic issues: the nature of the threat from the Soviet Union and the role of the United States in the world.

In handwritten notes he prepared for one session of the Georgetown group, Cheney laid out the two contrasting positions. The first held that "we are the last best hope of man on earth, the leader of the free world, a morally superior force for good in the world, justified in taking steps to promote democracy and defend freedom." The second was that "the U.S. is—from a moral standpoint—no different from the Soviets; that much of the troubles in the world stem from too much U.S. aggressiveness, too little U.S. understanding of third world developments and too little U.S. sympathy for the problems facing the Soviets; that U.S. defense spending and assertiveness [are] provocative and unjustified in spite of Soviet aggression" and that "the key requirement for U.S. congressional policy makers is to restrain [the] U.S. president—not Soviet leaders."[14]

In Congress, Cheney's leadership post enabled him to become more involved in national security issues. Although he had his differences with Gingrich, Cheney was impressed with the Georgia Republican's thinking about defense. Gingrich had been deeply involved in creating the Military Reform Caucus, an odd collection of liberals and conservatives interested in national defense policy and in changing the way the Pentagon conducted business. Although the caucus was in its infancy, it was unusually influential, owing in part to the stature of its members and their rejection of conventional thinking on important military matters. "It wasn't enough just to build up the military," Gingrich would later say, in summarizing the group's activities. "You also had to rethink what we were doing and how we were doing it."[15] Among the strongest advocates of reform were congressmen whose commitment to a robust national defense could not be questioned—Democrats like Sam Nunn of Georgia, Gary Hart of Colorado, and the future secretary of defense Les Aspin; in addition to Gingrich and Cheney, the Republicans included Senator Charles Grassley of Iowa and Nancy Kassebaum of Kansas.

Gingrich encouraged Cheney to join the young group and introduced him to Colonel John Boyd, a fighter pilot who was rapidly gaining a reputation as the leading intellectual of the military reform movement. Boyd was a character—he once wore a bright orange jacket and plaid pants to testify before Congress—but even those who disagreed with his views respected his brainpower. Cheney met privately with Boyd on a regular basis, discussing a range of issues that included military strategy and the peculiarities of the Pentagon procurement process. Pierre Sprey, a colleague of Boyd's who often sat in on his tutorials with the congressman, thought Cheney "did his homework" and "studied deeply the intricacies of Boyd's approach to strategy."[16]

Boyd's interest in military reform grew out of his experience as a fighter pilot. For years he watched the Air Force add high-tech equipment to the fighter planes he flew, expensive equipment of dubious value. Boyd was known as the "father of the F-15," so when he criticized excessive spending, he spoke with authority. Over time, Cheney would give great weight to Boyd's thinking.

As American politicians debated the future of Boyd's creations, those planes made news for another reason. On June 8, 1981, the Israeli Air Force used American-made F-15 and F-16 fighter planes to launch preemptive strikes on the Osirak nuclear reactor outside Baghdad, Iraq. Israel had gathered intelligence indicating that the Iraqis were using secret underground bunkers attached to the facility to build nuclear bombs, weapons it was certain were meant for Israel.

The attacks were widely condemned. The Reagan administration swiftly issued several statements denouncing the Israeli action and suspended plans to ship four more F-16s to Israel as it reviewed whether the strikes violated a 1952 agreement between the United States and Israel that restricted the use of American-made planes to defensive actions.

At the United Nations, the U.S. ambassador, Jeane Kirk-patrick, voted with the other fourteen members of the Security Council to condemn the attack. *Newsweek* magazine called this "one of the toughest rebukes Washington had ever delivered to Jerusalem."[17]

The reaction in Congress was divided. Many legislators warned their colleagues not to judge the strikes until they had more information, and a smaller group openly supported the attacks. Although some in the media predicted that the strikes against Osirak would provoke a broad discussion of preemptive war, that debate never materialized. Cheney did not comment publicly about the Israeli action, but he was impressed by the precision of the strikes, and he thought they were strategically necessary. "I can understand why they did it," he says. "They didn't want any nuclear powers in their neighborhood."[18]

Still, Cheney did not regard Saddam Hussein, distracted by his war with Iran, as a threat to the United States. "There wasn't a warm and friendly relationship by any means, but we didn't want the Iranians to win," he says, recalling that he once met with the Iraqi ambassador to the United States at the Iraqi embassy in Washington.[19]

Even as Cheney involved himself more deeply in national security issues, he remained attentive to "the Wyoming issues." He still spent the equivalent of two or three days each week on subjects that most Americans never contemplated: water rights, snowmobiles in national parks, wolves in Yellowstone, protecting livestock from wild predators, grazing rights on public lands, and energy exploration on federal government property.

Throughout his time in Congress, Cheney employed a state representative who lived and worked in Wyoming. The state rep, as he was known around the office, worked closely with constituents and handled local issues. When the congressman was back in the district, over a weekend or during a congressional recess, the state rep performed duties both mundane and important. He was a personal aide, driver, scheduler, press liaison, speechwriter, and policy adviser.

When the state rep's job came open in the spring of 1983, Cheney's office put out word that he was hiring and began to solicit résumés. On May 14, Cheney interviewed several candidates on the campus of the University of Wyoming in Laramie, in a large classroom with auditorium-style seating. Cheney sat at a desk at the bottom of the bowl as candidates filed in for their meetings.

Merritt Benson was young, tall, and athletic, with a mustache that made him look like the Marlboro man. He came to his interview with impressive credentials and good references. He had worked as the associate editor of *Outdoor Life* magazine before taking a job with the Wyoming Department of Fish and Wildlife; his wife worked for Malcolm Wallop, Wyoming's senior senator, like Cheney, a Republican.

When Benson walked into the room, Cheney stood, offered a brief greeting, sat down again, and then plopped his feet on the desk in front of him. He glanced at Benson's résumé. The conversation was short.

"I bet you know about a few fishing holes in this state," Cheney said.

"Yeah, I do."

"Well, I never travel anywhere without my pole in my trunk," said Cheney.

Benson had the job. Cheney's friends and family teased him about the outdoorsmen he chose to be his state representatives. Paul Hoffman would succeed Benson, and Cheney would unconsciously refer to him in an interview as "my outfitter and my state rep." Cheney makes no apologies. "Got to have your priorities straight," he explains with mock defiance.[20]

Fishing was such a priority that it was often included on Cheney's schedule, his appointment with cutthroat trout on the Wind River just down the page from his commencement speech in Thermopolis. Although Cheney had been fishing nearly all his life, he was in the early stages of becoming a serious fly fisherman when he met Benson. It was good timing.

"When we started, he really had no clue what he was doing," Benson recalls. "He had this bamboo rod and an automatic reel. It was way out of balance. I don't know how you could fish with that thing. Bamboo looks good in museums, but it stinks to fish with it. It's like a limp spaghetti noodle."[21]

Benson would work for Cheney for the next two years. Whenever the congressman returned home to Wyoming, they traveled the state together—sometimes in small charter planes, usually in cars—and fished together. "I'd pick him up, go with him wherever he needed to go, and I'd say, 'Hey, there's a creek nearby. Let's stop there after.'"[22]

On August 7, 1983, Cheney began a three-day staff retreat at Flat Creek Ranch, just north of Jackson. The Cheney family had visited there two years earlier, a trip that inspired Lynne Cheney to write a short history of the property for *The Annals of Wyoming*. The fishing was good, a fact that made it the perfect place for Cheney to hold a series of long-range planning meetings with his aides.

As often happened when those based in the district got together with the Washington-based staff, discussion centered on improving communications between the two groups. Benson noticed that Cheney did not look well. "He was chalky and gray as a ghost."[23]

Their cabin was not equipped with a telephone, and cell phones were not yet prevalent. Waiting for an ambulance would take too much time, so Benson loaded his boss into the backseat of his Suburban. Cheney's daughters were "freaking out," says Benson, and his wife, remembering her husband's first heart attack five years earlier, was "shaken."[24]

The fifteen miles of road from the secluded property into town are unpaved. The first five miles are actually a dried streambed, described by the ranch owners as "a bumpy jeep trail in the mountains." As the Suburban slowly jolted and jostled its way to the main road, Lynne, riding next to Benson in the passenger seat, was growing impatient. Benson knew that speeding was not an option; the last thing they needed was for the Suburban to bottom out in the middle of nowhere. But he shared her sense of urgency.

"Go faster!" she said, with panic in her voice. "Go faster!"

Cheney spoke up from the backseat. "Lynne," he said, "Merritt knows what he's doing. We'll be fine."

In the end, he was right. The hospital declined to classify the episode as a heart attack, but it was a sign of things to come.

When Cheney was in Washington, Benson spent much of his time meeting with constituents and interest groups on the Wyoming issues. The most contentious of these, exploring Wyoming's wilderness areas, had its roots in federal policies developed over the previous decade. In 1972, the U.S. Forest Service undertook a project known as RARE I—the Roadless Area Review and Evaluation—a study that sought to codify undeveloped lands across the country. Some of the land would be classified as "wilderness," a designation that would bring restrictions on the activities that could take place in those areas. "You can't use chain saws. You can't have any vehicles—you can only go in on foot or horseback," says Cheney. "There can't be any energy development or any dams built or any oil and gas production or mining and mineral activity. No grazing."[25] In 1979, the Forest Service conducted a second review, RARE II, that designated even more land as wilderness.

Cheney was concerned about the economic impact of the new restrictions on his state and said that Wyomingites resented what they regarded as the paternalism of outsiders, including federal regulators and environmental activists on both coasts. He summarized those feelings this way: "Folks in New York and San Francisco know what it's like in August in Jackson Hole, I mean, it's beautiful—but they don't have to live here year-round through the winters and make a living. And we do. We've got to raise our cattle and sheep; we've got to drill for oil and gas; we've got to mine our coal and our uranium and we're sick and tired of you guys telling us how to operate. . . . We've kept and protected our state and taken good care of it. If they're so smart and so interested in how we manage our business here—we're the guys with the pristine areas—what happened to their areas?"[26]

In an interview with Adam Clymer of the *New York Times*, Cheney said, "We don't have a lot of sympathy for the New Yorkers who belong to the Sierra Club."[27]

National coverage of wilderness issues in Wyoming was not common; most of the battles took place in the bowels of the U.S. Capitol, in the offices of the House Interior Committee's subcommittee on public lands. The chairman of that subcommittee was John Seiberling from Ohio, who was, according to Cheney, "a very liberal Democrat, all-the-way, balls-out, wilderness advocate and environmentalist."[28] For obvious, reasons they did not approach concerns about public lands from the same viewpoint.

Cheney thought Seiberling and the conservationists who backed him were so concerned with protecting the environment that they failed to account for the economic impact of the policies they were pushing. The environmentalists, in turn, claimed that Cheney simply did the bidding of Wyoming's business community. Cheney acknowledges that "a lot of my financial support came from people in the energy industry," but says his views represented a majority of Wyomingites. "Most of the people who voted for me in '78 were not wilderness advocates," says Cheney. This does not mean, however, that they were anti-environment. "It's important," Cheney says. "The wilderness is part of our heritage and you want to preserve and protect it; you don't want it all developed."[29]

Throughout the early 1980s, Cheney often found himself in the middle of disputes between environmentalists and Reagan's combative secretary of the interior, James Watt, a native of Wyoming. As a rule, Cheney was not sympathetic to the environmental lobby. Out of a possible 100, he earned ratings of twelve, six, and three from the League of Conservation Voters from 1980 through 1982. Still, Cheney fought off several plans by the Reagan administration to open up additional wilderness areas in Wyoming to energy development. His views were simple and uncompromising. "We want the northwest corner of the state left alone. Ninety percent of the state is up for development and we want the other 10 percent as it is."[30]

These occasional battles with Watt apparently didn't affect how the White House viewed Cheney. Watt resigned in September 1983 after describing members of a congressional panel investigating his tenure at Interior as "a black, . . . a woman, two Jews and a cripple." Almost immediately, the news media picked up on speculation that the White House was considering Cheney as a possible replacement.

On October 12, 1983, Cheney was meeting with constituents at the Niabora County fairgrounds in Lusk, Wyoming, when Benson, traveling through the state with Cheney, was beckoned to the telephone. "Someone came and got me and said, 'You've got a call from Jim Baker at the White House,'" Benson recalls.[31] Benson took the call in a small adjoining room. Baker told him that President Reagan was interested in considering Cheney for the vacancy at Interior. Benson explained that Cheney was finishing a meeting with constituents and promised that Cheney would be in touch when that session was over.

Cheney and Baker spoke briefly. The White House chief of staff told Cheney that the president was interested in offering him a job in the cabinet. Cheney asked for some time to consider the offer.

"We talked about it all the way to Moorcroft," says Benson, who drove Cheney to his next event. "The road from Lusk to Moorcroft is the kind of road where you can take your hands off the wheel for ten minutes and let the car drive itself. He was in the passenger seat drinking a beer and talking about it in an almost giddy fashion."[32]

Benson knew Cheney was excited about the opportunity because he didn't stop talking until they arrived in Moorcroft. "We spent the entire drive with him talking about the pros and cons and me asking him a question every now and then."[33]

Cheney was interested in the job, but he wasn't sure it would be a good career move. He had become an important member of the Republican leadership, and although he would be running a cabinet agency, Interior wasn't exactly in the middle of the major policy debates in the Reagan administration. In the end, the cons outweighed the pros and Cheney decided to remain in the House.

Back in the capital, the focus returned to foreign policy. On October 25, 1983, President Reagan dispatched some 2,000 U.S. troops to Grenada in an effort to put down a Cuban-backed Marxist coup that had taken place two weeks earlier. The coup targeted the Grenadian prime minister, Maurice Bishop, who was also a Marxist, and resulted in his ouster and later his death. Cold war tension in Central America and the Caribbean had been building for years, and denying the Soviet Union a foothold in the western hemisphere was an important component of Reagan's anticommunist foreign policy.

Leaders in the region who were friendly to the United States asked it to respond. In the months before the uprising Grenada had been building a new 10,000-foot airfield, helped by Cuban and Soviet workers and, the U.S. government suspected, backed by communist regimes in Cuba and the Soviet Union for possible military use.

Further complicating matters was the presence of approximately 1,000 American medical students taking courses at the country's St. George's School of Medicine. With fresh memories of the Iranian hostage crisis that terrified the nation—and crippled his predecessor in the 1980 elections—Reagan wanted to avoid a similar situation in Grenada. And given the aggressive hostility the new regime had shown toward America, Reagan was concerned for the students' safety. (They could be shot on sight if they were spotted in public areas after a government-imposed curfew.)

Reagan addressed the nation on the night "Operation Urgent Fury" was launched. America's objectives, he said, were threefold: "to protect our own citizens, to facilitate the evacuation of those who want to leave, and to help in the restoration of democratic institutions in Grenada."[34]

Many Democrats challenged Reagan's decision. The following morning, Speaker of the House Tip O'Neill said that Reagan's "policy scares me. We can't go the way of gunboat diplomacy. His policy is wrong. His policy is frightening."[35] Senator Lawton Chiles of Florida, a Democrat, asked rhetorically, "Are we looking for a war we can win?"[36]

The critics, including O'Neill, were angry that Congress had not been adequately consulted before the invasion. Senator Charles Mathias Jr. of Maryland, a Republican who was a ranking member of the Senate Foreign Relations Committee, complained that "congressional leaders were simply called to the Oval Office and told that the troops were under way. This is not consultation. The Prime Minister of Great Britain was advised about the invasion before the President told the Speaker of the House of Representatives or the Majority Leader of the Senate."[37]

"We weren't asked our advice," grumbled O'Neill. "We were informed what was taking place."[38]

On October 27, two days after the invasion began, television networks showed the return home of the American students who had been rescued by U.S. troops. "I have been a dove all my life," said one student who was rescued. "I just can't believe how well those Rangers came down and saved us."[39]

"I really loved Grenada, but I fully support President Reagan's move," said another student. "He really did save our lives."[40]

Despite some mistakes, the operation was popular with the American public. A telephone poll conducted by ABC found that callers favored the invasion by a margin of eight to one.[41] Opponents of the invasion fell silent.

President Reagan addressed the nation on prime-time television that evening. "Grenada, we were told, was a friendly island paradise for tourism. Well, it wasn't. It was a Soviet-Cuban colony, being readied as a major military bastion to export terror and undermine democracy. We got there just in time."[42]

Shortly after the hostilities ended, Cheney was one of fourteen members of Congress—Republicans and Democrats—dispatched by O'Neill to Grenada on a fact-finding mission. The evidence that he and others found confirmed Reagan's assertions. It was enough to persuade O'Neill and other Democrats that their initial misgivings had been wrong. One member of the delegation, Representative Michael D. Barnes, a Democrat from Maryland, confessed, "I came down here very skeptical, but I've reluctantly come to the conclusion that the invasion was justified."[43] After a two-hour meeting with members of the delegation, O'Neill admitted, "I believe that sending American forces into combat was justified under these particular circumstances."[44]

For Cheney, Grenada was an important lesson. His trip to the island following the invasion convinced him that a mission must be completed once it is started. After the fighting, he urged that some American troops remain on the island to restore order and to ensure Grenada's return to a constitutional government. "Most of us came down with a strong feeling that we should get our guys out as soon as possible," he said during his trip. "What happened was we met all kinds of Grenadians, including the Governor General, who implored us not to leave too soon during this period of transition."[45]

"The local people we saw made a strong case that since the United States mounted this mission, we have an obligation not to turn our backs on them now."[46]

A month after the United States' invasion of Grenada, Cheney continued his campaign on behalf of a strong executive at a panel discussion once again hosted by the American Enterprise Institute.

"I was present the morning of the Grenada event when the president met with a larger group and with the bipartisan leadership at the White House. This was at the time the invasion was under way. Not one member of either party in the room, of some thirty members of the House and Senate, raised a question about the operation. But three days later, the Speaker was publicly condemning the president for gunboat diplomacy. . . . How does the president deal with that kind of situation if he is trying to use some kind of reasonable, rational decision-making process?"[47]

That Cheney saw his colleagues' vacillation firsthand prepared him for many similar moments he would observe as vice president.

Cheney spent much of the debate challenging Senator Mathias, a fellow Republican who was a vocal proponent of enhanced congressional power. A strong executive branch, Cheney maintained, is not a partisan matter.

"When Jimmy Carter was president, I was not very pleased with his style of operation or the quality of the decisions made in his administration," he said. "But for the time a president is the constitutional president of the United States, he has the authority to make those kinds of decisions and judgments on behalf of all of us. If he makes a mistake, obviously we pay a price for it; but we have to trust him to make certain decisions. To keep coming back to the notion that every set of circumstances in which military force might be used lends itself to consultation and legal arguments is nice, but the world doesn't work that way."[48]

Although Cheney spent his days in the trade-and-barter world of Congress, he spent his nights pursuing his decades-long academic interest in the House of Representatives as a deliberative body. Although he never completed his doctoral dissertation on Congress, Cheney put his study to use in *Kings of the Hill:*

Power and Personality in the House of Representatives, a book
he wrote with his wife. *Kings of the Hill* examines the lives and
leadership of eight of the most powerful and colorful leaders in
the House of Representatives.[49]

The book included very little of Cheney's personal experience,
but the introduction gave readers some sense of how Cheney
looked at his job. "As one might suspect, the task of leadership
in such an institution has seldom been easy or predictable. Persuading ambitious and competitive individuals to follow another's lead is hard work under any circumstance, but in the House,
where there are so many members all pressured by intense and
diverse forces, it is an especially formidable undertaking."[50]

A review in the *Washington Post* described the book as "an
unabashed work of love," but criticized it for "inconsistencies of
narrative voice" and "confusing history."[51] Other reviews were
more positive. Dan Rather, in his third year as anchor of the *CBS
Evening News* after taking over for Walter Cronkite, endorsed
the book in a blurb for its back cover. "Superb history. Lively
writing. An excellent history book that doesn't read like one. I
loved it, and learned from it." *Kings of the Hill* also occasioned
a puff piece in *People* magazine's "Couples" column, under the
saccharine headline, "In Politics and Now in Print, Wyoming's
Dick and Lynne Cheney Go a Country Mile for Each Other."
Cheney, according to the story, "is regarded on Capitol Hill as
one of the GOP's rising stars."[52]

Twenty years after *Kings of the Hill* was published, it is still
in wide circulation in the Capitol. "One of the best books ever
written on the Speakers of the House was the book that he and
his wife had written," says a former speaker of the House, Dennis Hastert. "Almost everybody who ever comes into Congress
reads that book."[53]

Cheney's formal duties as chairman of the Republican Policy
Committee included presiding over weekly meetings for House
Republicans to hash out the formal policy positions of the caucus. Cheney served mainly as a facilitator, soliciting input from
conservatives and moderates alike, a role not unlike the role he
played in the Ford White House. It was an important job, but
over time, his colleagues regarded Cheney as a free-agent mem-

ber of the leadership whose actual power outstripped his formal title. Although he was the fourth-ranking Republican, he was widely considered Bob Michel's right-hand man. He had good relations with House Democrats, and he had strong ties to the Reagan White House.

Despite all the power Cheney was accumulating, his personal office remained low-key, even casual. Everyone on the staff called him "Dick," and Cheney would occasionally drop by "the Hindquarters," a nickname given to the collection of cubicles that housed the legislative staff. In the afternoon, the entire office suite would be filled with the smell of fresh popcorn, and Cheney would often take his helping back to his desk in a coffee filter. His staff worked hard, but kept reasonably normal hours. When Cheney worked on Capitol Hill as a staffer, he had a two-year-old daughter at home and a wife who was finishing work on her PhD. He often left at the end of the day to cook for his young family, so he understood that his staff would have priorities other than work.

As Cheney's influence in Congress grew, he faced continued difficulties with his health. On September 12, 1984, Cheney was working in his office when he began feeling queasy. Erring as he always did on the side of caution, Cheney paid a visit to the Capitol Hill physician to see about adjusting the dosage of his heart medication. The doctor recommended that he see a specialist at Bethesda Naval Hospital. Initial electrocardiogram tests did not show that Cheney had had a heart attack, but a subsequent blood test confirmed his suspicions. In a statement to the media, his press secretary Patty Howe said that Cheney believed the heart attack was not as severe as the one he suffered in 1978. Cheney remained hospitalized for a week as he underwent further testing.

Cheney stayed active after his second heart attack despite knee problems dating back to his days on the Natrona High School football team. He injured the same knee during a touch football game in Washington in the early 1970s when, in the tough-guy version of the story Cheney used to tell, "somebody clipped me from the side and the knee folded up." It turns out, however, that he only felt as if he had been hit.

"Somebody was making home movies and had it all on film,"

he says, shaking his head with exaggerated embarrassment. "And nobody was near me. I was all by myself in the middle of the field and it just gave out. So I couldn't even claim sports injury."[54]

The injury marked the end of Cheney's football career, but he continued to ski and remained an avid tennis player. Skiing on bad knees is a high-risk activity, but it was tennis that ultimately led to more trouble. The fateful injury came in a pitched battle on the public courts at Pineview Elementary School in Casper. Cheney was playing doubles with his wife, her brother Mark, and Mark's wife, Linda. Lynne had kept up with the game since taking it up on the trip to Eleuthera with the Rumsfelds. She had become pretty good, and as in everything she had ever done was an intense competitor. "We played like it was the championship of the world," recalls Mark.[55]

At one point in the middle of a long rally, the congressman stretched to reach a ball whizzing by his side. His knee buckled and he crumpled to the ground clutching his leg. The severity of the injury was obvious, as Cheney, never known for drama, screamed out in pain and rolled from side to side holding the damaged joint. His partner had returned a volley and Lynne Cheney, on the other side of the net, continued as if nothing had happened.

"They kept playing right over me," says Cheney.[56]

Back in Washington, the Republican leadership began to consider committee assignments for the next Congress. Cheney went to Bob Michel with a special request: a seat on the Intelligence Committee. "He thought it very important," says David Gribbin, Cheney's chief of staff. "He wanted it."[57]

Cheney had been a consumer of intelligence in the Ford White House, so he would bring a basic familiarity to his work on the committee. But oversight required much greater knowledge. So Cheney began to study. He bought a series of books that had been recently published by the Consortium for the Study of Intelligence and edited by a professor at Georgetown University, Roy Godson. Cheney scribbled his thoughts on the inside flap of one of those books, *Intelligence Requirements for the 1980s: Elements of Intelligence.* The list touches on the pro-

fessional, the personal, and the political, and begins with several
basic questions:

> *What is the product of the intel community? Who receives
> it? What do they do with it? How is an NIE put together?*

Cheney gave himself several assignments as he sought to im-
merse himself in intelligence policy making:

> *Review past NIEs and NIDs and see if they were accurate.
> How do they look in hindsight.*
>   *Evaluations of how well we're doing—does it work—
> what are our strengths and weaknesses—case studies.*
>   *Review of our relations w/foreign services.*
>   *Assessment of damage done to us by Soviet efforts over
> the years.*
>   *Do history of disclosures of 1970s—Church and Pike—
> Press accts, etc.*

Another assignment demonstrates Cheney's concern with
having an impact during his tenure on Intelligence:

> *Objectives over next 6 years—personal-institutional-
> legislative, etc. See old pros—ask their views—how to leave
> a positive legacy.*

Other subjects Cheney highlighted would figure prominently
in the controversies over intelligence at the center of the Bush administration some twenty years later:

> *Coordination between CIA, State and DoD on key areas*
>   *Consider security procedures for House Intel Commit-
> tee—look at downside potential if they fail*
>   *Develop long-term strategy for dealing w/Demo domi-
> nated Cong—importance of GOP takeover*
>   *FISA '78*
>   *FOIA details*

Intelligence Identities Protection Act—1982
Look at current weaknesses in counter-intel in light of
growing terrorist threat

Cheney's work on the Intelligence Committee consumed much of his time. His daily schedules from the mid-1980s include intelligence-related meetings nearly every day. When he first joined the committee, two of its staffers, David Addington and Steve Berry, visited Cheney's personal office to brief Jim Steen, a senior legislative assistant, on the requirements of committee work. After that session, however, Cheney kept his work on the Intelligence Committee strictly segregated from that of his personal office.

Cheney, as a member of the leadership, quickly became an important resource for minority staffers such as Addington, a lawyer with the CIA detailed to the Intelligence Committee. When they ran into problems they first consulted the senior Republicans on the committee and then went to Cheney. "He was the guy who could get things done," one staffer recalls.[58]

There were two intellectual forces on the minority side. Representative Henry Hyde, a veteran Republican from Illinois, was the "fire and brimstone, anticommunist" member who could be counted on to make passionate public arguments appealing to the ideals of freedom and justice. Cheney was a quiet, behind-the-scenes operator, who toiled away on the details of intelligence funding and operations. In committee meetings, he said little, occasionally jotting down a word or phrase on his yellow legal pad. But those notes often resulted in follow-up action directly with the intelligence community. Cheney developed a reputation as a committee member with lots of questions who would occasionally make trips out to Langley for an in-depth discussion of an issue that had caught his interest.

"He was a good, thoughtful critic, and that's exactly, from my perspective, what you wanted," said Richard Kerr, who served as deputy director of intelligence during Cheney's years on the committee. "You wanted someone who was interested, and he was very interested. He would take time out when he was on the House Permanent Select Committee [on Intelligence] to come to

the agency and spend a Saturday talking to analysts about a subject or problem. Not very many congressmen do that."[59]

"I never found in any situation that I dealt with of his trying
to push you to do things you didn't believe, try to argue you into
saying something that you didn't want to say," Kerr explained.
Cheney, he said, tried "to push you to be as precise and clear,
trying to get you to think about problems in a way perhaps you
hadn't thought about."

Cheney's understated style and his in-depth knowledge of issues won him admirers on both sides of the aisle. "He was always
a low-key operator," says Lee Hamilton, a Democrat from Indiana who served as chairman of the House Intelligence Committee when Cheney was a member. "He was highly respected
by both Democrats and Republicans. We obviously had a lot of
differences on policy matters, but we also agreed on a lot, particularly on the Intelligence Committee."[60]

Hamilton and Cheney grew to trust one another to such a
great extent that the two men began to hold private consultations aimed at forging consensus between their parties. The discussions were free of political posturing and allowed for a candid
exchange of views and strategies. "He would give me an honest
assessment of where the Republicans stood, and I'd give him an
honest assessment of where the Democrats stood. Then we'd see
where we could find consensus," says Hamilton.[61]

The sessions worked because both men understood that neither the fact of the meetings nor their substance would ever be
discussed in public. "I always felt that I could talk to Dick without him running to the cameras," says Hamilton.[62]

"Lee was a guy that you could reach across the aisle and talk
to, and didn't have to always agree by any means. He certainly
was a loyal Democrat, and I was a loyal Republican, but you
could work with him. He kept his word. I tried to operate the
same way."[63]

Still, ask his colleagues to describe Dick Cheney and few of
them would offer up "conciliator" or "mediator." Cheney believed
in legislative compromise, in working well with others, but only to
a point. And once he reached that point, he would not budge—no
matter how intense the pressure or who was applying it.

In the aftermath of Ronald Reagan's landslide victory in the elections of 1984, the White House wanted to do some legacy-polishing, and Democrats, recognizing the president's popularity, were suddenly eager to compromise. These objectives would overlap on tax reform.

But House Republicans were skeptical. And as tax reform moved from idea to policy, they liked it even less. The White House wanted to lower top marginal tax rates and simplify the tax code; and Cheney and his colleagues favored that. But the Reagan administration wanted to do this by eliminating deductions favored by fiscal conservatives and raising tax rates on capital gains, a plan that was anathema to pro-growth economists. Cheney made his views known to the White House.

A short time later, on June 19, 1985, Jim Baker wrote to Cheney. Following the election of 1984, Baker had swapped jobs with Donald Regan, secretary of the treasury during Ronald Reagan's first term. In his new position, Baker was responsible for guiding Reagan's tax proposals through Congress. His letter to Cheney did not seek support from his old friend. A clever vote counter, Baker sought to forestall public opposition from an important party leader in the House. "While I realize that you may have concerns about specific aspects of the plan," Baker wrote, "I strongly urge you to keep your options open and to avoid getting yourself locked into any specific position which could prevent you from supporting the enactment of tax reform."

Baker remembers that Cheney was confident. At a subsequent meeting, the congressman told his old friend: "We're going to oppose you on this, and we're going to beat you."[64]

The White House wanted its effort to be bipartisan and took great pains to win support from Democrats. It worked closely with Dan Rostenkowski of Illinois, the Democrat who was the powerful chairman of the House Ways and Means Committee. This had both practical and psychological effects. The emerging White House proposal more closely reflected House Democrats' thinking on economic policy than the views of House Republicans. And the House Republicans didn't like it. "Baker made a gamble he could date Rosty for six months and then take us on a weeklong stroll," said Newt Gingrich, as the plan appeared destined for defeat. "It didn't work."[65]

Actually, it did. Shortly after Gingrich delivered that assessment, the White House successfully split the House Republican leadership, one of the few times that this happened during Cheney's congressional career. Following intense lobbying from President Reagan himself, Jack Kemp—the lower chamber's leading supply-sider—and Bob Michel both told the White House they would at least allow a vote on Reagan's proposal.

"We all wanted to get the rates down," Kemp recalls. "That was the strategy. There was a dispute over tactics. And it got emotional for some. Cheney was never emotional about it. But Michel and I were roundly challenged and chastised by some in our own caucus. . . . There was a big dispute. I've got to admit it was more than just surface. There were some real harsh words said."[66]

"That was the only time I could really remember that Dick and I were at odds on a significant vote," says Michel.[67]

Reagan had argued that House Republicans should vote for the bill with the expectation that a better version would emerge from the Senate. Cheney didn't buy it. In a speech on the House floor he argued that the bill "discriminates against Wyoming" and worried that it would stifle growth. "The most critical consideration for me in evaluating any tax measure is its impact on Wyoming and the nation. In the short run, over the next two to three years, I believe the bill will result in lower levels of economic activity, slower growth, less investment, and fewer jobs for Wyoming."[68]

"It was a crappy bill," Michel recalled later. "As leader, I had to swallow that in the interest in furthering the procedure and the process that gets you to another end. We're not going to clean this thing up in the House, but let's not kill it."[69]

Cheney was stubborn. He wanted it dead, and the arm-twisting from Jim Baker and others in the Reagan administration did not persuade him to support it. "It is not reasonable for the president to expect me to vote for a tax bill that he himself would not sign," he said at the time.[70]

More than a decade later, when Cheney was named to the Republican ticket in 2000, Kemp credited him for his foresight. "While serving as the congressman from Wyoming, Cheney perceived the many flaws in the Tax Reform Act of 1986 and broke with his own party and voted against the bill," Kemp wrote. "It turns out he was right; it was the anti-supply-side aspects of tax reform that helped bring on the [1990] recession."[71]

Over the course of the 1980s, Cheney came to occupy one of the safest seats in the House of Representatives. He easily won reelection every two years, and Democrats in Wyoming had difficulty even recruiting a serious candidate to oppose him. This solid base of support allowed Cheney to devote large portions of his time to leadership and to the Intelligence Committee. It also meant that he did not have to be on guard when he spoke. Cheney was not going to be bounced from office because of a slip of the tongue or because he answered a question indelicately. So he was sometimes quite blunt even in addressing his constituents.

On one occasion, a Wyomingite wrote to Cheney to share his belief that there had been no Holocaust. Cheney's staff had drafted a bland response, thanking the correspondent for taking the time to write but leaving unaddressed the preposterous claims contained in the letter. Cheney, who frequently reviewed his mail—incoming and outgoing—happened to see the letter before it was mailed. In a handwritten note at the bottom of the page, he shared his thoughts on the original letter.

"Having visited Auschwitz, I can say that anyone who doesn't think there was a Holocaust doesn't have two oars in the water."

Later, Cheney took questions from voters in a series called "Ask Your Congressman" that ran in the *Green River Star,* a weekly newspaper in southwest Wyoming. Violet Smith wanted to know why Congress had failed to enact laws restricting food imports that hurt American farmers and cattlemen. Cheney responded with an unapologetic defense of free markets. "In the end," he said, "we'll lose if we erect trade barriers to agricultural commodities."

Another question came from a man named Sid Case. "I'm a smoker and people seem to treat us smokers like lepers. Smokers are taxed heavily for cigarettes, but we are told not to smoke in some public places we paid tax to build. Is there some way to keep the Congress from picking on the 'lepers' who smoke cigarettes?"

The photograph alongside his question left no doubt that Case was in fact a longtime smoker. Cheney, who had successfully quit smoking several years earlier, made no attempt to find common ground. "First of all, Congress has to annually appropriate funds for paying for all the heart disease, lung cancer and emphysema which smokers suffer from because of their refusal to recognize the damages to their health caused by cigarettes. Every year, through Medicare, we have to pay the cost for illnesses brought on by smoking. If anyone is discriminated against in this relationship, it's the taxpayers, not the smokers."[72]

Any risk that Cheney's frank and forthright style would alienate voters was mitigated when he appeared alongside Al Simpson, whose disarming candor made Cheney look cautious by comparison. Such appearances were common over the course of Cheney's decade in Congress, as Simpson campaigned with Cheney every two years, even when his own name wasn't on the ballot and when Cheney didn't have a serious challenge. "He enjoyed it," says Cheney. There were, however, occasions when it wasn't so pleasant. "In Wyoming, there are people who will sit at a town meeting for hours to nail your ass," says Simpson.[73]

One of those people was waiting to pounce at a town meeting at the VA Hospital in Sheridan attended by both Simpson and Cheney. Simpson was the chairman of the Senate Veterans Affairs Committee, and a veteran in the audience launched into an accusatory tirade about his inadequate benefits. Simpson was not amused, and Cheney, sitting next to him on stage, could see him steaming.

"Don't do it, Al," Cheney whispered to his friend. It was too late.

"How long did you serve?" Simpson asked his inquisitor.

"None of your business," the man responded. It was not the right answer.

"Yes, it is," Simpson thundered. "I bet you've never left Camp Beetle Bailey."

"I served eight months," stammered the man. Not good

enough, said Simpson. "I served overseas for two years and you don't know either end of a mortar from your asshole. In my opinion you've already taken too much of the government's money."

It was going to get worse before it got better. "He was on a roll," says Cheney, who remembers that Simpson ended his vituperation by suggesting that the veteran had "been disabled because he'd fallen off a bar stool."[74]

Finally, Simpson recalls, Cheney cut him off. "Old smoothie stood up," says Simpson, "and said, 'Al's not feeling well. He's off his meds.'"[75] Cheney's intervention took some of the sting from Simpson's counterattack, leaving the audience to wonder whether what they'd just witnessed had been a put-on all along.

Over the years, Cheney and Simpson had developed reputations as pranksters on the campaign trail, especially when they traveled together, as they often did. Cheney and Simpson had become good friends in the years since Cheney opted not to challenge him for Cliff Hansen's Senate seat. As the two traversed the state, each successive stop brought a prank or joke that amounted to an escalation over the other's last effort.

At one rally in the middle a string of campaign events, Cheney was determined to get even with Simpson for some transgression that neither can recall. Both men, however, remember the payback. Cheney spoke first; given his plans, that was crucial. He took the microphone and proceeded to deliver his remarks and, when he had finished, a nearly verbatim version of Simpson's standard stump speech. Cheney concluded his remarks with an admission: "I don't know what Al's going to say. I just used all his stuff," he said to roars of laughter.

"And he had," Simpson recalls. "He used all of my jokes."[76]

Cheney's favorite story, and one he would frequently tell on the campaign trail as vice president, involves a mix-up at a radio station in Riverton, Wyoming. Cheney was scheduled to participate in an interview at nine AM with KVOW radio and was running late. Riverton has a population of less than 10,000 people and the KVOW studios where the interview would take place were in a house on the outskirts of town. When Cheney arrived, he ran up the stairs and threw open the door. He found himself in what looked like a living room. When he stopped to get his bearings, he noticed a baby in a diaper and a woman vacuuming the floor in her nightgown.

"I'll be you're looking for the radio station, aren't you?" the woman said to Cheney.

"Yes, ma'am, I am indeed looking for the radio station."

"Well, it moved last week," she informed him. "This is now my private home and the new radio station's uptown."

Cheney thought for a moment, apologized for the intrusion, and introduced himself. "I'm Al Simpson, your United States senator."[77]

There have been rumors that this popular story is apocryphal. Cheney, asked by his staff about its authenticity, responded: "I'll never tell." It remained a staple of Cheney's speeches on the Washington, D.C., rubber-chicken dinner circuit.

The Cheneys were fast becoming something of a power couple. Dick's swift rise in the House made him a coveted guest at the regular dinner parties of sophisticated Washington. Lynne had developed a strong reputation of her own as a writer and thinker on big historical issues of the day. Her articles appeared regularly in such magazines as *American Heritage* and *Washingtonian*, and she had written two novels in addition to her work on *Kings of the Hill*.

In anticipation of the 200th anniversary of the Constitution in 1987, she was selected as a member of the Commission on the Bicentennial of the U.S. Constitution, led by the former chief justice Warren Burger. It was a group of distinguished Americans brought together for what was essentially a long public relations campaign on behalf of the Constitution.

A short time later, Ronald Reagan picked Lynne Cheney as chair of the National Endowment for the Humanities (NEH), the often controversial federal agency that provides grants to museums, libraries, universities, and scholars with the goal of improving cultural and historical literacy. She replaced William Bennett, a PhD and former Democrat who never shied away from a confrontation. Many observers in Washington assumed that Lynne Cheney would be a more reserved chair, perhaps even a caretaker. They did not know her well.

Cheney had consistently voted against funding the NEH

throughout the 1980s. He was not enthusiastic about its mission. Although he may have been tempted to vote against it after his wife took the helm, he never did. "I don't think I ever had that much nerve," he says.[78]

With Lynne running a federal agency and Dick a member of the congressional leadership, the Cheneys solidified their status as honorary—and temporary—members of the Washington establishment. Even if their modest backgrounds would prevent them from fully embracing establishment life—Dick's prize possession was a Mazda RX-7, not a BMW—the establishment certainly embraced the Cheneys. The columnist David Broder of the *Washington Post* called them "perhaps the most literate couple in town," and your Christmas party was a success if the Cheneys made time to attend.[79]

All this happened before Dick Cheney's stardom really took off in 1987, when he spent hundreds of hours on television screens across the country as a key player in the Iran-contra investigation. He had been a well-known personality inside the Beltway before those hearings, with regular appearances on his favorite show, *The MacNeil/Lehrer NewsHour*, and he would become a national presence once they were finished.

Cheney had maintained an active interest in the U.S. government's anticommunist efforts in Central America for a long time. He explained this interest in the November 1983 edition of his monthly newsletter to his constituents, *Congressman Dick Cheney Reports to Wyoming.*

"Vietnam taught us to think twice before again getting involved in the affairs of other nations, so some people question the justification for U.S. economic and military assistance to Central American countries and worry about getting sucked into their military conflicts," Cheney wrote. But avoiding involvement is not an option. "From a strategic standpoint, the United States cannot sit back and allow the Communists to take over Central America. They already control centrally located Nicaragua, and have vowed to export their war to other nations."

Cheney studied the political history of the region, learning about the significant figures on both sides of the struggle. Even before he won a seat on the Intelligence Committee, Cheney volunteered for a fact-finding mission that took him to Nicaragua. Cheney met with Daniel Ortega, leader of the Sandinistas, who were supported by the communists; and with Bishop Miguel Obando y Bravo, whose position in the Catholic church made him a powerful figure in the heavily Catholic nation. (Cheney has a vivid mental picture of his meeting with Obando y Bravo, who sat below a colorful portrait on the wall depicting Jesus Christ toting an AK-47.)

For years, the battle in the United States over aid to the anti-communist counterrevolutionaries in Nicaragua had been simple. The Reagan administration and its allies believed that the communist government in Managua, backed by Cuba and the Soviet Union, posed a threat to peace and stability in the Western Hemisphere. They sought U.S. funding to support the armed opponents of the regime with the goal of destabilizing or overthrowing it, and as a result, denying the Soviets another ally in a region of such strategic importance.

The opponents of the aid, which included a majority of the Democrats in Congress, were wary of military assistance, which they viewed as a precursor to potential military involvement by the United States in Central America. For them, the willingness of the Reagan administration to run from continent to continent chasing communists who posed no immediate threat to the United States was an indication that Republicans in Washington had failed to learn the lessons of Vietnam and the Bay of Pigs.

In the early 1980s, Democrats, led by Edward Boland of Massachusetts, chairman of the House Permanent Select Committee on Intelligence, offered a series of amendments to an emergency funding bill; these were meant to block support for the contras.

In October 1984, the Boland amendment was attached to an emergency funding bill. If President Reagan refused to sign it, the government would be shut down just three weeks before the presidential election, and the likely political costs were potentially huge. It was a shrewd move by the Democrats. Reagan could veto the legislation and risk being blamed for the closure, or he could sign it and assure congressional supporters of the contras that they would find other ways to continue the fight against communism in Central America. Against the advice of several senior Republicans in the House, Reagan put his pen to the bill. The Boland amendment became law.

The restrictions on aid did not end there. The following year, two leading Democrats in the House introduced legislation that would further restrict aid to the contras. Cheney grew increasingly exasperated by these efforts, because he disagreed with them substantively and thought they represented the worst kind of congressional micromanaging of U.S. foreign policy. One amendment, offered by Representative Michael Barnes and Cheney's friend Lee Hamilton, recommended targeting U.S. aid to humanitarian organizations. Cheney's frustration was evident when he made a rare appearance on the House floor on April 24, 1985, to mock the proposal.

"This Barnes-Hamilton amendment is a fascinating document. It really is," he said, derision dripping from each word. "I don't know why generations of American diplomats didn't think of this approach to halting and reversing the spread of communism. Think of what Harry Truman could have done when faced with the crisis in Europe after World War II. Instead of spending billions on the Marshall Plan and on building NATO, he could have made a donation to the International Red Cross, and instead of sending American troops to save Korea from communist aggression, he could have made a donation to the International Red Cross. This is a whole new doctrine in American foreign policy—we'll call it the Barnes Doctrine—and this is how it will work in Central America. Got a problem with communist government in Nicaragua? Are they censoring the press? Why, we'll give a donation to the International Red Cross. Are they persecuting the Catholic Church and other religions? Give a donation to the International Red Cross. Have they engaged in a massive military buildup that grossly distorts the balance of power in the region? Give a donation to the International Red Cross."[80]

The political dispute over funding the contras came to a head in the fall of 1986, when an aircraft carrying supplies to the rebels was shot down in Nicaragua. An American survivor of the crash, Eugene Hasenfus, told his captors that the plane was on a mission for the CIA. The CIA quickly denied the claim, as its involvement would have been a violation of the Boland amendment. The story was about to get more complicated.

In early November 1986, *Al Shiraa,* a Lebanese magazine, published an article that contained sensational allegations about the United States' counterterrorism practices in the Middle East. The U.S. government, according to this account, had provided arms to Iran in order to win the release of U.S. hostages kidnapped by Islamic Jihad.

It was a stunning claim. The policy of the Reagan administration on hostage-taking and other acts of terrorism had been firm. We will not negotiate with terrorists, the administration said. Period. If the story was true, it would represent an obvious and embarrassing violation of this fundamental principle.

Worse, the administration had aggressively pressed its allies to adopt the same policy, with no exceptions. The State Department had invested considerable diplomatic resources—mostly time, but money too—in an effort to make certain that its friends were not conducting back-channel negotiations that could jeopardize unity.

Adding to the embarrassment, the Reagan administration had been supporting Iraq and its secular dictator, Saddam Hussein, in its four-year war with the radicals in Iran. If the reporting was true, it would seem that the U.S. government had been caught supporting both sides in the regional war. Such double-dealing is hardly uncommon in the hidden world of tradecraft that lies behind the public conduct of international affairs, particularly in the Middle East. But the U.S. had been careful not to get caught.

On the surface, the events in Central America and those in Iran had nothing to do with each other. They involved different parts of the world, different interested parties, and different issues—on the surface.

The first hint of an overlap came well before the facts of the operation were known, when reporters observed, without understanding the meaning, that some of the same people involved in the Iran arms deal had responsibility for the contras. An article published in the *Washington Post* shortly after the story about Iran broke noted that Lieutenant Colonel Oliver North, a deputy director of the National Security Council, had helped devise the secret plan to send arms to Iran. "North, who supervises the White House counterterrorism operation, also has played a leading role in U.S. aid to the rebel, or contra, forces fighting the government of Nicaragua."[81] Reporters chased the arms-for-hostages story, unaware that those transactions were just part of a much larger scheme.

As the story unfolded on the front pages, members of Congress left Washington for the postelection recess. Cheney was scheduled to leave later than most, returning to Wyoming in mid-November to hunt elk with Al Simpson. But on the morning of November 12, Cheney received an urgent phone call from the White House. There would be an important briefing for congressional leaders later in the day. Because Bob Michel, Trent Lott, and Jack Kemp had already left town, Cheney would represent the House Republicans. Others present included the Senate majority leader, Bob Dole; the Senate minority leader, Robert Byrd; and the House majority leader, Jim Wright. They arrived at the White House "situation room" to find the president and senior members of his national security team: Secretary of State George Shultz; Secretary of Defense Caspar Weinberger; the director of the CIA, William Casey; Attorney General Edwin Meese III; the White House chief of staff, Donald Regan; the national security adviser, John Poindexter; and Vice President George Bush.

The president spoke first. He gave the congressional leaders a broad overview of the administration's dealings with Iran, insisting that there had been no negotiations with the terrorists and that the entire operation was undertaken to improve America's strategic position in the Middle East.

Poindexter filled in some of the chronology with a long explanation of the events and decisions that had led to this point. It could have been longer. Poindexter left out several important details. "He was vague to the point of prevarication," Shultz would later recall.[82] No one else in the room bothered to provide those details.

"We didn't get the whole picture," Cheney says, "but we knew something a little shady was going on when they summoned us down to the situation room in the West Wing basement, and we had the powers that be there in the administration saying, 'You guys need to know.'"[83] The seriousness of the matter required Cheney to remain in Washington.

"I never got to go elk hunting," he recalls. "But it was a miserable trip. They got snowed in and didn't get any elk."[84]

Two weeks later, Meese spoke at a hastily arranged press conference and disclosed what would become known as "the diversion"—a transfer of profits from Iranian arms sales to the contras in Nicaragua. A public relations mess quickly became a disaster. Not only had the Reagan administration taken a hit on its dealings with Iran; now there were serious questions about whether aid to the contras violated the law.

The day after Meese's press conference, Cheney told the Washington Post that Reagan's foreign policy was in "disarray," adding, "you can't run effective foreign policy if the man himself is unaware of something of that enormity."[85]

"It's pretty damn serious," Cheney told David Broder the same day. "I tried to defend him [Reagan] initially, but it's hard. You have to say it's a pretty fundamental flaw that would allow a lieutenant colonel on the White House staff to operate in defiance of the law."[86]

Cheney had another reason to be angry about the news. On August 10 he had been present at a briefing North gave to the House Intelligence Committee. North told the committee that his activities in Central America had violated neither the letter nor the spirit of the Boland amendment. It was not true.

The assessment Cheney gave Broder was notable not only because it contained a serious accusation, but because of who was making it. He was the fourth-ranking Republican in the House of Representatives, a close ally of the Reagan administration, a leading proponent of executive power, and an outspoken advocate of aiding the contras. That the White House was having difficulty keeping Dick Cheney on the team meant that it was in trouble.

An explosion of investigations quickly followed the disclosure of the potential scandal. There was an investigation by the Justice Department; the White House appointed the Tower Commission; the Senate had a Select Committee on Secret Military Assistance to Iran and the Nicaraguan Opposition; the House had a

Select Committee to Investigate Covert Arms Transactions with Iran; and eventually, an independent counsel, Lawrence Walsh, was appointed. Washington settled in for the scandal season.

Bob Michel named Cheney to serve as the ranking Republican on the special House committee set up to investigate Iran-contra, skipping over the higher-ranking Trent Lott and Jack Kemp to award him the coveted job. Michel thought Cheney's position as a leader and his seat on the intelligence committee made him an ideal head of the House Republicans on the committee. As he had done throughout his congressional career, Cheney would represent the minority: nine panel members were Democrats; six were Republicans.

The committee began its preliminary investigation, interviewing players and potential witnesses about their roles in the diversion. As Cheney learned more, he became increasingly sympathetic to the Reagan administration. He believed that the diversion was a serious error in judgment—he would later call it "a major mistake." But several senior officials in the Reagan administration had already paid for their error with their jobs, and Cheney saw the investigation taking a turn toward politics.

"I didn't want to see the president and his administration damaged," he says.[87] As the committee prepared to conduct public hearings into the matter, Cheney began to see himself not as an investigator but as the lead defender of the Reagan administration.

"I thought there were people, certainly on the [minority] staff, who were trying to damage the president, and I took it as my responsibility as the senior Republican on the House side to do everything I could to support and defend him," he says.[88] So Cheney prepared for a fight.

Cheney anticipated that one of the likely consequences of the Iran-contra scandal would be a concerted attempt by congressional Democrats to rein in the executive branch. He expected the kinds of legislation that had been proposed—and in some cases passed—following the Vietnam War and Watergate. It was more than an educated guess. Democrats on the panel had hired an expert in the separation of powers, Louis Fisher of the Congressional Research Service, as their research director. Fisher's writings had consistently challenged the authority of the executive branch. The post-Watergate erosion of executive power, which Cheney saw as a cause for concern, Fisher saw as a cause for celebration.

In the spring of 1987, as the panel members prepared for their first public hearings, Cheney met with Michael Malbin, an acquaintance from his first term in Congress who was seeking advice. The timing of Malbin's visit was fortunate.

Malbin had received a PhD in political science from Cornell University in 1973; and after four years as a reporter for the *National Journal*, he settled in as a resident scholar at the American Enterprise Institute. In 1979, Malbin helped coordinate the series of off-the-record dinner discussions among newly elected House members that had included Cheney.

Malbin had recently completed a yearlong fellowship at the University of Maryland, and he stopped in to see Cheney about career options. As Malbin explained his research in College Park—an examination of relations between the legislative and executive branches of the U.S. government—Cheney realized that he could offer more than advice. He had a job for Malbin.

Cheney explained the assignment to his new hire. Wherever the facts of this investigation lead, he told Malbin, let them lead. Your job is to preserve the institution of the presidency. Malbin was given a quiet office in the committee's annex, away from the chaos of the staff cubicles. As the drama of the contentious public hearings captured the attention of the American public, Malbin spent his days reading the voluminous legislative histories of the National Security Act of 1947 and war powers issues going back to 1789.

From the first moments of the hearings, Cheney understood that he did not have an ally in Warren Rudman, the ranking Republican on the Senate side of the investigation. Cheney thought that Rudman, a former prosecutor, enjoyed his time in the national spotlight and "flaked off" during the investigation. "The core of support for the administration in this political battle were the House Republicans," he recalls.[89]

Throughout the hearings, Cheney's goal was to put the mistakes in context. He took advantage of the prominent platform provided by the televised proceedings, making the case that the

United States should be supporting the contras in Nicaragua. When Democrats objected to Oliver North's plan to give the legislators—and the television cameras—the same presentation he had given to solicit funds for the contras from private donors, Cheney agreed to give North his time. It was a controversial decision. The phone lines in Cheney's personal office lit up. To some of the callers, he was enabling a criminal and a liar. To others, he was protecting a patriot and defending the country. Cheney defended the president to the end.

On August 3, 1987, Cheney delivered his closing remarks about the Iran-contra affair:

> *President Reagan has enjoyed many successes during his more than six years in office. Clearly, this was not one of them. As the President himself has said, mistakes were made.*
>
> *Mistakes in selling arms to Iran, allowing the transaction to become focused on releasing American hostages, diverting funds from the arms sale to support for the contras, misleading the Congress about the extent of NSC staff involvement with the contras, delaying notification of anyone in Congress of the transactions until after the story broke in Lebanese newspapers, and tolerating the decisionmaking process within the upper reaches of the Administration that lacked integrity and accountability for key elements of the process.*
>
> *But there are some mitigating factors; factors which, while they don't justify Administration mistakes, go a long way to helping explain and make them understandable:*
>
> *The need, still evident today, to find some way to alter our current relationships with Iran. The President's compassionate concern over the fate of Americans held hostage in Lebanon, especially the fate of Mr. William Buckley, our CIA station chief in Beirut.*
>
> *The vital importance of keeping the Nicaraguan democratic resistance alive until Congress could reverse itself and repeal the Boland Amendment. The fact that for the President and most of his key advisers these events did not loom as large at the time they occurred as they do now.*

Congressional vacillation and uncertainty about our policies in Central America. And finally, a Congressional track record of leaks of sensitive information sufficient to worry even the most apologetic advocate of an expansive role for the Congress in foreign policymaking.

Moments after Cheney finished his closing statement, Representative Henry Hyde passed him a note with the letterhead "United States Senate: Select Committee on Secret Military Assistance to Iran and the Nicaraguan Opposition." In barely legible handwriting, Hyde had scrawled: "A Superior Final statement—in fact—Churchillian!"

The word "superior" was underlined twice, and the note was signed simply "HH."

Not everyone agreed with Hyde, but most reviews of Cheney's work on Iran-contra were positive.

Mary McGrory, a liberal columnist at the *Washington Post*, criticized Cheney for his inability to see beyond his own experience in the Ford White House, but nonetheless said he was "by far the brightest of the Republicans chosen for the committee."[90]

If some Democrats were surprised and disappointed by Cheney's partisan defense of Reagan, Sam Nunn of Georgia was not one of them. In a letter of August 10, Nunn, a respected thinker on national security issues, commended Cheney's "excellent work" during the public hearings. "Our task was not an easy or a pleasant one," Nunn wrote. "Yet it had to be done and done in a professional, non-partisan manner. The issues were too important to be handled any other way. Your leadership, fairness, judgment and good humor were absolutely critical to the success of those hearings."[91]

Such plaudits did not persuade Cheney and several other Republicans to join their colleagues in producing a consensus congressional report on Iran-contra. The Minority Report—fourteen chapters taking up 155 pages—provides an overview of Iran-contra and includes a narrative that departs from the Majority Report on several points and, even more significantly, draws very different conclusions.

"Congress must recognize that an effective foreign policy

requires, and the Constitution mandates, the president to be the country's foreign policy leader," wrote the minority. If the legislative branch places excessive constraints on the executive, the report argues, the executive has a responsibility to ignore them. "Unconstitutional statutes violate the rule of law every bit as much as do willful violations of constitutional statues."[92]

The Minority Report argued that Congress had created the conditions for a scandal like Iran-contra by creating a "state of political guerrilla warfare over foreign policy between the legislative and executive branches."[93] The recommendations offered at the end of the Minority Report continue this theme. They make clear the authors' view that Congress, not the executive branch, was most urgently in need of reform. The report proposed a Joint House/Senate Intelligence Committee, a secrecy oath for those on the committee, and stiff penalties for those who violated the oath. It further recommended strengthening sanctions on national security leaks and limiting to four the number of congressional leaders the executive branch would be required to notify on sensitive issues.

Lee Hamilton remembers that Cheney's protectiveness of the executive branch made him stand out among his colleagues in the legislative branch. "As long as I knew him, he had a deep-seated conviction that the executive branch had suffered erosion since, I don't know, Watergate and that time frame," says Hamilton. "You usually think of a member of Congress as being of the opinion that Congress is a separate and co-equal branch. He leaned strongly toward executive power even as a member."[94]

The benefits of Cheney's work on Iran-contra were many. The hearings had propelled him to national recognition and raised his stature among congressional Republicans.

The editors of *U.S. News and World Report* included Cheney in the "Best of the Hill" issue. The article on Cheney described him as "the Republicans' most articulate, reasoned spokesman," and noted that "he solidified support with GOP colleagues, who almost certainly will choose him as their leader within a few years."[95]

Another reward came on December 8, 1987, at a state dinner President Reagan hosted for President Mikhail Gorbachev of the

Soviet Union. Not only had Cheney received a coveted invitation to the event; he was assigned a seat at a table with Nancy Reagan and next to Gorbachev. Cheney, who had worked on seating assignments in the Ford White House, believed that his proximity to the guest of honor was a thank-you from President Reagan.

The dinner gave Cheney an opportunity to question Gorbachev one-on-one. Cheney had been to the Soviet Union several times. He had taken Gorbachev's measure from a distance and in small groups. His most recent trip had come in April, over the Easter recess, when he traveled throughout the country as part of a congressional delegation led by the House speaker, Jim Wright. Cheney had never cared for Wright, and his experience with the new speaker in the Soviet Union did little to change his opinion. In one of the delegation's first meetings, Representative Les Aspin asked a tough question of a Soviet official. The exchange later prompted a reminder from Wright that the delegation was there to promote goodwill and common understanding, not to exacerbate tensions. That evening, after their meetings were done for the day, Cheney talked about Wright's admonition with Aspin and Thomas Foley, a Democrat from Washington. Both Foley and Aspin thought Wright's comments were out of line, but they were worried about confronting the leader of their party. They asked Cheney to make clear to Wright that the congressmen on the delegation were there to ask lots of questions — some of them friendly, some of them confrontational. Cheney asked Wright for a private meeting but never received a response. He took it on himself to ask difficult questions throughout the remaining days of the trip.

Although he returned from this trip pleasantly surprised by the seriousness of the Soviet Union about political reforms, Cheney remained skeptical that the changes would be lasting. In June, Gorbachev began to introduce a series of economic reforms known as perestroika. The initial changes were modest, allowing state-run enterprises to sell some of their goods at market prices. In the West, the reforms were greeted with great optimism and, in Cheney's view, hyped beyond their real significance.

Conversation at dinner was almost all substance. During an exchange on economic reform, the discussion turned to the economic success of the so-called Asian tigers—Hong Kong, South Korea, Singapore, and Taiwan. The question was whether the Soviet Union had anything to learn from an embrace of export-driven market reforms. The answer, according to Gorbachev, was absolutely not. "He gave you just the sort of standard Marxist line—the only reason they've achieved anything is the exploitation of the working class—that kind of standard Marxist BS," says Cheney.[96]

Cheney was not surprised. "He wasn't willing to say that anything resembling capitalism is going to be the answer. He thought the communists had it right. They may need to tweak the system a bit, but this was not a guy who was going to overturn Soviet communism."[97]

For Cheney, it was a conversation that confirmed his skepticism about the prospects for reform and would shape his thinking on the Soviet Union through its dissolution.

Back in Congress, after six years without changes in the makeup of the senior Republican leadership in the House, Jack Kemp stepped down to run for president in April 1987. Two months later, Cheney's colleagues unanimously elected him to replace Kemp as chairman of the House Republican Conference, the third-ranking leadership job. It was a welcome promotion, but Cheney had a bigger goal: to replace Bob Michel as minority leader when Michel decided to retire. Cheney was prepared to challenge the minority whip, Trent Lott, the second-ranking House Republican and a good friend, for Michel's job.

When Senator John Stennis of Mississippi announced that he would not run again in 1988, Lott, a natural candidate to replace him, quickly sought a meeting with Michel.

"Trent Lott came to me and said, 'Hey, Bob. Are you going to stick around here?' And at that time I said, 'Well, I'm not ready to retire yet,'" Michel recalled. "He always salivated to go over to the other body there. But he also salivated for my spot as leader."[98]

Michel told Lott to run for the Senate. He was comfortable with Cheney as his eventual replacement. And shortly after the 1988 elections, Cheney moved up another spot. "I thought, now that Cheney's won the doggone whip race there, unopposed, he's a natural follow-on, and someone whose philosophy mirrored mine, as far as I could see."[99]

As the 101st Congress convened, Cheney was well positioned to continue his climb up the Republican leadership ladder. "I was Bob Michel's understudy and, I think, logical successor," he recalls. "I certainly anticipated staying in the House, making that my career. And obviously it would have been a dream to become speaker, but I certainly had a respectable shot at becoming Republican leader."[100]

The election of George H. W. Bush as president would change his plans.

CHAPTER SEVEN

At War

Days after he delivered his inaugural address, President George H. W. Bush returned to Capitol Hill to meet with House Republicans. At the top of the agenda that day was the budget, but a casual conversation that he had on his way out the door would have a greater impact on the future of his administration.

On the other side of Capitol Hill from Room 227 of the Cannon House Office Building, the former senator John Tower, the man Bush had nominated to serve as his secretary of defense, was receiving a tougher grilling than expected at his confirmation hearings. Before the questioning had started, Sam Nunn, chairman of the Senate Armed Services Committee, promised that he and his colleagues would conduct a thorough review of the FBI's background investigation of Tower, who had a reputation as a heavy drinker with a keen interest in the ladies. Nunn, as one of a few members who had seen the full report, was one of the few people who might have known about the difficulties ahead.

Bush was another.

As he left his meeting with the House Republicans, he stopped to talk to Cheney. He asked specifically about Cheney's health. Cheney had suffered a third heart attack back in June. In August, he skipped the end of the Republican National Convention in

New Orleans in order to have a quadruple coronary bypass. The surgery had gone well and Cheney had resumed normal physical activities. Cheney took Bush's question as a friendly inquiry from a man he'd known for more than a decade. "Well, Mr. President," he said. "I was skiing at Christmastime."

Thinking back on the exchange, Cheney remembers that Bush seemed exceptionally interested in his answer. "It occurred to me that he may have been at that point thinking maybe if Tower goes down, Cheney's the guy I ought to go to."[1]

After more than a month of extraordinarily contentious debate, Tower's nomination indeed went down on Thursday, March 9, in a vote of 53 to 47. That afternoon, Cheney took a call from the White House chief of staff, John Sununu, who asked Cheney to come to his office to discuss possible replacements for Tower. Cheney told Sununu that he couldn't come immediately because he had to attend a previously scheduled taping of CNN's *Evans and Novak*, but that he would come as soon as it was finished. He suspected that the White House was interested in more than his advice.

Two of his good friends and longtime colleagues had high-profile jobs in the Bush administration. Brent Scowcroft had become national security adviser, and Jim Baker was secretary of state. The phone call from Sununu had much more to do with that fact than the desire of the White House to consult with the House Republican whip.

In the middle of the CNN taping, Bob Novak turned to Tower's nomination. He asked what his cohost, Rowland Evans, was hearing about candidates for secretary of defense. Evans had a scoop. He had heard from a high-ranking source that Bush would nominate Bobby Ray Inman, the former deputy director of the CIA. Cheney listened in silence.

After the show, he went to the White House, where he was ushered into his old office to meet with Sununu and Scowcroft. "They maintained the fiction for a few minutes that they wanted my advice," he says, asking him for suggestions of confirmable nominees.[2] Then the conversation got serious. Are you interested in the job?

He was, and he told them he'd like a night to think about

it. Sununu told him to call the next morning if he wanted to be considered.

Cheney returned to his office. He told no one about his meeting. That evening, he and Lynne met Tom and Marta Stroock for dinner at La Colline, a French restaurant on Capitol Hill. It had been thirty years since Stroock had recruited Cheney to go to Yale and more than a decade until they nearly faced off in the election to replace Teno Roncalio as Wyoming's representative in the House. They talked about old times and life in Washington—everything, it seemed, but the opening at Defense. On the way home, Cheney shared the news with Lynne, who was excited for her husband but not at all pleased that he had kept her in the dark throughout dinner.

When he returned to his house, he received a call from Baker. "Jim put the arm on me," says Cheney, "basically pitched me hard that I ought to take the job, secretary of defense, if it were offered."[3]

It was not a difficult sell. "It was a fairly easy call," says Cheney. "I didn't agonize over it. . . . The choice for me boiled down to whether I spent four more years in the House in the minority or went to be secretary of defense. In the House I was doing well, just been elected to the number two leadership post as whip. . . . I would have been perfectly happy, and up until this job came along, certainly anticipated staying in the House, making that my career. But the chance to go be secretary of defense, given my background in the executive branch, given the subject matter, given the chance to run a big operation, and—my God, secretary of defense, it's a big job and I didn't look at it on a long-term basis. I suppose if I were only concerned about the long term I would have stayed in the House."[4]

The following morning, Cheney called Sununu as promised. There was no time to waste. Sununu told Cheney to come to the White House as soon as possible. He quietly sneaked into the building, using the diplomatic entrance so that his visit would not generate any suspicion among the press corps or gossip-hawking White House officials. Once inside, Cheney was led upstairs to the president's residential quarters, usually off-limits to all but the most senior White House staff. He met with the president alone. "The conversation was conducted on the basis not of 'I'm offering you the job,' but 'Let's explore the possibility' and 'How do you look at the job?' and 'Here's the kind of thing I'm interested in' and 'What I'm looking for in a secretary' and so forth."[5]

There was one additional, unpleasant matter: the DUIs. Such a history might complicate any confirmation, and it had the potential to be a sticking point given the attention paid to the issue during the battle over Tower. Bush, however, was not concerned. "He knew the full extent"—Cheney pauses and laughs—"of my youthful transgressions. . . . He didn't think it'd be a problem."[6] After an hour, Sununu rejoined them and ended the meeting without a resolution. But Cheney thought the interview had gone well.

He raced back to Capitol Hill for a previously scheduled interview with a first-year assistant professor from the University of Georgia, John Maltese. Such meetings were common for Cheney, who had a soft spot for academics: his schedule was cluttered with speeches to student groups, interviews with professors, and think-tank conferences around the country, in addition to his congressional duties.

The congressman was relaxed as the interview began. Maltese explained that he was researching a book on news management at the White House and was interested in Cheney's experience as Ford's chief of staff. For sixteen minutes, they discussed the ins and outs of the Ford White House and the campaign of 1976, their conversation occasionally interrupted by the loud thud of Cheney's boots pounding the desk as he sat back in his chair. As Cheney explained that he had asked a senior staff member to work on the transition throughout the summer of 1976, he was interrupted by the high-pitched beep of the telephone intercom. He was in no hurry to answer it, speaking for another eleven seconds.

"He laid the plans for the transition on my desk on Election Day"—beep—"and I never did anything with it because we lost"—beep—"so you know, it became irrelevant after that. I'm not sure I even read it."[7]

Maltese clicked off the tape recorder and Cheney picked up

his phone. The congressman sat up in his chair slightly and asked Maltese to step outside. One of Cheney's staffers noticed Maltese standing near the photocopier and thought it odd that he was not allowed to remain in the office. Behind the door, Cheney took the call from President Bush and accepted his formal offer to become secretary of defense. It was a quick conversation, and moments after he had been ushered out, Maltese took his chair in front of Cheney's desk and asked about Michael Duval, who worked for Cheney on Ford's campaign.

"Mike had been, when he originally came to town, he was an attorney," Cheney responded, his even manner betraying no hint of the life-changing conversation he had just completed. "He came to town in 1967."

Maltese remembers that Cheney's deportment was the same before and after the interruption. "There was no change in tone," he says. "No change at all."[8]

Bush had instructed Cheney to return to the White House at four PM. When Cheney left Capitol Hill that afternoon, the phones in his personal office lit up as word began to leak out that he would be introduced as the new nominee. The reporters calling the office had more information than Cheney's staff.

The congressman was already on his way to the White House when Pete Williams, Cheney's press secretary, and Patty Howe, the chief of staff, returned to the office from a meeting on acid rain. Others in the office had gotten word of the appointment and assumed that both Williams and Howe had been told. In fact, neither of Cheney's top advisers had any idea what was about to happen. Williams went to sit at his desk, looking puzzled by all the activity. Jim Steen asked him, "You mean you didn't know Dick was picked to be secretary of defense?" Williams glanced quickly at Steen and dashed out of the room.

In an announcement in the Rose Garden, Bush introduced Cheney as "a thoughtful man, a quiet man, a strong man— approaches public policy with vigor, determination and diligence." Bush said that he'd gotten to know Cheney in the Ford administration as a "government manager," noted his service on the Intelligence Committee, and reported the substance of their discussion a few hours earlier. "I've heard his thinking on arms control, Central American policy, strategic defense posture, and on the difficult challenges that he knows he faces of reforming procurement process in the Pentagon."

Cheney thanked the president for asking him to serve. "I look forward very much to working with him, and especially also with Brent Scowcroft, who's an old friend of many years' standing, and with Jim Baker, in the difficult assignment ahead."

No one, of course, had any idea how difficult. The relationships would prove valuable as Bush's foreign policy team navigated the choppy waters of international affairs. The long tradition of rivalry—sometimes animosity—between the diplomats at the State Department and the war fighters at the Pentagon would be eased by the close personal friendship between Baker and Cheney. The men would not always agree, and battles at the staff level were as fierce as ever. But with rare exceptions the differences on policy would not became personal. At a time of great potential volatility in the world—with turmoil in the Persian Gulf, the continued rise of China, volatility in Central America, and the crumbling of a rival superpower—the familiarity of the three men would bring a measure of stability to the conduct of U.S. foreign and defense policy.

The announcement was carried live on CNN. Cheney's staff watched in stunned disbelief, wondering what this would mean for their immediate future, but also how Cheney had managed to keep it from them. As the anchor reappeared after the brief ceremony, the office descended into chaos, with a phone call from France from someone wanting to know how to pronounce Cheney, a fax from an eager job-seeker looking for work at the Pentagon, an inquiry from Representative Paul Gillmor's office about whether Cheney would be able to speak at the local Lincoln Day dinner later that month, and a photographer from the *New York Times* snapping pictures of the office. Within an hour, investigators from the FBI were interviewing Cheney's staff, asking the most personal questions about their boss and their colleagues.

The entire process—from the initial White House inquiry to the announcement—had taken little more than twenty-four hours. And the only reason it hadn't been completed earlier—the

CNN taping—had been rendered obsolete by the news. "Evans and Novak had to do a rerun that weekend—could not use that tape," says Cheney. "There I am sitting right in front of them and they say it's going to be Bobby Inman."[9]

Reaction to the nomination was positive and in some cases effusive. At the *Washington Post*, the reporter and columnist David Broder ran out of superlatives. "Here are three things to understand about Rep. Dick Cheney which will explain why President Bush made such a superb choice in naming him as secretary of defense following the John Tower fiasco. Cheney is smart, he is tough and he is totally trustworthy," he wrote.[10] It was high praise from the man known as the dean of the Washington press corps, whose columns would often embrace views diametrically opposed to Cheney's.

Broder told of a an interview in 1987 at which Cheney said that he would rather the Democrats nominate Sam Nunn than someone like Michael Dukakis in the upcoming presidential race. Although Nunn would certainly make the race more difficult for Republicans, Cheney said, he would be better for the country if he were elected. Cheney's "brain is as good as anyone's in town, and he is totally unafraid to voice his convictions in any company. . . . Though he is a conservative by conviction and a staunch partisan Republican, he is admired by dozens of Democratic colleagues."[11]

Like the nomination, the confirmation was expedited. Senators from the two parties had not agreed on much during the battles over John Tower. But they knew that the hearings had damaged not only Tower's reputation, but theirs too, as an institution. They were unanimous in their desire to avoid a repetition.

As the confirmation hearings began, Senator Alan Simpson introduced Cheney to his colleagues as "one of the most steady and unflappable men I have ever known anywhere, any time, under any circumstances." Simpson told the panel that although the two men took their work seriously, they had fun together. "He has unceremoniously stolen some of my finest stories. He has invented new ones and blamed them on me. And I hope and pray that the elevation to this national post will deter him from ever again relating the dog-eared old saw about the Riverton radio station."[12]

Moments later, Cheney, at the invitation of Nunn, opened his testimony with a light touch that suggested a much easier process than Tower's nomination. "I suppose the best way to proceed would be to open with the Riverton radio station story."[13] So he did.

The hearings shifted to substance and remained focused on issues until they ended three days later. Cheney fielded questions on missile defense, budgets, strengthening alliances, and the nature of the threat from the Soviet Union. His remarks suggested a softening of his views on the political and economic changes under Mikhail Gorbachev, though he cautioned that significant cuts in the defense budget in reaction to developments in the Soviet Union would be premature. After watching Gorbachev throughout the late 1980s and talking to him directly about the changes taking place in his country, Cheney had concluded that Gorbachev is "convinced that the status quo in the Soviet Union is unacceptable" and that "he is committed to a policy of reform." Gorbachev, he explained to the panel, "is serious about trying to improve the overall efficiency of the Soviet economy" and understands that "there are going to have to be some fundamental political changes as well." Still, he continued, "I think it would be a grave mistake for us to assume at this point that it is appropriate for us to in any way reduce our own military capabilities or lessen our own defense posture. I am frankly somewhat skeptical about the likelihood of his success in terms of his reform drive."[14]

Cheney's testimony not only made clear that he harbored doubts about the prospects for change but, in a response to a question from Senator Carl Levin, also revealed his deep ambiguity about the wisdom of those reforms.

"Obviously, if you had a Soviet Union that was revitalized, one with a strong, healthy economy and the same kinds of policies toward the rest of the world that have dominated during the postwar period, that would not be good from the standpoint of the United States," Cheney explained. "It would mean that we were faced with an adversary far more capable than the one we're faced with today. On the other hand, I'm inclined to think that if you buy the argument that economic change and progress in the Soviet Union requires and will only occur with fundamental political change as well, less central control, greater freedom for individual citizens, more elements of a market economy involved and a less onerous hold on eastern Europe, then I think you could conclude that in fact his success would be a benefit to the United States."[15]

In the aftermath of Tower's nomination, Cheney thought that the best way to handle his DUIs would be to disclose them publicly, and to do so early. But the committee wanted to handle his DUIs privately—along with a ticket he had gotten for fishing without a license—in a closed session. Senator John Glenn, a Democrat from Ohio, asked Cheney about his arrests for drinking and wondered why there had never been a third incident. Cheney had a simple explanation: "I got married and quit hanging out in bars."[16]

Support for the nomination was enthusiastic and bipartisan. Democrats who would years later count themselves as Cheney's most outspoken detractors urged his confirmation. Levin, who would become a relentless critic of Cheney more than a decade later, was the first senator to second the call to report Cheney's nomination favorably to the Senate. Senator Al Gore congratulated President Bush for his choice, and Ted Kennedy declared simply: "America's defense policy is back on track."[17]

He was confirmed on Friday, March 17, 1989, on a vote of 92 to 0.

Cheney led the Department of Defense at a critical time in America's history. Under his leadership—which was often tested by the department's unwieldy bureaucracy—the Pentagon began to reposition itself away from its cold war traditions and move toward newer emerging threats, whatever they may be.

But as much as Cheney made a mark on the Pentagon, his experience there made a mark on him. In the lead-up to the Gulf War that forced Iraq out of Kuwait, Cheney was on the receiving end of U.S. intelligence estimates that were flat-out wrong—intelligence that, for instance, failed to anticipate Iraq's invasion of Kuwait and later grossly underestimated how close Saddam Hussein was to acquiring nuclear weapons. Those failures of intelligence in the early 1990s would shape his views on the interplay between policy decisions and intelligence—a subject that would become highly controversial a decade later.[18]

Cheney wanted to get to work immediately, and this would require him to take the oath of office at a small private ceremony in his own office. He had planned to have Jim Ford, the House chaplain, administer the oath, but the Pentagon had other ideas. He was told that David "Doc" Cooke was on his way to Capitol Hill. Doc Cooke, with his Mr. Clean head and bassett-hound jowls, was known as the mayor of the Pentagon. He had sworn in nearly a dozen previous secretaries of defense. Cheney, whether he liked it or not, would be next.[19]

The tug-of-war with the permanent Pentagon had started early. Cheney was taken aback—the first decision of his tenure as secretary of defense was not his to make. But he wanted to get off to a good start, so he acquiesced.

Cooke came to 104 Cannon to administer the oath. Lynne, in a bright aqua suit, held the Bible as her husband took the oath. His parents and children looked on, along with senior members of his staff. When the ceremony was over, he walked down the long marble hallway to the stairs that would take him outside to the secretary of defense's blue Cadillac limousine. Trailing close behind was the phalanx of security personnel that would follow him everywhere he went for the next four years. Cheney climbed into the car, and someone slammed the door behind him. "Boom, your House career is over and you're embarked on this new adventure," he says. "I always felt a little sad for what I left behind and the staff."[20]

There was little time for nostalgia. For years, Cheney had studied defense issues and had been a strong advocate of military reform. Now, he was in a position to effect some serious changes. He was full of big-new-job enthusiasm, and it was time to get started. As he stood behind the oversize desk that once belonged to General John Pershing, the legendary commander in World War I, he ordered an aide to fetch the Pentagon organization chart.

The aide returned and flopped the mammoth diagram in front of his boss. "It sort of fell off both ends of the desk," says Cheney. "And I rolled it up and stuck it in the trash and never looked at it again. I decided right then and there that I wasn't going to spend a lot of time trying to reorganize the place."[21]

Cheney would nevertheless have tremendous influence on the Pentagon. Because the Bush administration had a unique place in history as the midwife of a post–cold war world, Cheney would have an extraordinary role in shaping the national security of the United States for what his boss would famously call "the new world order." He would fight the prevailing political consensus that America was due a "peace dividend" and fight his own impulse to fund virtually every defense program that crossed his desk.

The day after he was sworn in was a Saturday. Cheney was in early. He had important things to do—"you know, figuring out where the restroom was and all that stuff"—and wanted to get a head start on the deluge of paperwork that would greet him Monday morning. But shortly after he arrived, he was summoned to the White House for a meeting with the president. It was urgent.

Cheney made his way to the elevator located directly in his new office. He punched the button for the garage and waited to be delivered to his limousine. He stepped out of the elevator. There was no car; there were no people. He was alone in the basement—not where he was supposed to be. He turned to board the elevator to try another floor and faced a new challenge. There was no call button—a reasonable security measure to keep would-be troublemakers from riding directly from the basement into the secretary's office, but a real problem for Cheney. He grew impatient as he thought about his predicament. "It's a Saturday morning, nobody's around, and the president is waiting for me over in the Oval Office." After a quick search, he found the stairs that would take him up. He burst through the doors out of the garage, and through a set of glass doors he could see a chaotic scene outside his limousine: agents from his security detail were barking questions at one another—"Where the hell is he? What happened to the secretary?"—and trying to locate the man they were supposed to protect with their lives.

Cheney, amused, wanted to project calm. "I tied my tie and opened the door and walked out like I knew exactly what I was doing," he says. "And I got in the car and nobody ever had the nerve to ask me, 'Hey, where the hell were you?'"[22]

"That was my first day on the job. Lost in the basement of the Pentagon."

It did not take long for Cheney to get his bearings. He held his first press conference one week after he started. He opened by declaring that he did not have any special announcements, didn't want to make "any major news," and viewed the briefing as a simple get-to-know-you session with the Pentagon press corps.

Reporters on the Pentagon beat weren't interested in friendly chitchat. The first question to Cheney concerned reports in the morning papers that the Pentagon had reached an agreement with members of Congress on a controversial plan for missile defense. Cheney said such reports were "premature." The follow-up question was more specific.

"Mr. Secretary, General Welch, the chief of staff of the Air Force, apparently has been up on the Hill working this program himself. Is that a change of policy for the Defense Department to have a service chief negotiate his own strategic system?"

"General Welch was freelancing. He was not speaking for the department. He was obviously up there on his own hook, so to speak," Cheney said.

"Do you accept that?"

"No, I'm not happy with it, frankly. I think it's inappropriate for a uniformed officer to be in a position where he's in fact negotiating an arrangement. I have not had an opportunity yet to talk to him about it. I've been over at the White House all morning. I will have the opportunity to discuss it with him. I'll make known to him my displeasure. Everybody's entitled to one mistake."

Cheney's public rebuke of a senior military officer may not have been major news outside the defense establishment, but within minutes word of the scolding was bouncing throughout the miles of marble corridors in the Pentagon. Such a tongue-lashing was unusual. That it came in public during the new secretary's first press conference made it extraordinary.

The castigation of Welch set the tone for civilian-military relations for the next four years. "He wasn't looking to make an example out of anyone," says a senior adviser to Cheney at the Pentagon. "But frankly, it was important."

"The entire building said, 'Holy shit,'" says another.

"In fairness to Welch—I liked Larry Welch," says Cheney. "He was one of the better service chiefs that we had under my watch. Very bright guy. He was under the impression he had been authorized by Will Taft, who was the outgoing, acting guy, to go up and deal with the Congress on the missile program and nobody had bothered to tell me. . . . It was probably a little unfair to Welch, but on the other hand it was a gift to me, in terms of my putting down a marker in my first few days on the job that you've got to be very careful, make sure the secretary's plugged in, and that I didn't have any qualms about taking on the uniformed military."[23]

In his confirmation hearings, Cheney said that his first priority would be personnel. The White House had provisionally filled two of the top positions at the Pentagon in anticipation of John Tower's confirmation. Don Atwood left his position as vice chairman of General Motors to serve as Tower's deputy secretary of defense, a job that typically involves day-to-day management of the sprawling Pentagon bureaucracy. When Bush picked Cheney to replace Tower, he asked Cheney to meet with Atwood and to consider keeping him. "He didn't say keep him," Cheney explains. "He said take a good look at him. So that's what I did."[24]

Cheney regarded the position as more important than most cabinet-level jobs. He had observed the work of deputy secretaries under Nixon and Ford—David Packard and Bill Clements—and knew what he wanted in his own number two. The meeting went well, and Cheney hired Atwood.

Cheney drew up a list of things he would delegate to his deputy. The secretary would focus on big policy questions, use of the force, chain of command, personnel, the budget, relations with Congress, and intelligence. Atwood would handle management of the service secretaries, procurement, contracts, and the military in domestic crises such as hurricanes.

The other senior position nominally filled was undersecretary of defense for policy. Paul Wolfowitz had served as U.S. ambassador to Indonesia under Ronald Reagan and before that as the director of the State Department's policy and planning office. His résumé also included work on defense policy issues, as assistant secretary of defense for regional programs in the Carter administration and as a staffer in the Arms Control and Disarmament Agency under Ford and Nixon. Wolfowitz had another attribute that appealed to Cheney: he was an academic, having received a PhD in political science from the University of Chicago. After a get-to-know-you meeting, Cheney agreed to keep Wolfowitz on as his top policy adviser.

Even before Cheney arrived at the Pentagon, he knew he would need a new chairman of the Joint Chiefs of Staff. The current chairman, Admiral William Crowe, was scheduled to retire on October 1, 1989. The position, filled by a four-star general or admiral, coordinates the activities of the chiefs of the four services—Army, Navy, Air Force, Marines—and the chairman is the highest-ranking uniformed soldier in the country.

Three years earlier, on a trip to Europe, Cheney was on a congressional delegation that made a brief stop in Germany to refuel and receive a briefing. The young general with oversized glasses seemed personable and knowledgeable. It was a short meeting, Cheney thought, and a good one. Cheney had known Colin Powell—by reputation, anyway—through the general's service on Reagan's NSC staff.

Three years after their first meeting, on a trip back from the headquarters of U.S. Central Command in Tampa, Florida, he took a detour to Atlanta for a follow-up. Powell was now running U.S. Forces Command (FORSCOM) at Fort McPherson, which deploys and supports conventional Army forces around the world. The two men got reacquainted, and Cheney once again left impressed.

He knew then that he wanted to hire Powell as the new chairman of the Joint Chiefs of Staff. But there was a potential problem. Powell was the youngest of the fifteen four-star generals eligible for the position. Cheney was worried that elevating him would cause jealousy and resentment among those who were passed over for the job. Cheney was not the only one concerned. When he took his idea to the president, Bush expressed the same sentiments. He thought Powell was a good pick, but he worried about possible fallout from the selection.

Cheney decided to raise the issue with Powell. The general had suspected that he was being considered. An article in the *New York Times* reported that he had been writing letters to Cheney with the hope of endearing himself to the new boss at the Pentagon, a report Powell would later dismiss as erroneous.[25] Still, when Cheney summoned Powell to the Pentagon in early August, the general knew why.[26]

There was no small talk. Cheney had dispatched a helicopter to Baltimore, where Powell was attending a conference, to shuttle Powell to the Pentagon. They met in Cheney's office.

"You know we're looking for a chairman," Cheney told Powell. "You're my candidate."[27]

Powell was Bush's candidate, too, and the next day he stood at the president's side as Bush made the announcement in the Rose Garden.

In truth, Cheney did not mind once again ruffling the feathers of the military leaders. He had come to the Pentagon with the goal of reestablishing civilian control, and jolting the world of the uniformed military was a good way to do that. After the censure of Larry Welch, this was the second time in six months. It would not be the last.

Cheney brought with him senior staff members from his offices on Capitol Hill. Pete Williams would run the Pentagon's public affairs operations; Dave Gribbin and Jim Steen worked on congressional relations. Cheney also tapped Michael Malbin, then working in the whip's office, as a speechwriter, and brought David Addington to the Pentagon as his special assistant.

These connections to Congress would pay immediate dividends. One of Cheney's first priorities was to shepherd a defense budget through Congress—no easy task under normal circumstances and more difficult because he had come to the job late, after the protracted fight over Tower's nomination. Further complicating matters, the prevailing political winds—and a Democratic Congress—were pushing cuts, in some cases major cuts. Cheney understood this.

"I am one of those members of Congress who, as a member of Congress, voted for every single defense program they ever put in front of me. You couldn't find a weapon system I didn't like as a member of Congress," he explained at a public forum in Pittsburgh. "My problem is, now that I'm Secretary, I've got to figure out how to pay for all that stuff."[28]

He was open to some reductions, but wary of cutting too much. It was the posture he would take for his entire four-year tenure. Determined to resist what he viewed as irresponsibly deep cuts, Cheney publicly warned about putting too much faith in a "kinder and gentler Soviet Union."[29]

Behind the scenes, however, Cheney was open to negotiating with congressional Democrats. His flexibility won him plaudits from his former colleagues. "You will only hear praise for Mr. Cheney from the Congress," said Representative George J. Hochbrueckner, a Democrat from New York, who worked with Cheney on his first budget as secretary. "The guy, he's good, he's honest, he has high integrity, he works hard, a very smart man. . . . We were very impressed when after only 39 days [in office], he came up and went through three hours of grueling testimony."[30]

Cheney faced his first crisis less than eight months after being sworn in. On October 3, 1989, Cheney was touring the Civil War battlefield at Gettysburg with his counterpart from the Soviet Union, Minister of Defense Marshal Dmitry Yazov, when he received an urgent call on his oversize late-1980s cell phone. It was Powell. There was trouble in Panama.

Political unrest in this small Central American nation had been growing. For years, Panama had been a convenient, and sometimes important, ally of the United States. Its nominal leader, General Manuel Antonio Noriega, had worked with the United States—sometimes in public, usually in the shadows—as it waged its low-level war against communists in his neighborhood.

Lately, though, the agendas were increasingly in conflict.

As the U.S. government stepped up its campaign against illegal drugs in the final years of the Reagan administration and the early years of the Bush administration, authorities found that many of their leads took them to Panama—which borders Colombia, the world capital of the drug trade—and to Noriega. The Panamanian strongman found himself a victim of shifting U.S. priorities, and as Republican administrations turned their sights from communism to drugs, he was no longer a useful friend.

"We'd inherited a mess in Panama," Cheney recalls. "It got more and more difficult to ignore this legacy that had been left behind by the Reagan folks."[31] With potentially historic changes taking place in eastern Europe and the Soviet Union, Panama was not a high priority when George Bush moved into the White House. The CIA explored ways to weaken Noriega's grasp on power, but the matter was not considered so urgent that it was worth taking any great risks to unseat him.

Over time, he became more than a mere nuisance. Noriega's sympathizers in the Panamanian Defense Forces (PDF) regularly pestered American military and civilian personnel, most of whom were in Panama to work on the canal. Much of the antagonism was harmless. Soldiers were taunted verbally or stopped as they drove in the streets of Panama City. But as relations between Noriega and the American government soured, the frequency and seriousness of the harassment worsened. The instability grew because Noriega's support among Panamanians was dwindling. His countrymen were living with the effects of economic sanctions imposed by the United States designed to strangle his regime; and word of Noriega's often violent suppression of any political opposition spread quickly among the population of some 3 million.

In May, less than two months into Cheney's tenure at the Pentagon, Panama held national elections. They were rife with corruption. Noriega simply nullified the results he didn't like. Even Jimmy Carter, in Panama to monitor the voting, declared it a "fraud."[32]

Cheney had received regular reports on the election and military maneuvering in Panama from General Frederick Woerner, the commander of SOUTHCOM, in charge of U.S. forces in

Panama. But Cheney and his top advisers regarded Woerner as weak and as unwilling to challenge Noriega. One of these advisers recalls Woerner as an "apologist" for Noriega. Cheney did not go so far, but he worried that Woerner always "had an excuse why we couldn't do anything in Panama." He had his eyes open for a possible replacement. Over lunch in June with a former colleague in the Ford administration, Jack Marsh, now serving as secretary of the Army, Cheney found one. Marsh had brought with him Maxwell Thurman, a four-star general who was retiring from his post running the Training and Doctrine Command. Marsh thought the two men would get along. He was right.

Cheney had another source on developments in Panama: Representative John ("Jack") Murtha, a Democrat from Pennsylvania. Cheney and Murtha knew each other well from their time in Congress together. Murtha was an old-school anticommunist and, compared with most Democrats in Congress, a hawk. Cheney and Murtha disagreed on plenty but agreed on the big things — the threat from the Soviet Union, aid to the contras, and the importance of maintaining a strong national defense.

Others knew about Murtha's good relationship with Cheney. When the speaker of the House, Jim Wright, had found himself in trouble for publicly confirming classified intelligence, he asked Murtha to make things right with the House Republicans. "Jim asked me to intercede with Dick Cheney," says Murtha. Cheney told Murtha that the problem could be solved if Wright would simply apologize. After several days of back-and-forth, Wright refused. But the negotiations brought Cheney and Murtha closer. "Those are the kind of things we dealt with all the time," says Murtha.[33]

Murtha had made himself something of an expert on Panama. He visited regularly and sought to divide the PDF from Noriega. In meetings with PDF officials — and occasionally on television — Murtha would welcome cooperation from the PDF but make clear that the U.S. government was wary of any relationship with Noriega. Murtha was in Panama for the election in May, and like everyone else he dismissed the results as bogus.

Over the summer, reports that the PDF was harassing Americans increased. The incidents were sometimes violent, and

Washington took them as acts of deliberate provocation. Behind closed doors, senior Bush administration officials spent more time discussing how they might solve "the Noriega problem." Cheney and his colleagues regularly heard that coups were being plotted by Panamanians, but most of the reports were vague and few plots were attempted.

Late in the evening on October 1, the Pentagon received word that a senior Panamanian general, Moises Giroldi, had reached out to the CIA for help in overthrowing Noriega. The CIA was wary. The details of the plot were fuzzy, but the CIA reported that General Giroldi planned to install himself as Noriega's replacement. There were few indications that he was the kind of democrat who would make the potential benefits of American help worth the obvious risks. So the United States declined to play a major role. Giroldi decided to proceed without significant operational help from the United States.

His forces took the PDF headquarters in Panama City and captured Noriega. It was a surprisingly successful coup, but only for a matter of hours. Before he was captured, Noriega had called in PDF soldiers loyal to him—and in time, he talked his way out of captivity. Giroldi was shot and killed. The coup was over.

The botched coup presented the U.S. government with unwelcome complications. As news of the uprising filled newspapers, politicians on Capitol Hill blamed the Bush administration for doing too little, and Noriega criticized it for doing too much. As Noriega's paranoia grew, so did the frequency and seriousness of incidents against Americans in Panama. And as critics, mostly Democrats, took shots at Bush for being a "wimp," he resolved to solve the problem.

Politics in Washington and the skirmishes in Panama made escalation a virtual certainty. All it would take would be an incident or two—something that would leave Bush little choice but to act.

Such a provocation came on December 15, 1989. Panama's National Assembly declared Noriega "maximum leader for national liberation" and also declared that Panama was in a state of war against the United States.

The next day, PDF forces shot and killed a Marine lieutenant, Robert Paz, who was a passenger in an American car that had been stopped at a PDF checkpoint. In a moment of panic, the driver sped off before the examination of his vehicle was finished, and PDF soldiers began firing at the car. The shooting was not directly related to the assembly's declaration, but the high tension between the PDF and the U.S. government was certainly a factor. As it happened, a Navy lieutenant and his wife had been detained at the same checkpoint and saw the shooting of Paz. They were taken into custody by PDF troops and physically abused as they were interrogated.

In Washington, Cheney met with Powell and several top civilian aides. Cheney had been worried about possible terrorist attacks on American soldiers and civilians. There had been reports of anti-American plots to be conducted by guerrilla armies of Colombian drug lords and the PDF itself. Still, Cheney wasn't ready to commit the United States to an invasion. Two of his top advisers—Colin Powell and Paul Wolfowitz—made arguments that would run counter to the reputations each would develop in the coming years.

Just three months into his tenure as chairman of the Joint Chiefs of Staff, Powell, who would later become known as a reluctant warrior, argued for a quick and forceful military response.[34] Wolfowitz, who would later be considered one of the world's preeminent hawks, was hesitant. He thought the incidents were unfortunate, but not reason enough to launch a full-scale counterattack. The other civilians nodded their heads in agreement. Cheney listened to the debate intently. Powell won.

The invasion happened quickly. Civilians at the Pentagon had been frustrated by a lack of creativity in the existing plans for military operations in Panama. Several believed that the uniformed military submitted unworkable plans on purpose, as a way to prevent the Bush administration from using force to oust Noriega. But after some tweaking, Cheney thought the plans could work.

As he reviewed the plans, he noticed something odd. "I see in there F-117 stealth bombers going into Panama and I thought: 'Wait a minute. What kind of air defense system have they got in Panama? Why do we need this?'"[35]

Cheney had seen the stealth bombers up close as a member of the House Intelligence Committee. He secretly flew at night to Tonapah, Nevada, where the stealth bomber was being put together in what was at the time a highly classified program. Cheney thought it looked like something out of Buck Rogers. "Weirdest plane I'd ever seen in my life," he says.[36]

The Air Force, he says, "was dying to use the darn things." He acquiesced, but limited their participation to two bombing raids. "It was hard to justify stealth bombers over Panama City."[37]

By January 3, 1990, Noriega was in the custody of the United States. The short war was over. "It was interesting because it was a trial run to some extent," Cheney recalls. "It was the first time the president and the administration used force, exercised our authorities to launch a military operation. And it was good practice, in effect, for what you have to go through to do that kind of thing."[38]

And a year later, they would do that kind of thing again. The trouble started much earlier, with a destitute Saddam Hussein making noises about Kuwait's stealing Iraqi oil. Iraq was left impoverished by the Iran-Iraq War of 1980–1988. Kuwait had supported Iraq throughout that conflict, both diplomatically and financially, and when the war ended the Kuwaitis wanted Iraq to repay its debts. Iraq did not have the means to do so, and in an effort to force Kuwait to forgive some of those liabilities, Saddam began a public campaign against the Kuwaiti royal family. Among other alleged transgressions, the Iraqis accused Kuwait of slant-drilling to gain access to oil on the Iraq side of the border. The Kuwaitis denied the claims and sought to force Iraq to repay its debts.

By the summer, as Saddam Hussein openly threatened to take action in Kuwait, he backed up his belligerent rhetoric by massing Iraqi troops on the border with Kuwait. Although Saddam had provoked the war with Iran a decade earlier, the consensus among U.S. intelligence analysts was that the Iraqi troop movements were meant to do nothing more than scare Kuwait into a weaker negotiating position. It was bluster, they said, not a sign of an imminent invasion.

There were dissenting voices. Pat Lang, an analyst at the Defense Intelligence Agency, had been watching the Iraq-Kuwait border closely and was becoming increasingly concerned about the likelihood of an attack. He made his views known, but others at the DIA were not convinced. Analysts at the CIA, too, downplayed the Iraqi maneuvers.[39]

Cheney did not know what to make of the situation. But as he had in the past, he reasoned that the safest course was to assume—and prepare for—the worst. On July 19, he offered a public warning to Saddam Hussein.

"We take very seriously any threat that could risk U.S. interests or U.S. friends in the region," he told reporters. "That's a part of the world you've got to be concerned about and where you have to take into account the possibility that at some future date U.S. forces could become involved in conflict."[40]

But Saddam was getting mixed messages from the U.S. government. Even as Cheney talked tough, the State Department took a decidedly less confrontational line, at one point assuring Saddam that the United States would not interfere in regional disputes.

Captain Michael McConnell of the Navy had been at the Pentagon for less than a week as the top intelligence official for the Joint Chiefs. He was new and less experienced than many of his colleagues and was not eager to make waves. But as the border activity intensified, he felt strongly that the intelligence analysts and regional experts had to give Cheney their best educated guess. Were the Iraqis going to invade or not?

Look, he told his colleagues, if there's going to be an invasion, let's call it. Let's be bold here.

"There's pretty strong resistance in the intel community to taking a risky stance," says McConnell. "And being an intel professional my whole life, I'd watched us go through many other crises and not make the call. I didn't want that to happen on my watch."[41]

CENTCOM was reluctant. It had never done anything like this before and told McConnell that it did not want to begin with Iraq. In an interview, McConnell says he decided to force its hand by consulting with Powell. "I don't know if I should admit this," he said, "especially with the tape recorder running." He paused to think about it. "OK."

"I'm going to see the chairman," he told his colleagues. "I went up to his office on the second floor, which is the ground floor of the Pentagon. My office is down below. So I went up to the second floor and went into the executive assistant's office, just outside the chairman's office. There's a little peephole—you can look in to see if he's busy so you don't interrupt him, that sort of thing. So I looked in and he was sitting at his desk. I came out and I walked down the hall and had a Coke and went back down. And I said: 'I've seen the chairman and here's what we're going to do.' I never said that I'd talked to the chairman. 'I've *seen* the chairman.'"[42]

It was late, but CENTCOM made the call. Iraq would invade.

Cheney arrived at the Pentagon at eight AM on August 1, 1990, later than usual. That morning, he met with his speech-writers and attended an honor cordon for President Gnassingbé Eyadéma of Togo before an intelligence briefing with Assistant Secretary Duane Andrews and a budget meeting with the Pentagon comptroller, Sean O'Keefe.

Cheney kept an eye on developments in the Gulf in frequent discussions with Powell, Wolfowitz, and one of Wolfowitz's top aides, I. Lewis "Scooter" Libby. With little to do but watch and wait, Cheney spent much of his day wrangling with Congress over the budget. The issue of the day was the future of the B-2 bomber. Democrats on the House Armed Services Committee were pushing Cheney to eliminate funding for the bomber, arguing that such weaponry was unnecessary in light of a reduced threat from the Soviet Union. It was exactly the kind of cut Cheney had warned about at his confirmation hearings. Cheney maintained that although the conventional threat was on the decline, the strategic threat remained.

He made his case in an appearance that evening on the *MacNeil/ Lehrer NewsHour*. "There are two things that we focus on when we talk about the possible threat, if you will," Cheney said. "One is intentions, the other capability. Intentions can change overnight. You can get new leadership in the Soviet Union, some crisis develops, those intentions change. Capability takes a very long time. We have been working on the B-2 for some twelve or thirteen years and the danger is that we will make the mistake of assuming that current Soviet intentions are improving [and] then we can get rid of capability that we may need at some future point."[43]

It was an important debate, but one that would be eclipsed by events later that evening. At eight PM in Washington—two AM in the Persian Gulf—Iraqi troops poured over the southern border into Kuwait. Admiral Bill Owens, Cheney's military aide, who had been feted at a going-away party that afternoon, informed his boss about the invasion at nine PM.

The Bush administration, guided by the consensus of the U.S. intelligence community, had misread Saddam Hussein's aggressive intentions. Through his public rhetoric and his provocative actions, Saddam had given abundant warning of the coming invasion. "Instead of seeing Iraq's war preparations for what they were," wrote Michael Gordon and Bernard Trainor in the definitive account of the Gulf War, the Bush administration "had embraced the most benign possible explanation of the Iraqi moves."[44]

Cheney's skepticism had been vindicated.

By morning, the Iraqis controlled most of Kuwait City and the few Kuwaitis who had initially resisted had either given up or been killed. Cheney reported to the White House at eight AM for a meeting of President Bush's foreign policy team; only Secretary of State James Baker was absent—he was in Siberia, leading a delegation to Russia.

Cheney decided early that the United States should be prepared to use force. He thought it important to make clear to the world that the U.S. government would not tolerate acts of naked aggression. His colleagues were decidedly less enthusiastic about the prospects of sending the U.S. military to roll back the Iraqis.

In a brief photo op before the meeting, President Bush was asked about a military response to the Iraqi invasion. He said he was "not contemplating such action." It was exactly the wrong message to send, Cheney thought.

After reporters were escorted out of the room, General Norman Schwarzkopf, the barrel-chested chief of CENTCOM, told the gathering that the United States had two military options: conduct air strikes on Iraqi military targets or implement Plan 1002-90, an off-the-shelf plan to defend Saudi Arabia that required the deployment of 200,000 American troops.

Cheney's days following the invasion were a flurry of briefings and contingency planning meetings all with the purpose of providing answers to two questions: What is the United States' proper response to the Iraqi invasion? And is Saudi Arabia next?

The national security team met with the president at Camp David to review their options. When Cheney returned to Washington, he received a call at home from Scowcroft, who told Cheney that he and Powell would probably be leaving for Saudi Arabia to meet with King Fahd the following day.

Cheney wanted to go instead. From early on, Cheney thought that it would probably be necessary to use the U.S. military to expel the Iraqis from Kuwait. The very boldness of the invasion of a sovereign country suggested that reasoning with Saddam Hussein was not likely to produce results. And if the U.S. military was going to be involved, Cheney wanted to shape that involvement.

"I thought it was important from the standpoint of the department as well as my ability to run things that they get used to—over at the White House—having me do that sort of thing. It was easy to look and see Scowcroft and Powell going off and doing their thing and I'm back manning the fort. I didn't want that to happen."[45]

Bush agreed to send Cheney.

The White House was awaiting word from Prince Bandar bin Sultan, the Saudi ambassador to the United States, who was in Saudi Arabia seeking approval from King Fahd for the trip. Cheney had already met with Bandar to share the broad outlines of the United States' possible war plan, an operation that would include more than 100,000 U.S. troops. Bandar had told Cheney that he understood the stakes and would fly back to Saudi Arabia to prepare King Fahd for a forceful presentation.

The American delegation literally waited on the tarmac at Andrews Air Force base for word from Riyadh. Pete Williams, who had come to the Defense Department from Cheney's congressional office, was on the phone with his office back at the Pentagon. President Bush was on television from the White House making a statement about Iraq. A member of the public affairs staff held the phone to the television so Williams could hear the president.

Bush was talking tough. "This will not stand, this aggression against Kuwait," he declared.

When Williams told Cheney about Bush's statement, his boss answered with one word.

Good.

The Saudis had agreed to receive the American delegation. The trip had a simple mission: to obtain permission from King Fahd to deploy U.S. forces to Saudi Arabia. On its face, it was an obvious next step. With Iraqis continuing to move toward the border of Saudi Arabia, the Saudis would be foolish to reject the offer. But there were domestic political considerations for the Saudis that made the picture more complicated. The Saudi population was not likely to welcome the presence of American troops. It was Cheney's job to convince the king that the potential political consequences were far preferable to an Iraqi invasion.

Cheney had taken Schwarzkopf; Wolfowitz; Williams; Robert Gates, a former deputy director of the CIA, now working as Scowcroft's deputy at the CIA; and Charles Freeman, the U.S. ambassador to Saudi Arabia, who had been back in the United States. The men got together in Cheney's quarters on the plane to run through the meeting. Cheney thought the CIA's briefing was weak. "The caliber and quality of his presentation was too hesitant—'We didn't know this and we might know that,' and so forth. What we wanted to show the king was: 'Look, here are Iraqi tanks lined up along the border.'"[46]

Cheney also rejected the guidance provided by Freeman and his colleagues from the State Department, who suggested a gentle, patient approach to the Saudis. They told Cheney to be prepared to wait several days for an answer. "Standard, traditional State Department advice," says Cheney.[47]

When Cheney arrived, he took a short nap and then met with Bandar again. Bandar told Cheney that the biggest problem he would face was the Saudis' skepticism about the intentions of the United States—an attitude left over from the Carter administration.

Following the fall of the shah of Iran, the Saudis had requested a demonstration of support by the United States, something to show that any aggression toward Saudi Arabia would be met with American might. President Jimmy Carter agreed to help. "His response to all of this was to order up a squadron of F-15s to fly to Saudi Arabia. When they were halfway there, he announced they were unarmed," says Cheney. "Now, I suppose in the minds of the State Department types, this was crafty diplomacy."[48] From the Saudis' perspective, it showed weakness. Bandar extended the analogy to make his point to Cheney: we don't want unarmed F-15s.

The Americans made their presentation to King Fahd. Cheney did most of the talking, with assistance from Schwarzkopf on the Iraqi army and the details of the proposed deployment. Bandar translated. After about two hours, Bandar stopped translating as the Saudis conferred among themselves. Only Charles Freeman, who spoke Arabic, understood the conversation.

"We have to do this," Fahd told the group. "The Kuwaitis waited; they waited too long and now there no longer is a Kuwait."

Crown Prince Abdullah disagreed. "There is still a Kuwait."

"Yes," said Fahd, "and all the Kuwaitis are living in our hotel rooms."[49]

The king agreed to the plan and asked Cheney for two assurances. First, he wanted the United States to send enough troops to do the job; second, he wanted the troops to leave when they were asked to leave. Cheney gave his word.

On the return to Washington, Cheney stopped in Egypt to see Hosni Mubarak. At the airport in Alexandria, Cheney's plane parked next to a large jet belonging to the Iraqi government. Saddam Hussein had sent his trusted deputy, Izzat Ibrahim al Duri, on a mission designed to undercut the Americans. Izzat Ibrahim, a devout Muslim, explained the Iraqi invasion to leaders in the region and asked them to reject American overtures.

Cheney won this round of his battle with Izzat Ibrahim, as

the Egyptians supported the decision of the Saudis to accept U.S. troops and later sent two divisions to fight alongside the coalition forces. But Izzat Ibrahim would prove more successful after the United States invaded Iraq a second time, more than a decade later.

Cheney's last stop before his return to America was Morocco. President Bush had asked Cheney to brief King Hassan on the United States' agreement with the Saudis. When the group arrived in Rabat, in the middle of the night, a small welcoming party was waiting. After an exchange of diplomatic pleasantries, Cheney explained to King Hassan that the information he was about to share was highly sensitive and asked for a private audience.

The two men met in a small office attached to a large conference room. A Moroccan translator joined them. Before the meeting started, King Hassan placed a small silver box in the palm of his translator's hand and began speaking to him in Arabic. A moment later, the translator responded, then turned to Cheney and told him the meeting could begin. Curious about the exchange, Cheney asked King Hassan what the ritual meant. The silver box, the king explained in halting English, contained a piece of the Koran. He made his translator swear to secrecy. "I told him that what he was about to hear he should not reveal to anybody on pain of death," explained King Hassan. Cheney, a fanatic for secrecy, had a predictable reaction. "I thought: Damn, I need one of those."[50]

Cheney landed at Andrews Air Force Base at three-twenty AM on August 8. At seven-fifteen he met Baker and Scowcroft for breakfast at the White House. The three men met each Wednesday at the same time whenever they were all in Washington. They updated one another on the latest developments in the Gulf and elsewhere and tried to resolve any differences over policy before they became problems.

Even as they worked on issues of war and peace, their friendships continued. In late October, as these men were pointing the country toward a war with Iraq, Baker and Cheney stole away for a long weekend of fishing in Pinedale, Wyoming.

Even more frequent than his meetings with Baker and Scowcroft were Cheney's late-afternoon sessions with Colin Powell.

The two men often got together or spoke several times each day before a session, but the standing meeting allowed them to compare notes on military planning and interagency discussions.

Cheney would take a much more active role in war planning for Iraq than he had taken for Panama. He had been disappointed by the existing contingency plans for fighting in the Persian Gulf region, designed to respond to an attempt by the Soviet Union to invade the oil-rich countries there.

"The first plan we got back from Schwarzkopf was not very attractive," Cheney says. "It basically involved sort of straight up the middle into Kuwait, which we didn't think was a very good idea."[51] Cheney understood that Schwarzkopf was trying to be realistic about their options. Still, he thought, with clear air superiority it made little sense to send the ground troops into what was almost certainly the heart of the enemy forces.

Cheney asked Wolfowitz and his policy team to come up with an alternative plan. He wanted an answer to a basic question: "How you could use what you had basically, along with some modifications, to go at the Iraqis without having to go straight into their strength?"[52]

At the same time, Cheney went back to Powell and Schwarzkopf and asked them what they would need to be successful. "And they started ticking off troops. 'Well, we want another Marine division. We want Big Red One out of Fort Riley, Kansas. We want to bring down the Seventh Corps out of Europe. You know, sort of the ultimate wish list. . . . I always felt that part of what was happening was that the military was testing us. They weren't absolutely certain that we were going to give them what they needed. And there was still—you know, this is the aftermath of Vietnam—questions of the relationship between the military and civilians and the sense that the civilians had sort of hamstrung them in Vietnam. And from our perspective, we didn't want to be in a position where we denied them something they thought they needed."[53]

So they granted virtually every wish. "We said: 'OK, you got it. Now, what are you going to do with it? How are you going to use it?"[54]

For Cheney, giving the generals what they wanted solved another problem. "The odor of defense cutbacks is in the air," he recalls. "So we were in the midst of budget negotiations as well that were going to lead to big cutbacks of the military with the cold war ending. . . . I was perfectly prepared to spend whatever they thought was needed because frankly it would make it easier to defend against what I thought was going to be unwise pressure to cut the defense budget on the grounds that the Soviet Union was going out of business."[55]

Still, there was a sense among some of Cheney's staff that Schwarzkopf and the uniformed military were offering inadequate war plans in an effort to delay the onset of war. The civilians had seen this foot-dragging before, during the planning for Panama.

Henry Rowen, assistant secretary of defense for international security affairs, had done some thinking about the war plan. Rowen previously served in government as chairman of the National Intelligence Council. In the private sector, he had been president of the Rand Corporation and a fellow at Stanford University's Hoover Institution. He was, in short, a man whose ideas were taken seriously.

Rowen was concerned that the Iraqis might try to draw Israel into any war with the United States. The Iraqis had missiles capable of reaching Israel from western Iraq; and earlier that year, Saddam Hussein had threatened to "burn half of Israel" in response to any perceived provocation. If Israel became involved, the fragile Arab coalition could splinter or break apart altogether. Rowen and Wolfowitz, his boss, thought that any good war plan would account for Iraqi missiles and protect Israel.

Rowen went to Wolfowitz with his idea, and the two men went together to brief Cheney. The policy planners wanted to flood Iraq's western desert with American troops at the outset of the war. Doing so, they reasoned, could prevent the Iraqis from firing SCUD missiles at Israel and would draw Iraqi troops from Kuwait so that they might defend Baghdad. "I put it more or less this way," said Rowen. "When our forces have occupied the Western Desert up to the Euphrates, some of them will be within sixty miles of Baghdad. Saddam might pull his troops out of Kuwait to try to defend Baghdad but they will never get there because our air will destroy them; Saddam might be overturned; he might flee. Something will happen in Baghdad.'"[56] Rowen thought Cheney seemed particularly intrigued by that possibility. Cheney instructed the small team to continue planning and insisted that they keep their activities a secret. The civilians recruited Lieutenant General Dale Vesser to help refine the plan.

Schwarzkopf and Powell did not like the idea. Powell, in fact, was increasingly skeptical about using force at all. He thought sanctions should be given a greater opportunity to succeed and went to Cheney to share his concerns. Cheney didn't agree with Powell, but he nonetheless arranged for Powell to make his case to the president.

"I thought it was important that the principal military adviser to the president get a hearing," says Cheney. "It was all amicable. Colin didn't say, you know, obviously you can't do this or I won't work for you, like that. He was more of a 'Go slow, maybe we can get them out without having to use force' approach. My argument was: you're less likely to have to use force—diplomacy has a better chance of working—if he thinks you will use force. And the best way to prepare for either eventuality is to deploy the force. Whether he hangs it up and withdraws, so you don't have to use it, or whether he basically tells you to stuff it. Then you've got them there and you're ready to go." In the end, says Cheney, "George Bush wasn't interested in sanctions."[57]

So the buildup continued and Cheney pressed on with "the western option," presenting an updated version of the plan to Powell and Schwarzkopf. Cheney thought the alternative plan could be useful even if it contributed little to the final war plan.

"It also was a spur to the chairman, Colin Powell," Cheney recalls, "a signal, if you will, that clearly said: 'Look, we're deadly serious about this business. And we want a plan that will work. And it created quite a stir in the building.'"[58]

As planning for the war was becoming increasingly serious, the media coverage increased. On Sunday, September 16, Cheney awoke to an unwelcome headline, above the fold on the front page of the *Washington Post*. "U.S. to Rely on Air Strikes If War Erupts."[59]

"The Joint Chiefs of Staff have concluded that U.S. military air power—including a massive bombing campaign against Baghdad that specifically targets Iraqi President Saddam Hussein—is the only effective option to force Iraqi forces from Kuwait if war erupts, according to the Air Force chief of staff, Gen. Michael J. Dugan."[60]

Cheney had grown accustomed to chest-thumping from the services in more than a year at the Pentagon. But this time it was worse.

The Israelis had suggested taking out Saddam Hussein, Dugan told reporters, and it was advice the United States was inclined to heed. Cheney was not happy. These were exactly the kind of internal deliberations that he had long insisted remain internal. That a high-ranking uniformed officer would speak to reporters about prospective targeting before a potential conflict was one of the worst violations of protocol he could imagine.

Cheney had become acquainted with Dugan as a member of the House Intelligence Committee. Dugan was then the Air Force's assistant deputy chief of staff for plans, and his responsibilities included briefing members of Congress on tactical intelligence and budget matters. In 1986, according to Dugan, the two men clashed on funding for SR-71, a reconnaissance aircraft known as the Blackbird. Cheney, then still a military reformer, wanted to cut funding for the plane; Dugan believed he was there not only to argue on behalf of the Air Force's requests for funding but to defend President Reagan's budget. The tension that resulted resurfaced when Cheney was secretary of defense, says Dugan, and Cheney only reluctantly promoted him to Air Force chief of staff in July 1990.

Dugan was three months into his new position when he invited three reporters to travel with him to the Persian Gulf as the Air Force prepared for possible fighting in Iraq. Dugan had come to the job eager to improve relations with the press and ignored several warnings from the Pentagon's public affairs shop against taking the press with him. Throughout their travels, Dugan spoke at length to the journalists who had accompanied him on the trip. They were surprised at his candor and dutifully amplified his words for their readers.

"The cutting edge would be in downtown Baghdad. This

[bombing] would not be nibbling at the edges," Dugan told the reporters. "If I want to hurt you, it would be at home, not out in the woods someplace.

"If and when we choose violence," Dugan explained, Saddam Hussein "ought to be at the focus of our efforts."[61]

Cheney immediately called Pete Williams at home. It was early and Williams was still groggy. Williams optimistically raised the possibility that the *Post* had just mischaracterized the comments or inflated their importance. A reporter from the *Los Angeles Times* had also traveled with Dugan, he told Cheney. Maybe its story will be better. Williams called a Pentagon liaison office in California, where it was still the middle of the night, and told the desk officer who answered the phone to fax a copy of the article in the *Los Angeles Times* to him at home.

But that story was not significantly different in tone or content from the article in the *Post*. Williams called Cheney with the bad news and reminded him that public affairs had cautioned Dugan about taking the press on the trip. This was a detail Cheney had forgotten, and it made him even angrier.

Williams met with Cheney later that afternoon. Cheney recorded his views on his yellow legal pad. He thought the comments created problems for three practical reasons. First, they potentially compromised intelligence provided by an ally. Second, they came at a time when the United States was trying to win support for the use of force from Arab countries hostile to Israel; whatever popular support there was for rolling back the Iraqi invasion was sure to disappear if these populations came to believe that their governments were taking marching orders from Israel. And third, the remarks could be read as a violation of the executive order banning assassination of foreign leaders.

Dugan was called in to see Cheney at his office on Monday morning, shortly after Cheney adjourned an eight o'clock meeting with Powell and Don Atwood. Cheney read Dugan a list of grievances from the yellow legal pad. The secretary of defense showed no anger, says Dugan, and did not raise his voice. "He was just very cold." When Cheney finished reading his brief, he told Dugan that he had discussed the issue with Senator Sam

Nunn, chairman of the Senate Armed Services Committee, and then matter-of-factly asked Dugan to resign. Dugan refused, but said he would retire. "You will have a letter by eleven o'clock," he told Cheney.[62]

At a press briefing later that day, Cheney said that Dugan "took the news as I expected him to take it, as a gentleman." Cheney explained the decision. "There are some things we never discuss." He listed several: "future operations, such as the selection of specific targets for potential air strikes, . . . targeting of specific individuals who are officials of other governments," and "classified information about the size and disposition of U.S. forces." Dugan's statements, he concluded, "showed poor judgment at a very sensitive time."

Reporters pressed Cheney on whether the firing of Dugan represented a tightening of the media's access to senior Pentagon officials. "It's not a question of whether or not we make ourselves available to the press," he said. "Clearly, I think that's expected of many of us in senior positions in the Department, and I expect that of the chiefs. I do expect them to exercise discretion in what they say. And I would come back again to the proposition that sort of wide-ranging speculation about those matters that were discussed in the interviews that were granted by the general is what I felt was inappropriate, not the fact that he granted interviews."[63]

Nunn and John Warner, the ranking Republican on the committee, released a joint statement that echoed Cheney's criticism and supported his decision to retire Dugan.

Mike McConnell remembers the firing and its impact at the Pentagon. "You talk about forceful decision making," he says. "It was—boom! There were no ifs, ands, or buts. . . . Boy, that sent a wake-up signal to the United States military."[64]

The issue soon disappeared from the front pages. Behind the scenes, though, Cheney kept it alive in a surprising way. If Dugan's forced retirement went into effect after ninety days, as was required by law, he would have narrowly missed a pay raise, and so his pension would have been cut by 25 percent annually. More important to a career military officer, Dugan would have lost two of his four stars. Cheney thought Dugan's dismissal was punishment enough for his comments. He asked his congressional relations staff to request that the Armed Services Committee make an exception for Dugan.

"Nunn and Warner were so outraged by Dugan that they were determined to retire him as a two-star," says David Gribbin, Cheney's top legislative liaison at the time. "He went to battle privately with the committee."[65] After weeks of political wrangling, Nunn and Warner signed off on Cheney's decision. They did not agree with it, however, as they made clear in a statement announcing their position. "The committee is concerned not only about the circumvention of the generous 90-day transition period, but the effect that such a favored decision for one senior officer could have on the perception of all other military personnel of the fairness and integrity of the personnel system."[66]

The firing and its aftermath did not hurt Cheney at the Pentagon. "The services watched that very carefully," says Gribbin. "He didn't want them to be confused about who was running the place, but also wanted them to know how he felt about them. They knew it and they got it."[67]

The dispute over Dugan was one of many issues that required Cheney to spend time with his former colleagues on Capitol Hill. In internal deliberations, Cheney made clear his opposition to seeking congressional approval for using force in the Persian Gulf. He made arguments based on principle and practicality. The Constitution gives the president, as commander in chief, powers that provide wide latitude to use the U.S. military. Approval from Congress was unnecessary, he said.

On a practical political level, he said, the risks of seeking congressional approval outweighed the possible benefits. Cheney was particularly concerned that a vote against authorizing the use of force would mute the strong message the massive troop buildup was sending to Saddam Hussein. And he thought that a negative vote could have a disastrous impact on domestic politics. Congressional approval, he said, might provide a short-term political boost, but it was virtually meaningless in the long term. Even if members of Congress authorize the use of force, they will have no hesitation in disowning their votes if the war goes badly.

Cheney's strongly held views on executive power occasionally made their way into the public debate. On one Sunday show appearance, Cheney dismissed Congress as a "great debating society," but a body unequipped to make urgent decisions about when and where to apply U.S. military might.[68]

Still, Cheney thought it important to keep Congress informed of the administration's thinking, and he made several trips to Capitol Hill to brief his former colleagues. In one briefing for House members, held in the ornate ceremonial office adjacent to Minority Leader Bob Michel's office, Cheney provided an update on recent developments in the Gulf. He went on to describe the diplomatic efforts by the Bush administration to persuade Saddam Hussein to leave Kuwait: international sanctions, more United Nations resolutions, pressure from erstwhile allies of Iraq, strong warnings about the consequences of defiance, and a continued buildup of U.S. troops to suggest that the Iraqis take such warnings seriously. Each stage of diplomatic escalation, Cheney said, would make it more difficult for the Iraqis to maintain what was clearly an untenable position.

"He was giving us the lines in the sand on the Gulf War," recalls Porter Goss, then a first-term representative from Florida. Goss, a former CIA officer, was impressed with the briefing, but thought Cheney had left two questions unanswered. So he asked them.

"What if all of that doesn't happen? What if they still refuse to leave?"

Cheney glared at Goss, who couldn't decide if the look was meant to convey surprise or disgust. "He was either saying: 'You are really stupid' or 'You just asked a question you shouldn't have asked.'"[69] Cheney left unspoken the one-word answer: "War."

On January 12, 1991, Congress voted to authorize President Bush to use force in Iraq. The Democratic House of Representatives voted 250 to 183. The Senate vote was 52 to 47, with Tennessee senator Al Gore casting one of the decisive votes at the final hour.

The reality of a likely war weighed heavily on Cheney. As the imminent conflict approached, his staff sensed that Cheney was growing more resolved and yet more contemplative. Outwardly, Cheney carried out his duties with his characteristic deportment: "I have a job to do." But there were signs of the burden of impending war. For one thing, the walks he took along the C & O canal—sometimes with his wife, sometimes alone—increased in frequency.

News reports were filled with horrific projections of American dead. Saddam Hussein had used chemical weapons before and was thought to be working on both biological and nuclear weapons. Just how close he was to these other weapons of mass destruction was a matter of considerable debate.

Shortly after Iraq had invaded Kuwait, Cheney met with an Israeli delegation that presented him with deeply disturbing intelligence on Iraq's nuclear weapons program. The Israelis' assessment suggested that Saddam Hussein might be less than a year away from developing a nuclear weapon. Estimates from the CIA and others in the U.S. intelligence community were far less alarming, suggesting that the Iraqis were perhaps a decade away from nuclear capability.

A two-man team at the Pentagon undertook an alternative analysis, designed to challenge the emerging consensus in the U.S. intelligence community. "We took ten years of information, intelligence, even open-source information, and from that we were able to conclude that in ten identified areas that are essential for nuclear weapons development, all the boxes had been filled. We were able to point to intelligence or information in every one of those areas," said Michael Maloof, one of the analysts.

"We then took that information up to the [under]secretary, Paul Wolfowitz, at that time. Paul then sent the information to Cheney, as secretary of defense. Cheney immediately responded and sent our entire report off to the agency and asked for a reevaluation. Our estimate was they were not five years out; they were eighteen months out."[70]

Cheney had discussed Saddam's weapons programs in an appearance on CBS's *Face the Nation* on November 25, 1990:

We've known for a long time from public sources that he was trying to acquire nuclear weapons. The British intercepted devices intended to trigger a nuclear weapon in shipments to Iraq just last spring.

We also know from intelligence sources that he's working on this problem very hard. The president made reference to it, but it shouldn't be a surprise to anybody. This man has developed ballistic missiles and used them, developed chemical weapons and used them, is trying to develop biological and nuclear weapons, has tested missiles at longer ranges than just the short-range SCUD. It's only a matter of time until he acquires nuclear weapons and the capability to deliver them. And that has to be, I think, an element of concern and—as we decide how to deal with the problem.[71]

Asked specifically about the status of Iraq's nuclear program, Cheney shared the assessments he'd gotten, including the one provided by the Israelis:

There are a lot of estimates. They range through worst-case assumption—a matter of a year or less to having some kind of a crude device—to one to five to ten years in terms of having a deliverable weapon. The experts are all over the lot. What we do know is, he's doing everything he can to acquire the capability.[72]

Cheney would learn soon enough which one was the most accurate. It was a lesson he would not forget.

In the late afternoon of January 16, 1991, Cheney's military aide walked into his office with a blanket and a pillow, then left without saying a word. One day earlier, Cheney had signed the "execute order" that would launch operation Desert Storm.

The war began. As the bombing continued, Cheney and Powell took the podium at the Pentagon. President Bush had just spoken to the nation. Cheney opened the briefing by warning that he would have little to add to the president's remarks. He then gave a succinct description of the mission.

"At seven tonight, as you all know by now, Eastern Time, three Thursday morning in the Gulf, the armed forces of the United States began an operation at the direction of the president to force Saddam Hussein to withdraw his troops from Kuwait and to end his occupation of that country."[73]

When he was done, Cheney returned to his office and turned on CNN. He was alone, watching green tracers light up the night sky over Baghdad. In time, David Addington wandered in and they watched the fireworks together.

Bush had given the military a clear objective, and a limited one. Cheney agreed with him. He often did. But even when they had different views on an issue, he respected and liked the president. He told his friends that Bush was "a good man"—high praise from Cheney. He liked Bush personally, and their management styles were similar. Each man believed in delegating responsibility. "George Bush would say, 'Go liberate that country. Check back with me when you're finished,'" says one longtime adviser of Cheney's who worked closely with him at the Pentagon. "Cheney was just like that one level down." The secretary of defense had played a crucial role in shaping the war plan, but once the war started he largely left its conduct to the generals.

Cheney spent the early days of the war involved in high-stakes diplomacy. In the months before the war, Cheney had stayed in close contact with his Israeli counterpart, Minister of Defense Moshe Arens, sometimes speaking by phone several times a day. The Israelis had better intelligence on Iraq than anyone else did. Several months before the invasion of Kuwait, Saddam Hussein had threatened to "burn half of Israel." Most observers took this as a threat indicating Saddam's willingness to use chemical weapons. Even for the bellicose Iraqi leader, the statements were incredible. But the threat itself was nothing new. Israel had long been concerned about Saddam's potential use of weapons of mass destruction (WMD), especially because he had used them before against the Kurds and Iran.

The United States did not want Israel involved in the fighting. In the buildup before the war, officials at the State Department urged the administration to keep Israel at a distance. Consulting Israel too closely—and too publicly—could create political problems for the Arab countries that had contributed money and troops to fight Saddam.

The State Department strategy created its own problems. The Israelis came to resent how they were being treated, and several Israeli officials made this clear—sometimes in very colorful language—to their counterparts in the Pentagon.

One thing was certain: military involvement by Israel would complicate matters considerably. The consensus at the State Department was that no amount of diplomacy would keep the Israelis from responding if they were attacked. It would be better to keep them at arm's length so that the United States could credibly deny coordinating with the Israelis when they did strike back.

Cheney and his top advisers at the Pentagon thought it better to keep Israel close, in the hope that by doing so they might limit the Israelis' military involvement. The Pentagon set up a secure hotline between Cheney and Arens known as "Hammer Rick," and the United States dispatched a planning cell to Tel Aviv so that Israel might benefit from real-time knowledge of the unfolding war. Cheney also offered to have the United States set up Patriot missile defenses, with American teams to operate them, in order to defend Israel from SCUD missiles launched from Iraq's western desert. The Israelis rejected that offer.

On the second night of the war, Iraqi SCUD missiles began to fall on Israel. Cheney spoke on the secure line to Arens, who had changed his mind. He now wanted American help to defend against the SCUDs. In interagency discussions, representatives of the State Department recommended against sending them, pointing out that the Israelis had already once rejected help from the United States.

Cheney, meanwhile, asked Schwarzkopf about American air cover in western Iraq. He was told, in effect, that there was no U.S. air cover in western Iraq. Schwarzkopf had never taken the SCUDs very seriously. He reasoned that any American aircraft involved in suppressing SCUDs was not protecting American forces on the ground.

Cheney told Arens that the United States would take care of the Iraqi SCUDs and said that the United States could do a better job of protecting Israel than the Israelis could. The Israelis were skeptical, but reluctantly agreed to hold off on responding to the attacks. Cheney ordered Schwarzkopf to send U.S. Special Forces troops into western Iraq to find and destroy the SCUD launch sites. It was a risky move that would have a dramatic effect on the course of the war.

As Iraqi SCUDs continued to fall on Tel Aviv, the Israelis grew increasingly impatient. Israel wanted to respond. It had an obligation to protect its population, and whatever assurances it was getting from the Americans were not enough to get that done. The Israelis wanted to start bombing.

The Pentagon once again told them no, this time more emphatically. The United States had Special Forces soldiers on the ground in western Iraq searching for SCUDs, the Israelis were told, and any offensive action you take in Iraq would put American soldiers at risk. Unacceptable.

The plan worked. The Israelis couldn't join the war, because doing so would put at risk the lives of Americans trying to protect them.

The Gulf War lasted six weeks—a sustained aerial bombardment followed by an overwhelming application of ground force. The campaign would go down as one of the most successful military operations in American history. American casualties were low—only 148 combat deaths—and contributions from members of the broad coalition of countries that had supported the war largely covered the cost of the conflict.

It was a quick war and, for the Americans, a clean war. Returning soldiers, along with their military and civilian superiors, were celebrated at ticker-tape parades throughout the country.

There was only one major problem: Saddam Hussein remained in power.

Although there had been some debate about continuing the march of American troops to Baghdad in an effort to capture or kill the Iraqi leader, Cheney agreed with Powell and President Bush that the United States had met its limited objective of expelling Iraqi forces from Kuwait. Removing Saddam Hussein by force risked the dissolution of the strong coalition the United

States had put together and almost certainly would have meant more American casualties. Also, Cheney was getting reports from intelligence agencies that Iraqis themselves would likely topple their leader, making any American military effort to do so unnecessary.

On February 14, two weeks before the end of the war, President Bush had encouraged Iraqis to overthrow Saddam Hussein.

"There's another way for the bloodshed to stop," Bush had said, "and that is for the Iraqi military and the Iraqi people to take matters into their own hands," and remove their despotic leader.[74] His words were broadcast into Iraq. After years of Saddam's oppressive rule, some Iraqis were so moved by Bush's admonition that they committed the words to memory.

Many Iraqis, both in the largely Kurdish north and the Shiite south, took this advice. American pilots bombed Iraqi weapons depots, allowing the rebels to arm themselves. As the Iraqi army withdrew from Kuwait and retreated toward Baghdad, the rebels made significant gains. The numbers are disputed, but at the height of the uprising, opposition forces may have controlled as many as fourteen of Iraq's eighteen provinces.

Just as the pressure on the regime intensified, however, American and Iraqi military leaders met near the Iraq-Kuwait border at Safwan to sign a cease-fire. As the negotiations drew to a close, the Iraqi representative, Lt. Gen. Sultan Hashim Ahmad, had a request, recorded in the official transcript of the meeting. "We have a point, one point. You might very well know the situation of the roads and bridges and communications. We would like to agree that helicopter flights sometimes are needed to carry some of the officials, government officials, or any member that is needed to be transported from one place to another because the roads and bridges are out."[75]

General Norman Schwarzkopf, representing the United States, playing the generous victor, told his counterpart that so long as no helicopters flew over areas controlled by U.S. troops, they were "absolutely no problem." He continued: "I want to make sure that's recorded, that military helicopters can fly over

Iraq. Not fighters, not bombers." Lt. Gen. Ahmad pressed the issue. "So you mean even helicopters that is [sic] armed in the Iraqi
skies can fly, but not the fighters?"

"Yeah, I will instruct our air force not to shoot at any helicopters that are flying over the territory of Iraq where we are
not located," Schwarzkopf replied, adding that he wanted armed
helicopters to be identified with an orange tag.

This moment of magnanimity would prove to be costly.
"Along Highway 8, the east-west route that ran from An Na-
siriyah to Basra, the American soldiers could tell that Saddam
Hussein was mercilessly putting down the rebellion," wrote Michael Gordon and Bernard Trainor in *The Generals' War: The
Inside Story of the Conflict in the Gulf.* "The tales at the medical
tent had a common theme: indiscriminate fire at men, women,
and children, the destruction of Islamic holy places, in which the
Shiites had taken refuge, helicopter and rocket attacks, threats of
chemical weapons attacks."[76]

Saddam's soldiers used the helicopters to stop the insurrec-
tion, spilling the blood of tens of thousands of Iraqis to do so.
On the ground, allied troops had reversed course and were now
taking weapons from any Iraqis who had them, including the
rebels. In the end, it was a massacre, with conservative estimates
of 30,000 dead. Nearly everyone in southern Iraq lost a family
member or knew someone who did.

Not surprisingly, the American abandonment of the Shiites
shaped the way many Iraqis viewed the United States and its rulers. The decisions of the Bush administration in 1991—to leave
Saddam Hussein in power and to allow his regime to snuff out
the rebellion—would have consequences that Cheney felt acutely
more than a decade later, in the next Bush administration.

Although Iraq required most of Cheney's attention in the
first half of 1991, the Soviet Union would soon be back to its
prominent place in his thinking. On August 18, 1991, Cheney
was fly-fishing in the spectacular Bella Coola Valley, in British
Columbia, when he took a call from Washington. There had been
a coup in the Soviet Union.

"We were just getting into the swing of things," says Merritt
Benson, Cheney's former state representative, who had accom-

panied him on the trip. "He was cooking and doing the dishes, just like the rest of us. A call came about the Gorbachev overthrow and he was called back to D.C. by Mr. Atwood."[77]

Hard-liners from the Communist Party, including Cheney's counterpart, Dmitry Yazov, had objected to Mikhail Gorbachev's willingness to share more power with the increasingly independent republics. Their troops stormed Mikhail Gorbachev's villa in the Crimea and reported that Gorbachev was too ill to perform his job. Boris Yeltsin, the newly elected president of Russia, called for a popular uprising against the plotters and for Gorbachev's return to power.

There was more trouble back in Washington. While Cheney was fishing, his daughter Liz had been driving his car. Cheney's friends remember the sleek, two-seat sports car as an odd departure for someone so consistently low-key. The inexpensive but flashy Mazda RX-7—with flip headlights—was the closest Cheney would come to a midlife crisis.

Liz had parked it on the street during a day trip to Georgetown. She returned to find that it had been hit by another car. It was still drivable, but the damage was significant. Among other problems, one of the two headlights was stuck, jutting out from the aerodynamic contours of the front hood. She took it to the local garage and was told it would take two or three days to fix. She left it.

When her father returned unexpectedly from Wyoming, he had a lot on his mind. How had they managed to pull off a coup? What was the proper U.S. response? Where are my clean shirts?

Cheney gathered an armful of dirty dress shirts and asked his daughter where he could find his car. "It's in the garage," she said, truthfully. She watched nervously as he punched the button to open the garage door and looked down as it opened to reveal her sleigh. Her father was perplexed.

She feigned confusion—Oh, I meant the fix-it garage, not our garage—and explained to her father that his prized possession had been damaged. After twenty-five years, Liz knew that her father would be disappointed, not angry. He was almost never angry. Disappointment was worse.

At the White House, Cheney told the president that the

United States should denounce the coup plotters in the strongest terms, and reiterate its support for democratic reforms in the Soviet Union. Scowcroft and Baker advised Bush to proceed with caution, telling the president that an unequivocal condemnation of the coup could complicate U.S.-Soviet relations if it were to succeed.

Bush sided with Cheney. "The unconstitutional seizure of power is an affront to the goals and aspirations that the Soviet people have been nurturing over the past years," he said at a press conference on August 21. "This action also puts the Soviet Union at odds with the world community and undermines the positive steps that had been undertaken to make the Soviet Union an integral and positive force in world affairs."

Following overwhelming opposition from the Russian people, the coup failed and Gorbachev returned to power in Moscow. But his power had been weakened by the four-day crisis. And Boris Yeltsin, by encouraging opposition to the coup plotters, had enhanced his own image and hastened the formal end of the Soviet Union.

On December 25, 1991, Gorbachev resigned and declared that the Soviet Union was no more.

Looking back, Cheney, a skeptic about Gorbachev dating back to his days in Congress, believes Gorbachev played an important role in ending the cold war—more for what he chose not to do than for what he did.

"To give Gorbachev his due, I think he deserves a lot of credit for not calling out the troops when the Berlin Wall came down," Cheney says. "Ample opportunity there for him to do what his predecessors had done—in other words, to respond very forcefully and violently to the developments in eastern Europe in '89. And he didn't do it."

But Cheney thinks the lionization of Gorbachev is overdone. "I think he deserves some credit, but I think he discovered that the place was coming apart, that they couldn't compete with the West economically, and they couldn't even keep up militarily. And part of that was driven by what Reagan had done."

Cheney reflected on the dramatic changes in a speech in Seattle, Washington, one year after the failed coup. "There's a revolution going on in the former Soviet Union and its influence is likely to extend far beyond its borders," he explained. "At present, democracy appears to be in control in the former Soviet Union, and we welcome and support it. Certainly. But we cannot make the mistake of assuming that there's going to be a peaceful, orderly progression of developments and that somehow the future of democracy in that part of the globe is assured. There is every reason to hope that they make it. We must do everything we can to support them. But there is no guarantee that it's going to work out that way."

After his remarks, the secretary of defense took questions from the audience.

One audience member wanted to know about "the many lives of Saddam Hussein," and told Cheney that the country had expected a different outcome. "We were—the American public was kind of led to believe he'd be out of the picture by now, and he's still causing trouble. How do you view all of that?"

Well, the question often comes up about Saddam. My own personal view continues to be one that he is not likely to survive as the leader of Iraq. I emphasize that's a personal view. You can get all kinds of opinions. That's based on the fact that he's got a shrinking political base inside Iraq. He doesn't control the northern part of his country. He doesn't control the southern part of his country. His economy is a shambles. The UN sanctions continue to place great pressure on him. We've had these reports of an attempted coup at the end of June, early July, against him. I think he—I think his days are numbered. But that's, again, my personal view.

The question that is usually asked is why didn't we go on to Baghdad and get rid of him? And let me take just a moment and address that if I can, because it is an important issue. Now, as you think about watching him operate over there every day, it's tempting to think it would be nice if he weren't there, and clearly we'd prefer to have somebody else in power in Baghdad. But we made the decision not to go on to Baghdad because that was never part of our objective. It wasn't what the country signed up for, it wasn't what the Congress signed up for, it wasn't what the coalition was put together to do. We stopped our military operations when we'd achieved our objective—when we'd liberated Kuwait and we'd destroyed most of his offensive capability—his capacity to threaten his neighbors. And no matter what he may say today, he knows full well that he lost two-thirds of his army, about half of his air force, most of his weapons of mass destruction, a lot of his productive capability. His military forces were decimated, and while he can try to regroup and reorganize now, he does not at present constitute a threat to his neighbors.

If we'd gone on to Baghdad, we would have wanted to send a lot of force. One of the lessons we learned was don't do anything in a halfhearted fashion. When we committed the forces to Kuwait, we sent a lot of force to make certain they could do the job. We would have moved from fighting in a desert environment, where you had clear areas where we knew who the enemy was. Everybody there was, in fact, an adversary—military, and there was no intermingling of any significant civilian population. If you go into the streets of Baghdad, that changes dramatically. All of a sudden you've got a battle you're fighting in a major built-up city, a lot of civilians are around, significant limitations on our ability to use our most effective technologies and techniques. You probably would have had to run him to ground; I don't think he would have surrendered and gone quietly to the slammer. Once we had rounded him up and gotten rid of his government, then the question is what do you put in its place? You know, you then have accepted the responsibility for governing Iraq.

Now what kind of government are you going to establish? Is it going to be a Kurdish government, or a Shia government, or a Sunni government, or maybe a government based on the old Baathist Party, or some mixture thereof? You will have, I think by that time, lost the support of the Arab coalition that was so crucial to our operations over

*there because none of them signed on for the United States
to go occupy Iraq. I would guess if we had gone in there, I
would still have forces in Baghdad today, we'd be running
the country. We would not have been able to get everybody out and bring everybody home.*

*And the final point that I think needs to be made is this
question of casualties. I don't think you could have done all
of that without significant additional U.S. casualties. And
while everybody was tremendously impressed with the low
cost of the conflict, for the 146 Americans who were killed
in action and for their families, it wasn't a cheap war. And
the question in my mind is how many additional American
casualties is Saddam worth? And the answer is not very
damned many. So I think we got it right, both when we
decided to expel him from Kuwait, but also when the president made the decision that we'd achieved our objectives
and we were not going to go get bogged down in the problems of trying to take over and govern Iraq.*[78]

The decision to end the war without removing Saddam Hussein, and the Bush administration's handling of Iraq policy more
generally, became an issue in the waning days of the 1992 campaign for president. Al Gore, chosen by Arkansas governor Bill
Clinton as his running mate, attacked President Bush for not taking the threat posed by Saddam Hussein seriously enough. Iraq,
he said, was a well-known state sponsor of terror.

Gore lamented that the Iraqi "tyrant remains firmly in power,
resisting by every means the will of the international community,"
and accused the Bush administration of "a blatant disregard for
brutal terrorism, a dangerous blindness to the murderous ambitions of a despot."[79]

Again and again, the speech returned to Saddam Hussein as
a terrorist pariah. Gore listed a series of terrorist attacks that
were launched from Iraqi soil or conducted with Iraqi complicity and claimed that the Bush administration ignored "evidence
that he was not only promoting terrorism, but was also pursuing a nuclear weapons program." He cited a study by the Rand

Corporation that reported "an estimated 1,400 terrorists were operating openly out of Iraq." In all, Gore made more than a dozen specific references to Iraq-sponsored terrorism.[80]

Saddam Hussein, Gore said, "had already conducted extensive terrorism activities, and Bush had looked the other way. He was already deeply involved in the effort to acquire nuclear weapons and other weapons of mass destruction, and Bush knew it but he looked the other way. Well, in my view, the Bush administration was acting in a manner directly opposite to what you would expect with all of the evidence that it had available to it at the time."[81]

The obvious problem, Gore said, was the Iraqi leader himself.

"Saddam Hussein's nature and intentions were perfectly visible."[82]

CHAPTER EIGHT

President Cheney?

Once again, Dick Cheney had nothing on his schedule. George Bush had lost his bid for reelection, and Cheney was out of a job.

His tenure as secretary of defense had been busy: two wars; an abrupt realignment of global power; domestic pressure to dramatically shrink the U.S. military. But the four-year grind of ceremonies, meetings, overseas diplomacy, and crisis management had come to an end.

This time, he didn't need a trip to the Bahamas to discover that he wanted to explore a run for elected office. One week after he left the Pentagon, in an appearance on *Larry King Live*, he hinted broadly at a possible bid for the presidency.

After praising his handling of the conflict in Panama and the Gulf War, King noted that there was already buzz about Cheney's running for president and went directly to his guest.

"Let's get right to it. What about '96? I mean, you know how quickly we react. Ready to throw it into the ring?"

"Well, the new administration hasn't quite got a week under their belt yet, Larry," Cheney replied. "And obviously, it's something I'll take a look at, but I think it's far too soon at this point to make any decisions."

The host, unsatisfied, pressed Cheney about whether he had considered running.

"I've worked for three presidents and watched two others up close," Cheney reminded him, "and so it is an idea that's occurred to me."[1]

At first glance, the prospect of acting on that idea was daunting. Cheney's former boss had won just 37 percent of the vote in a three-way race with Bill Clinton and Ross Perot. Although the Democrats lost nine seats in the House, they maintained a solid majority there and in the Senate. The country was generally optimistic about President Clinton and the new brand of Democratic centrism he had promised throughout his campaign.

That reality started to change almost immediately.

The Clinton administration began with a series of missteps and policy reversals. None would be so costly as the prolonged public debate on gays in the military. By the time of Cheney's appearance on CNN, King was calling it "the biggest story of the day."

Clinton had promised throughout his campaign to end the practice of banning gays from service in the armed forces. A week before the presidential election of 1992, a spokeswoman for the campaign said that Clinton "has a sense of urgency" about the issue and "considers it one of his top priorities."[2]

On January 30, 1993, just ten days after his inauguration, Clinton set aside objections from his Joint Chiefs of Staff and issued a memorandum to Cheney's successor at the Pentagon, Les Aspin. "I hereby direct you to submit to me prior to July 15, 1993, a draft of an executive order ending discrimination on the basis of sexual orientation in determining who may serve in the Armed Forces of the United States."[3]

Whatever goodwill the new president earned for sticking to a campaign promise soon evaporated. Clinton faced resistance from the military and from Republicans, as expected. But some of the fiercest opposition to the proposals came from his own party and, in particular, Senator Sam Nunn, chairman of the Senate Armed Services Committee. King asked Cheney about the controversy.

"I am one of those people who believes that people's sexual preference and orientation are a private matter. It's something that is a personal matter for them, and no one else's business," Cheney said. "That's the way I ran the civilian side of the Pentagon."[4]

The military side was different, he explained. There is no privacy in the military. The most important function of the military is to fight and win wars. Policies that make the military more likely to succeed are adopted, and policies that might compromise its effectiveness are rejected. "I've reviewed the policy with respect to gays. I, basically, don't believe in discrimination, but I did conclude, as secretary of defense, that the ban on gays in uniform was appropriate. It was the best advice I could get from our military commanders, and it clearly reflected the majority sentiment of those who were serving in uniform, and I felt it was sound policy. I think it's inappropriate to want to repeal it."[5]

For Cheney, the issue of sexual preference hit close to home in a way that few people outside his immediate family understood. Several years earlier, the younger of his two daughters, Mary, then a junior in high school, had told him that she was gay. His response was just what she had hoped.

"You're my daughter and I love you and I just want you to be happy."[6]

Nonetheless, Mary's sexuality had become a political issue in the summer of 1991, when activists launched a campaign to out gays and lesbians close to high-ranking Republican officials in an effort to overturn the ban on gays in the military. Both Dick and Lynne Cheney had gotten calls threatening to expose Mary unless they worked to change the policy. Her parents had also heard that reporters were getting phone calls from some of the same activists, hoping to generate media interest in their daughter.

It wasn't just Mary. Cheney's highly regarded press secretary, Pete Williams, was also gay, and he had been getting similar calls. Williams turned to Steve Herbits, an openly gay man who had helped Cheney staff the Pentagon when he arrived.

Williams wanted to quit, so as not to put his boss in the awkward position of having to answer questions about a subordinate's sexual preference. But Herbits told him that the decision wasn't his alone. He'd have to talk to Cheney.

When Herbits told Cheney that he'd been counseling Williams,

Cheney said that he supported Williams unconditionally, and that President Bush had endorsed this position. Cheney had already told Williams about Bush's support, but Williams had been skeptical. When Herbits relayed Cheney's message, he began to be reassured.

On July 31, 1991, Cheney testified before the House Budget Committee. Barney Frank, a frequent crusader for gay rights, asked him about gays in the military. Cheney emphatically noted that it was a policy he had inherited and called the notion that gays pose a security risk "a bit of an old chestnut."

In August, a writer in a gay magazine, *The Advocate*, attacked Williams for not publicly disclosing his sexuality. The story prompted dozens of follow-ups, but only a handful of publications reprinted Williams's name; most simply referred to "a senior Defense Department official" or "a prominent Pentagon spokesman."

In an interview on ABC's *This Week with David Brinkley* on August 4, Sam Donaldson asked Cheney about the outing.

"I take it, Mr. Secretary, that this individual who must defend department regulations as a spokesman is not going to be asked to resign?"

"Absolutely not," said Cheney.

Cheney once again noted that this was a policy he had inherited. "I have operated on the basis over the years with respect to my personal staff that I don't ask them about their private lives. As long as they perform their professional responsibilities in a responsible manner, their private lives are their business," said Cheney. "I would also argue that it's none of your business."[7]

Barney Frank praised Cheney for taking a "half step" toward rejecting the policy and said he deserved credit for repudiating the notion that gays pose a security risk. Still, Frank thought Cheney was uncomfortable with the subject. "He left me with the distinct impression that he wished that nobody would bring it up to him, because I think he finds himself in a dilemma, to be honest, where politically he's not ready yet to have the president abolish the policy, but intellectually, being an intelligent man, he can't think of anything to say in its defense. . . . Dick Cheney, having addressed this issue twice now in public forums, has yet to say a word in its behalf, really."[8]

The issue simmered for another few weeks and then died down, until it was resurrected in the early days of the Clinton administration.

After months of public hand-wringing, Clinton eventually settled on a solution designed to please everyone: the policy regarding homosexuals would be "Don't ask, don't tell." The military thought this policy went too far, and gay rights groups who had supported Clinton in the election thought it did not go far enough. Politically, it was not a win. At precisely the time a president should be strongest—his first 100 days—the fight over gays in the military made Clinton look weak.

The confrontation marked the first step in Clinton's difficult transition from the promise-the-world euphoria of a political campaign to the roll-up-your-sleeves reality of governance. The new president would later blame the Republicans for his troubles on the issue, charging that his political opponents had played to the media in order to damage him.

But there were other problems, too—difficulties with the transition, with his White House staff, with nominees to the cabinet, with Congress, and with the budget. Some of these problems were brought on by Clinton himself; others were not. The cumulative effect was devastating. The Democrats whispered their concern about the future of their party. The Republicans saw an opportunity to capitalize on the errors of the first Democrat to occupy the White House since Jimmy Carter, more than a decade earlier. News columnists, including some who were normally friendly to the Democrats, wondered about Clinton's competence.

In early June, less than six months after Clinton had arrived in Washington, *Time* magazine published a cover story calling him "The Incredible Shrinking President." The cover art was a small black-and-white photograph of Clinton, and the headline for the article wondered, "Is Clinton Up to the Job?"[9]

Not coincidentally, Cheney's thinking about a bid for the presidency was getting serious. That summer, he set out on an adventure which would afford him plenty of time to consider his prospects: a road trip across America.

His journey was a mix of paid speeches, Republican fund-raising, and fly-fishing. It gave him an opportunity to do two things he had not done as secretary of defense: drive and be by himself.

"I was alone," he says with pride. "It was fantastic. It was about 8,000 miles all together. And it was a hoot."[10]

The car was a new 1992 Cadillac STS, black. This was the first year Cadillac had made the STS, and Cheney had seen one at the Washington auto show. He needed a new car. "So I splurged." Recalling the trip, Cheney uses language that, for him, approaches effusiveness: "I love to drive. I loved that car. I loved getting out on the open road. I had my—fish along the way and make some money. Speak."[11]

Cheney gave speeches for Henry Hyde in Illinois and John Kasich in Ohio. He gave a talk to the New Hampshire Federation of Republican Women; an address to the Young Republicans' national convention in Charleston, West Virginia; a corporate speech at a dude ranch in Colorado; a lecture in New Mexico.

He had engagements to keep, but in between he was free to choose his routes and allow himself diversions. Sometimes he drove on the interstates; at other times he chose local roads. He did many things he had not been able to do for four years at the Pentagon—he ate at McDonald's, stopped at truck stops, pumped his own gas.

At one filling station along a highway in the West, a man at the next pump studied him intently and then wandered over to inquire about the familiar face. When Cheney said who he was, the other driver was skeptical. "He said, 'Naw, you can't be Dick Cheney.' He didn't believe me." So they argued. Cheney had the time.

Cheney stopped in Casper to see his parents and then headed up through the Pacific Northwest to British Columbia to spend a week fly-fishing on the Dean River. He drove back down along the Canadian Rockies, through Glacier National Park, and ended up in Wyoming, at his new home in Jackson.

In Washington, Clinton would recover from his early missteps, only to encounter more trouble with his plan to reform the U.S. health care system. Clinton delegated this, his most important first-term project, to the first lady, Hillary Clinton, his most important adviser. The first lady put together a task force of health care advocates and experts to consider options for a new federal role in ending the "health care crisis" in America.

Cheney assailed the Clintons' health care plan in a December 1993 speech at the American Enterprise Institute, where he accepted the Francis Boyer Award for academic contributions to good government. He challenged the Clintons' underlying diagnosis, asking, "What health care crisis?" Cheney's main criticism, however, was that the Clinton administration, with its neglect of foreign policy and national security, had its priorities backward. "We have turned inward as a nation and signed on to the proposition that the only truly important matters on the public policy agenda are domestic issues," he said. "It is more important than ever that our president be a foreign-policy president, that he be actively engaged in the business of shaping world events, that national security issues be right at the top of our national agenda, and that we constantly remind ourselves that peace and freedom in the world will continue to depend upon U.S. leadership backed up by U.S. military capability."[12]

The prevailing political winds were blowing in the Republicans' direction. Cheney, after spending a year as a senior fellow at the American Enterprise Institute contemplating a bid for the presidency, plunged into this favorable political environment in early 1994.

He established a political action committee called the Alliance for American Leadership, an entity that would allow him to raise money as he began to explore his potential candidacy more seriously. The name highlighted what would become Cheney's chief criticism of the Clinton administration: a failure of leadership. It was an obvious strategy for Cheney that allowed him to emphasize Clinton's difficulties while reminding voters that they had come to know Cheney himself first as a leader in Congress and then as a wartime defense secretary.

"Leadership" was also a convenient way to spin Cheney's twenty-five years in Washington. There was little question that if

Cheney were to run for president, his primary opponents would paint him as an inside-the-Beltway politician out of touch with average Americans. Cheney seemed to accept this eventuality.

"I don't think the way to get elected to anything at this stage in my life is to start trying to repackage myself," he said. "I am what I am, and if people aren't willing to accept me on that basis then, hell, vote for somebody else."[13]

In the early handicapping of the Republican race, Cheney was a second-tier candidate behind Bob Dole and Jack Kemp. Cheney usually scored only single digits when Republican voters were asked about their preferred choice in 1996. On the stump, he was solid but not spectacular. Still, many observers, especially those in Washington, thought he would be a strong contender if he decided to run.

Morton Kondracke of *Roll Call* offered this assessment: "One of Cheney's strengths, all GOP pros agree, is that he has no detractors. He's a pro-lifer who's acceptable to pro-choice Republicans. And he's a deficit-cutter who would be acceptable to supply-siders. He's also well-liked by the press and the bipartisan Beltway establishment, which is unusual for a Republican."[14]

The columnist Marianne Means wrote that Cheney was "much admired in Washington."[15] David Broder of the *Washington Post* was also bullish on a run by Cheney. "His assets in that fight are largely a byproduct of the public and private persona that he has developed—a sense of gravity, seriousness, competence and self-confidence, unmarred by either pomposity or obvious self-promotion. As one former House colleague, a Democrat, put it, 'I trust Dick, and so does damn near everyone who knows him.'"[16]

The Alliance for American Leadership was a four-person PAC led by David Addington. The mood in the office was efficient and businesslike, not chummy. Its employees were well aware that their jobs with this organization would be over the moment Cheney bowed out.

The communications director was Joseph Duggan, a young but experienced wordsmith who had worked for the UN ambassador Jeane Kirkpatrick during the Reagan administration and as a White House speechwriter under George H. W. Bush.

For Duggan, the job, although tenuous, had a significant upside. If Cheney decided to run for president and went on to win, Duggan would almost certainly end up with an important position in the White House. If it didn't work out, he would still learn what it takes to put together a presidential campaign, experience that would be valuable to someone who planned to stay in Washington.

When he first showed up for work, he learned that his job handling press for the PAC came with an odd restriction. He wasn't allowed to talk to the media.

Duggan spent most of his time on the phone with prospective big donors, but he didn't ask them for money. Instead, he solicited their views on policy for the campaign. The goal was to get major Republican donors and corporate citizens to feel as if they were part of Cheney's team. The financial contributions would come later.

The PAC raised funds to pay for Cheney's cross-country travels; whatever was left over was parceled out to Republican congressional and gubernatorial candidates in tight races. Cheney gave money to the congressional campaigns of two future Bush administration cabinet members: John Ashcroft and Spencer Abraham. He also contributed to a first-time politician from Tennessee, Bill Frist; to Senator Slade Gorton, later of the 9/11 commission; to the future Speaker of the House Newt Gingrich; to a congressional classmate, Olympia Snowe; and to Jim Jeffords, who would later reward the support in an unusual way. Cheney gave $5,000 to a band of midwestern governors-to-be, stretching north from Oklahoma to Iowa and across to Tennessee.

The list of contributors to the Alliance for American Leadership read like a yearbook of important figures from Cheney's past and his future: former secretary of state Henry Kissinger; former secretary of the treasury George Shultz; former White House counsel C. Boyden Gray; former deputy secretary of defense Donald Atwood; the CEO of Halliburton, Tom Cruikshank; Scooter Libby, from Cheney's policy staff at the Pentagon; Paul O'Neill, a colleague from the Ford administration; and Stephen Hadley, a Washington lawyer who had been on Cheney's staff at the Pentagon.

Hadley, who would become national security adviser for George W. Bush, helped Cheney draft a scathing critique of Bill Clinton's national security policies for *CommonSense,* a journal published by the think-tank arm of the Republican National Committee, known as the National Policy Forum (NPF). The incoming president of the NPF was John Bolton, a foreign policy intellectual who had worked with Cheney in the first Bush administration and would work with him again in the second. Hadley drafted the article "Inadequate Strategy, Inadequate Resources: Bill Clinton's National Security Deficit" with Joe Duggan.

This article, published in the issue of fall 1994, echoed the notes Cheney had written to himself a decade earlier about the capacity—and the obligation—of a strong America to shape the world.

"As the presidency passed into Mr. Clinton's hands, America had the tools and the credibility to quell crises and capitalize on opportunities. And world peace and freedom depended on it: Nothing was more vital to hopes for peaceful transformation to democracy and free markets in Russia, Eastern Europe and China than the credibility of American leadership. No efforts to shield the civilized world from the spread of ballistic missiles and nuclear, chemical and biological arms could succeed without American preparedness and American leadership. No deterrent was more effective against powerful tyrannies like North Korea than the clarity and consistency of American leadership. Today no less than before, peace and freedom in the world depend on American leadership—and American leadership depends upon maintaining the world's finest fighting force."

The bottom line: "In just 20 months, Bill Clinton has squandered the legacy he inherited—and the world has become a more dangerous place."[17]

Cheney sounded the same theme in more than 150 speeches and campaign appearances across the country that year. He portrayed Clinton as an inexperienced foreign policy naïf whose inability to make decisions had disastrous consequences. "Bill Clinton entered office with gaping holes in his understanding of international economic, political, and military dynamics," Cheney said. "More fundamentally, he exhibits no coherent sense of America's purpose and role in the world. Those on his national security team who are not novices are returnees from the Carter administration."[18]

At the same time as Clinton was cutting the defense budget, Cheney argued, he was overextending the armed forces with humanitarian missions in Somalia and dubious military adventures in Haiti. Clinton's focus on side issues of foreign policy meant that the larger issues—the real threats, in Cheney's view—were not getting the attention they deserved.

"The most perilous immediate threat, of course, is from North Korea," Cheney wrote. "President Clinton has used bold words to assert that under no circumstances will North Korea be allowed to acquire nuclear weapons. But North Korea's dictators go right along pursuing their nuclear program. . . . A president's most important commodity as Commander-in-Chief is his credibility. Bold talk that is never followed up by bold action leads our adversaries to conclude that we do not have to be taken seriously. The cost of reclaiming that credibility once it is lost is likely to be paid in terms of American lives. In these dangerous times, a president must always say what he means and mean what he says."[19]

As interesting as what was said in Cheney's article was what was not said. Although Republicans on the stump around the country were beating up Clinton over gays in the military, Cheney did not once mention this subject in his long critique of Clinton's national security policy.

The Republicans dominated the midterm elections of 1994, picking up fifty-four seats in the House and eight in the Senate. Cheney saw his friendly rival Newt Gingrich ascend to the speaker's chair, which he himself had once coveted.

Cheney understood that he would soon have to decide whether to run for president. He had been well received throughout the election cycle in 1994 and had collected dozens of political chits from candidates to whom he had doled out campaign visits or money from his PAC. Still, he was reluctant. Despite his best efforts, the campaign was being waged almost exclusively on issues of domestic policy. Cheney was experienced in this arena, but his strength was foreign policy and national security. His alarms about cutting defense budgets too deeply might have appealed to Republican voters in the primaries, but would they excite the rest of the country? His talk of threats such as North Korea and ballistic missiles seemed incongruous to a nation enjoying the post–cold war "peace dividend."

Cheney began telling friends privately that while he was willing to serve as president, he did not want to do the things one has to do in order to win a presidential election. His family was ambivalent. Of the three Cheney women, only one—his elder daughter, Liz—enthusiastically supported a run. And just days before Cheney would announce his decision, his father told the *Dallas Morning News*: "To tell you the truth, I kind of hope he doesn't do it."[20]

Cheney spent Christmas in Jackson, thinking and talking to his family. He knew exactly what kind of commitment the race would be, and he understood the costs. "The question I asked myself is, am I ready to do this, to go do what I would have to do over the next two years to mount an effective presidential campaign, and when I asked myself that question, kicked it around with the family and so forth, the basic conclusion was, I really don't want to go do it. It was a serious look at it, but the answer was pretty unequivocal," Cheney says.[21]

Finally, on January 3, 1995, after months of crisscrossing the country in his Cadillac and in the air, after shaking thousands of hands and raising millions of dollars, after dozens of media appearances and even more speeches, Cheney faxed a three-sentence statement to news organizations: "After careful consideration, I have decided not to become a candidate for the presidency in 1996. I appreciated very much the kind words of encouragement and support I received from the many Americans who had urged me to seek the Presidency. I look forward to supporting the Republican nominee for President in 1996."[22]

Joe Duggan, the muzzled communications director for Cheney's PAC, says that David Addington informed the staff. "Addington told us he decided not to run for family reasons," Duggan recalls.[23] And though Cheney believed he could run and win, a poll released the day before the announcement showed that he was the choice of just 3 percent of Republicans surveyed, two points behind Rush Limbaugh.[24]

Later, much of the speculation about his decision would focus on his daughter Mary. Asked whether there was truth to suggestions that protecting her privacy was an important consideration, Cheney responded sharply. "It's exactly what I said at the time: I was not prepared to do those things I would have to do to run an effective presidential campaign." He later added: "It wouldn't be fair to put it on Mary."[25]

Perhaps the best way to understand Cheney's decision is to consider what he said before he started seriously exploring a bid. As Cheney rolled across the country on his solo road trip in 1993, he explained how he would come to a decision. "I want to have a sense that I can see a way to campaign and then govern that feels right to me."[26]

It simply never felt right.

Cheney's decision to pass on a presidential run had longer-term implications. He was done with electoral politics.

After leaving the Pentagon, Cheney had joined the boards of major American companies, including Union Pacific Railroad, Procter and Gamble, and Morgan Stanley. He kept his affiliation with AEI. And as soon as he dropped out of the race, he had a flurry of inquiries from people and companies seeking to use him in one capacity or another. Cheney listened to them all. But he was in no hurry to clutter up his schedule. Fishing remained a priority.

In August 1994, in the midst of campaigning for other candidates and considering his own run in the election of that year, Cheney took time for a trip to the Miramichi River with friends. What was meant to be a relaxing break from a hectic campaign would be the most important fishing trip Cheney would take.

Tom Cruikshank, then chairman and CEO of Halliburton, was among the anglers. He and Cheney chatted several times during the trip. Cruikshank was impressed. He could understand why Cheney had been chief of staff at such a young age and why he had emerged from the Pentagon with a strong reputation. Cruikshank didn't think much more about his interactions with

Cheney until the following spring, when Halliburton was struggling in its search to find someone to replace him as he eased into retirement.

Halliburton was looking for an executive who could manage a $5.7 billion company with nearly 60,000 employees in hundreds of countries across the world. This massive, multifaceted corporation provides services to the oil industry at nearly every stage of energy development. Halliburton also includes an engineering and construction component that accounts for a large segment of the company's revenues. The successful candidate would replace Cruikshank in his dual role.

A four-person search committee had interviewed several people from the oil industry and had considered promoting from within, but hadn't been enthusiastic about any of the possibilities. In the effort to broaden the search, Cruikshank reminded the committee members of one talent pool they hadn't even looked at: former government officials. He offered some examples of people who had successfully made that transition and others who had failed. Then he threw out a name.

What about Dick Cheney?

Another member of the search committee knew Cheney and seconded the nomination. The committee studied his background and experience. "A lot of the things he was overseeing as secretary of defense would be similar to the things he would be overseeing at Halliburton," says Cruikshank. "Big infrastructure. A lot of foreign operations. Long-range planning. Logistics planning. Budgeting. Personnel matters. Technology management."[27]

The most serious potential impediment was Cheney's lack of corporate experience. He had been in government almost continuously since graduate school. And although Cheney had dabbled in consulting during the gaps in his government service, this was not necessarily the kind of experience that would prepare him to take over a Fortune 500 company whose revenues were comparable to those of Continental Airlines and Coca-Cola.

Also, although Cheney had worked on energy issues as a member of the Interior Committee and had grown up in Casper, the "Oil Capital of the Rockies," his knowledge base would not compare to that of someone who had made a career in the oil services industry.

But the committee was looking for a leader, not just someone to oversee operations. The members believed that Cheney had a good reputation with their customers in the oil industry and, importantly, he had good contacts in the oil-rich Middle East. And over the course of the week together in Canada, Cruikshank, who had respected Cheney enough to contribute to his presidential PAC, had come to like him personally.

"When you're isolated for about a week, fishing and sitting around the dinner table at night, you get to know people a little better," he says.

Cheney agreed to let Halliburton fly him to Dallas for a meeting with the search committee. The group met on a Sunday afternoon at Cruikshank's house, over lunch prepared by Cruikshank's wife. Cruikshank and his colleagues laid out their plan. "We explained to him what we saw that he could bring to the party," he recalls. Cheney flew back to Jackson and the next day took a follow-up call from Cruikshank.

The two men discussed details of the proposed arrangement, and later that week Cheney called back.

"Let's do it," he said.

"One of the things that was clear was he was through with Washington," says Cruikshank. "He wasn't going back to Washington. And if you thought about it, once he'd made his mind up not to run for president—you've been chief of staff, you've been in Congress, you've been secretary of defense. Where do you go back to?"[28]

"I'd had a great career in politics," Cheney recalls, "but it was through. I'd done twenty-five years and it was time to do something else."[29]

As he had done in Congress and then in the executive branch, Cheney sought to learn everything he could about the oil industry. He studied trends in oil services, read books, and consulted leading authors and industry analysts. "I made a very conscious decision: I'm going to go be a CEO, do my job. It's a whole different world out there. I'm going to go get totally immersed in that." Cheney's official starting date was October 1, 1995. He arrived in Dallas in August and began to sit in on meetings before he took over.

As Halliburton's CEO, Cheney stayed in touch with some of his Washington friends and kept track of legislation that affected Halliburton, but his friends got the impression that he did not miss political life. Still, Cheney was often asked about his views on both policy and politics.

When he traveled back to Wyoming for two public appearances in Casper, a story ran in the *Casper Star-Tribune* under the headline "Cheney Out of Politics—Mostly." The article is an object lesson in political irony. In the course of two speeches and an interview with the newspaper, Cheney called the vice presidency a "cruddy job" and insisted that his political career was "over with." He backed Colin Powell as a running mate for Bob Dole. He defended the decision to leave Saddam Hussein in power in Iraq, saying again that Saddam was not worth the lives of American soldiers. He pushed to lift the sanctions on Iran—"The good Lord didn't see fit to put oil and gas resources only in democratic countries friendly to the U.S."—and warned that a failure to do so might upset the United States' allies in Europe.[30]

Despite his repeated insistence that he would not accept a spot on the Republican ticket—"How direct do I have to be?" he asked in Casper—Cheney's name stayed in the mix. He was never approached by Dole's campaign and says he would have turned the offer down if he had been. Cheney did meet with Elizabeth Dole, who wanted to involve him in the campaign. Cheney liked Dole, but declined any active role in the campaign in order to focus on his work at Halliburton.

Even the selection of his longtime friend Jack Kemp as Dole's running mate was not enough to entice him to participate in the Republican National Convention in San Diego in 1996. Lynne Cheney spoke, but her husband did not even make the trip. It was the first Republican convention he had missed in two decades.

Asked what he was doing, Cheney says he doesn't remember, but offers the most likely answer.

"Probably fishing."

Among the few political issues Cheney did follow closely

were the debates over unilateral sanctions in places where Halliburton did business, something he called "my favorite hobbyhorse." As his comments in Casper suggest, Cheney opposed unilateral sanctions even against countries with abysmal human-rights records or a history of involvement with terrorism.

His argument was practical: unilateral sanctions don't work. In most cases, he argued, they hurt American companies and have little effect on the government they are intended to punish.

Cheney thought unilateral sanctions were an easy way for members of Congress and the Clinton administration to get political cover by signaling their opposition to bad regimes.

Although Cheney had occasionally voted in favor of unilateral economic sanctions as a member of Congress, his view that they are ineffective predates his tenure at Halliburton. The sanctions imposed on Haiti by the Bush administration when Cheney was secretary of defense were "stupid" and "a mistake," he said shortly after leaving office.[31] And later, after he became vice president, Cheney said that the unilateral sanctions against Pakistan, favored by Congress, were one reason the United States' relations with Pakistan soured in the 1990s.[32]

In May 1997, Cheney called on the oil and gas industry to fight Congress over unilateral sanctions, suggesting that further restrictions lay ahead unless businesses took a "proactive approach" to ending sanctions.[33] He followed his own advice.

On June 23, 1998, Cheney delivered a speech at the libertarian Cato Institute decrying the deleterious effects of sanctions. He cited two examples, both having to do with Iran. Cheney complained that U.S. companies operated at a significant disadvantage in the Caspian Sea region because of its proximity to Iran, which was under American sanctions at the time.

"As a result," he said, "American firms are prohibited from dealing with Iran and find themselves cut out of the action, both in terms of opportunities that develop with respect to Iran itself, and also with respect to our ability to gain access to Caspian resources. Iran is not punished by this decision. There are numerous oil and gas development companies from other countries that are now aggressively pursuing opportunities to develop those resources. That development will proceed, but it will happen without American participation."[34]

Cheney further pointed out that the Clinton administration was attempting to rally Arab states in the region to align with the United States against Iran. Doing so, he claimed, raised questions about "the wisdom of U.S. leadership."

Cheney later explained how, exactly, the sanctions affected Halliburton:

> *There was a limit, a dollar limit on how much you could do there, and our best customer at the time, most years either Exxon or Shell, was either number one or number two customer worldwide. But Shell was often up there. Shell, of course, was a British and Netherlands, Hague, based company. They'd go in to operate in Iran and we couldn't go with them as a service company. Our competition could, because that was Schlumberger. And they were domiciled in the Netherlands Antilles. Although they were listed on the New York exchange, they weren't subject to sanctions. So they'd go into Iran, service Shell, charge an arm and a leg because there was no competition. We weren't there, kept out by our own government. And then turn around and go back someplace say like Algeria where we could operate, did go head to head, they would use the profits they'd earned in Iran to undercut you financially in a place like Algeria. They had an unfair advantage and it was all because of the policies of our government. That was the source of the frustration as much as anything.*

Cheney says he advocated lifting sanctions on Iran "clearly when I was speaking with my business hat on as a chairman and CEO" of Halliburton.

From a national security perspective, didn't the United States have a pretty good idea about the nature of the regime?

"I wasn't arguing a national security issue," says Cheney. "I was saying, in effect, what was being done here wasn't having the desired policy effect on the other end, and it was damaging American companies."[35]

Did you think about national security issues when you made these kinds of decisions?

"I was more focused on the energy," he says. "But also, I mean, it's something I repeated periodically was that the good Lord didn't see fit to put a lot of gas resources only where there are democratically elected regimes friendly to the United States. If you're going to go get oil and gas that the nation needs to run its economy, you're going to have to deal with some unsavory characters in some potentially hostile parts of the world."[36]

One of the reasons the Clinton administration may have been pushing Arab governments in the region to stand against Iran is the alleged involvement of Iran in the bombing of Khobar Towers, in Dhahran, Saudi Arabia, on June 25, 1996. That attack, near a U.S. air base, killed nineteen American servicemen. The Saudi regime had conducted an investigation of the attacks and reported back to the FBI that responsibility lay with Islamic terrorists who were backed by Iran.

The fact that the Saudis had fingered Iran had been reported in the American media, but the Clinton administration said little about the charges. "I didn't know it at the time," says Cheney, now ten years after those attacks.

Nonetheless, Iran's record as a prominent state sponsor of terror was well known, especially to someone who followed regional issues as closely as Cheney did. Three months before Cheney advocated a "proactive approach" to fighting sanctions, the State Department reported that Iran was the "most active" state sponsor of terrorism in the world. "Iran remains foremost among the states which sponsor terrorism. . . . Iran continues its support for terrorist organizations such as Hizballah and Hamas and continues to host offices of other terrorist groups in Tehran."[37]

Indeed, the sanctions Cheney had fought to overturn, the Iran-Libya Sanctions Act of 1996, begin with a statement about the legislation: "The efforts of the Government of Iran to acquire weapons of mass destruction and the means to deliver them and its support of acts of international terrorism endanger the national security and foreign policy interests of the United States and those countries with which the United States shares common strategic and foreign policy objectives."

And the target of the policy was Iran's oil industry. "The Congress declares that it is the policy of the United States to deny Iran the ability to support acts of international terrorism and to fund the development and acquisition of weapons of mass destruction and the means to deliver them by limiting the development of Iran's ability to explore for, extract, refine, or transport by pipeline petroleum resources of Iran."

Halliburton nonetheless did business legally in Iran by using subsidiaries owned and operated on foreign soil. But one month before he joined George W. Bush on the Republican ticket, Cheney was still frustrated. "We would like to do more than we're able to do in Iran at present."[38]

By the time Cheney left Halliburton in July 2000, he was earning a salary of $1.3 million and had accumulated stocks and options worth an estimated $44 million.

But the new money didn't significantly change his lifestyle. "I didn't buy a yacht," he says.

Cheney drove the same Cadillac STS that he had bought while he was at the Pentagon. He was given a free membership at the prestigious Dallas Country Club; but he never played golf there, and he ate there only once, as a guest.

When gas prices rose, he got grief from his own father. "He knew the price of regular gasoline at every filling station in Casper, Wyoming. He could literally tell you what they were charging at the local Chevron station because he used to keep track of it," Cheney recalls. "I'd say, 'But Dad, don't you understand? Now I'm part of the industry; you want them to do well.'

"He never bought that."[39]

CHAPTER NINE

Another George Bush

Cheney was content in Dallas. In many ways, life was considerably less complicated as the CEO of a massive oil services company than it had been as a member of Congress or even as secretary of defense. A Congressman is a public figure, accessible to all—at the airport, at the barbershop, at the grocery. And while Cheney had not sought the spotlight as secretary of defense, it had found him. For all but six months of the previous two decades, Cheney had held jobs that required him to be on duty twenty-four hours a day.

At Halliburton, despite an intense schedule and huge responsibilities, he could usually leave his work at work. It was possible for Cheney to come home, kick his shoes off, and just be Dick Cheney. If this wasn't quite a return to the anonymity he had enjoyed as a midlevel bureaucrat in the executive branch, it was close.

"I was enjoying very much the boards I was on, being chairman and CEO of Halliburton, and making good money," Cheney says. "For the first time in my life I hadn't been a government employee or an academic. I had no desire, or aspiration, or interest, really, in returning to government."[1]

Aside from the policies that directly affected Halliburton, Cheney had largely avoided the political scene in Washington. He paid more attention to things closer to home, and had developed a friendly relationship with Governor George W. Bush of Texas.

The two had crossed paths occasionally while Cheney worked for Bush's father, but they were not close. The younger Bush worked on his father's campaign for reelection; Cheney had to stay out of politics while he was at the Pentagon. When Cheney returned to electoral politics as a potential presidential candidate in 1994, his PAC gave the younger Bush's gubernatorial campaign $2,500.

When Cheney accepted the job at Halliburton, Bush had been in office a little less than a year. Cheney's new home in Dallas was a short flight from the governor's mansion in Austin. For Cheney, having a good relationship with the governor was smart business. For Bush, who was already being mentioned as a national political candidate, an experienced political hand willing to give free advice was invaluable. He began to consult Cheney on a wide variety of subjects, inviting him to Austin on several occasions to talk about business, Texas politics, and world events.

Cheney contributed $2,000 to Bush's reelection campaign for governor in 1998, telling the *Dallas Morning News*: "I'm a Texas voter, and I'm a Republican, and I know him personally. I've known a lot of presidential offspring over the years. I would not have automatically supported them just because I knew the father or worked with the father. He has established a legitimate career in his own right."[2]

Bush was reelected that year amid rumors that he had his eye on the presidency. As those reports grew in number and specificity, Joe Allbaugh, Bush's longtime chief of staff, went to Dallas for a meeting with Cheney. Allbaugh wanted to talk about how to run a national campaign with someone who had done it. Cheney drew on his own experience with Gerald Ford and as a potential presidential candidate to caution Allbaugh about the pitfalls of the campaign trail.

Allbaugh was impressed. When he returned to Austin to report to Bush on the meeting, he opened with a piece of advice: if you run for president, this man should be your vice president.

It wasn't yet time to make that decision, but Bush knew that he wanted Cheney to play some role in his political future. A few weeks later, he invited Cheney to Austin as part of a group of national security experts, to share ideas with the prospective candidate. Cheney recommended that the group include Paul Wolfowitz, his policy chief from the Pentagon. They joined several influential thinkers who focused on foreign policy but who had been displaced six years earlier by the change in the administration. The group gathered around the dining table in the governor's mansion included Richard Armitage, Condoleezza Rice, Richard Perle, and Dov Zakheim. Together, they would form the core of a foreign policy advisory group known as "the Vulcans."

When asked what he and Cheney had discussed during the visit, Bush was evasive. "What did we discuss? Guess. Dick Cheney is a friend of mine. It's not the first time he has been down here. It won't be the last time he is down here. He is a person whose judgment I rely upon a lot."[3]

But Cheney's growing involvement with Bush's campaign didn't stay secret for long. Shortly after the meeting, Bob Novak reported on CNN that Cheney had become "a key figure" in the Bush operation, "in line to become the national campaign chairman."[4]

The frequency of Bush's consultations with Cheney increased. They spoke by phone regularly and often got together when Bush was in Dallas or Cheney was in Austin. Cheney was a generalist, dispensing advice on politics, national security policy, campaign personnel, and a variety of other issues. He was not meant to be a day-to-day adviser, but an elder statesman. Whenever his schedule permitted, he joined Bush's growing policy team for strategy sessions in Austin.

Bush, who considers himself a keen observer of body language, liked what he saw in Cheney. "Cheney came down to a lot of our policy meetings and he was the kind of guy that—he didn't speak a lot," says Bush. "But then when he spoke, everybody—like, the energy level in the room kind of shifted a little bit. . . . It was a very impressive group of people and impressive presentations. But the one guy that pretty much commanded, I felt, the respect of everybody around the table during these meetings was Cheney. And so he got my attention."[5]

News reports of Cheney's involvement with the campaign began to appear more frequently, but Cheney kept a low profile, refusing most requests for interviews about the campaign and the candidate. Then, on September 23, 1999, he made his first appearance as a surrogate for Bush on the Fox News Channel's *Special Report with Brit Hume.*

The timing was deliberate. Earlier in the day, Bush had given his first major address on foreign and defense policy. "If elected, I will set three goals," Bush told an enthusiastic crowd at the Citadel in South Carolina. "I will renew the bond of trust between the American president and the American military. I will defend the American people against missiles and terror. And I will begin creating the military of the next century."[6]

In his appearance on Fox, Cheney helped Bush make this case, hewing close to the campaign's talking points. When Hume asked him about the greatest threats to the United States, Cheney didn't answer. Instead, he highlighted the three main points in Bush's speech, only one of which had anything to do with threats. He spoke generally about the need to ensure that the United States would maintain "the preeminent military force in the world for the next hundred years," and refused to answer a question about whether he would have advised sending troops to Kosovo.[7]

It was not a great performance. But from the perspective of the campaign, what Cheney said mattered less than the fact that he said it as a representative of George W. Bush. It was also a hint about how the campaign would deploy him in the future. Foreign policy was Bush's obvious weakness. As a former businessman and governor, Bush lacked the stature to speak with authority about national security. Voters seemed to care about his views on education, health care, and tax cuts because there was a reasonable assumption that he had handled those issues as governor. The same could not be said for missile defense, military readiness, or international finance. Cheney could fill that void.

The buzz picked up almost immediately. *U.S. News & World Report*'s Paul Bedard included Cheney in a brief item, "Veep Watch," in October. The magazine reported that Cheney was "topping the list" of potential vice presidential candidates.[8]

Bush may have been thinking about Cheney as a possible running mate, but in the near term he had another position in mind. On November 2, 1999, the two men spoke at a dinner for Barbara Bush's literacy foundation in Arlington, Texas, just outside Dallas. Bush took Cheney aside and asked him to chair his election campaign. Cheney turned him down. "I said, 'Look, I'm really committed here at Halliburton. I'm eager to do what I can to be helpful.' Without any question, I wanted to support him but just—I couldn't walk away from what I was doing at Halliburton."[9]

But Bush was persistent. "It became apparent to me that Cheney was the kind of guy that would be a good fit for a two-term governor from Texas who, while he had a pretty good political pedigree, didn't have a lot of what they call 'Washington experience,'" says Bush. "And so I began to think about him."[10]

In early 2000, Bush was the front-runner for the Republican nomination, and he once again dispatched Allbaugh to Dallas for a meeting with Cheney. "Allbaugh talked to me about whether I'd consider being vice president," Cheney recalls. "He was there just to explore that possibility. We may have talked about other subjects, too. It wasn't an offer by any means, but it was whether or not I would be willing to consider."[11] Once again, Cheney said no and reiterated his reasons.

"I focused on health, and I focused on the fact that I had obligations at Halliburton, that both the president and I had backgrounds in the oil business," says Cheney. Further complicating matters was "the fact that we were both living in Texas." Cheney insists he wasn't playing hard to get. "It was a firm no."[12]

On March 14, Bush swept to victory in six southern Republican primaries and ended any remaining chances that John McCain could win the nomination. The national media, with a sudden paucity of campaign news, turned to speculation about potential running mates. Cheney was considered a "dark horse" whose selection would indicate a preference of "substance over style." Other possibilities included Elizabeth Dole, Senator Bill Frist, Governor Frank Keating of Oklahoma, Representative John Kasich, Senator Connie Mack, and Governor Tom Ridge of Pennsylvania.[13] Senator John McCain, who had defeated Bush in New Hampshire only to lose the nomination in a bitter and sometimes personal fight, said repeatedly that he did not want to be considered. Colin Powell had said the same.

In early April, Bush asked Cheney to work with the campaign in yet another role: as head of his vice presidential search committee. This time, Cheney concluded that he could handle the job running the search despite his other obligations. "It was the kind of thing that would take a piece of my time, but then was definable," says Cheney. "It would be over and done with and then I could get on with my life."[14] When the two men talked about the most important qualities of prospective candidates, Bush's answer in private was the same one he gave in public: someone who could help him govern. He was looking for a vice president, not a running mate.

It was a distinction that made an impression on Cheney.

Cheney told Bush he would begin the process by looking at obvious prospects: Republican congressmen, former cabinet members, current and former governors. Cheney says Bush "had a special feeling for governors," many of whom he'd gotten to know through the Republican Governors Association.[15]

Cheney quietly put together a small team of his most trusted advisers to vet the candidates. In addition to Cheney, only three people would be privy to the search process: Liz Cheney, David Addington, and Dave Gribbin. The search team met infrequently but stayed in close contact by telephone and e-mail. At one early session, Addington joked that if Cheney did a good job on the search, Bush might ask him to be his running mate. Cheney did not mention his conversation with Allbaugh and dismissed the comment with a laugh.

On April 24, at a campaign stop in Dayton, Ohio, Bush formally announced that Cheney would lead his vice presidential search. In the course of the announcement, Bush said three things that might have given a hint of his eventual choice. First, he said that Cheney's position leading the search wouldn't disqualify him from being considered. Second, Bush said that the most important criterion was the ability to serve as an effective vice president. Third, Bush said that he would continue to consult his father about possible candidates.

"He will not be a part of the formal process, but he does not need to be," Bush told reporters. "He's my dad, I talk to him all the time. He's got a few suggestions to make on occasion on different subjects, and sometimes I listen to him and sometimes I don't."[16]

As Cheney began the search process, he heard from friends and colleagues who had advice and in many cases a favorite candidate. Several people jokingly suggested that he should choose himself. Shortly after Bush announced that Cheney would head the selection team, Trent Lott, Cheney's former colleague in the House leadership, told him: "You da man! You don't need to go through all this process. You're the one that ought to be in that position."[17]

After several brainstorming sessions, the search team developed a long list of prospective candidates. "What happens when you start looking for vice presidents, frankly, is you start out with a notion that it's going to be a real plus, that you're going to get somebody that's really going to add a lot to the ticket," says Cheney. "My experience over the years has been that you usually end up with the least worst option. . . . It's hard to find many examples in history where the vice presidential candidate has really made much difference in the final outcome. People don't vote for vice president. They vote for president."[18]

Cheney called the prospects to gauge their interest and got very mixed responses. When he called Connie Mack, an old friend from Congress who was a former senator from Florida, Mack told Cheney that they would never speak again if his name were added to the list of prospective candidates.

To guard against leaks, the list was sprinkled liberally with decoy candidates, individuals who were never under serious consideration. Cheney hired lawyers with Latham and Watkins to provide exhaustive research dossiers on each of the candidates and then supplemented those files with his own investigations of the real candidates. Addington reviewed each of the files and produced final, comprehensive memos to Cheney on each of the serious prospects. At regular intervals, Cheney spoke with Bush by phone to brief him about the candidates; less frequently, he traveled to Austin to see Bush. Bush listened patiently and often asked questions about the candidates Cheney presented. The entire time, however, he was evaluating Cheney.

"As he came and gave me briefings on his vice presidential searching, I was really in my own mind interviewing him—he just didn't know it."[19]

Cheney had his suspicions. Bush often ended these sessions by joking that Cheney was the solution to his problems. Cheney shrugged off the suggestions by telling Bush that he took them as a threat, a motivational tool designed to make him work harder to find a suitable candidate.

Addington and Liz Cheney studied questionnaires provided to vice presidential prospects in previous campaigns and developed their own. A small group completed those surveys and even fewer got an interview with Bush. In the end, says Cheney, "we weeded it down to a very short list."[20]

Cheney continued to spend most of his time on his duties at Halliburton. He traveled extensively and gave speeches that one would not expect from a man on the verge of joining a presidential campaign. In one talk at the World Petroleum Congress in Calgary, just two weeks before telling Bush that he was willing to be considered as the running mate, Cheney offered a full-throated defense of the oil industry's social and environmental policies.

"What we do—finding, developing and producing petroleum products—is in itself an enormous value to our society," he said. "Without the energy resources we provide there would be no global economy. The world is far better off at the beginning of the twenty-first century because of all we have accomplished over the last 150 years in the oil and gas industries. Any discussion of social responsibly has to take note of the enormous contribution made by this industry."[21]

He continued Halliburton's attempts to lift unilateral sanctions on oil-rich countries, urging the U.S. government to seek rapprochement with Iran. "There's been enough aggravation on both sides, whether you consider the Iranian occupation of our embassy in 1979 and holding hostages for over a year, or the shoot-down of the Iran airliner by a U.S. naval vessel in 1988. It's been a tragedy in terms of the relationship."[22]

The United States, he said, should take steps to improve relations. "One of the ways I think is to allow American firms to do the same thing that most other firms around the world are able to do now, and that is to be active in Iran," he argued. "We're kept out of there primarily by our own government, which has made a decision that U.S. firms should not be allowed to invest significantly in Iran, and I think that's a mistake."[23]

The comments would cause a public relations issue for Bush's campaign. Although Cheney was only an adviser and had made the remarks in his capacity as CEO of Halliburton, reporters wanted to know whether Bush agreed with them. Two months earlier, Bush had written to the heads of major Jewish organizations in the United States and vowed that Iran would be kept at a distance until the Islamic Republic freed more than a dozen Israelis it was holding on espionage charges. A spokesman for the campaign said Bush's views had not changed.

On July 3, Cheney traveled to Bush's ranch in Crawford to review the remaining candidates. In the press, speculation focused on Governor Frank Keating of Oklahoma and Governor Tom Ridge of Pennsylvania. Bush liked them both. But he had not given up on Cheney.

Bush and Cheney spent the morning reviewing candidates. When they finished a leisurely lunch with Laura Bush, the two men retired to the porch. It was a steamy day, and Cheney couldn't figure out what they were doing outside, away from the air-conditioning.

Bush got serious. In their time working together, he had gotten comfortable with Cheney. He trusted Cheney's instincts, admired his loyalty, and valued his discretion. "You know," Bush said, echoing his many jokes over the past three months, "you're the solution to my problem."

For the first time, Cheney said he was willing to look at the steps he would need to take to place himself under consideration. It was an abrupt change of heart, influenced by two factors: a sense of duty and a growing belief that he was the least worst option.

"If the president of the United States asks you to do something, you really have an obligation to try to do it, if you can," says

Cheney. "And the other thing that I suppose affected it to some extent was that I'd been through the process with him, and when he said to me, now, 'You're the solution to my problem,' it was harder to argue with because we'd looked at the available—and I don't mean to be negative on the other candidates out there—but as I say, when you go through one of these processes, you nearly always end up in a situation, it's a very short list."[24]

Cheney told Bush that he would ask his doctor to provide the latest health information to the campaign and asked for an opportunity to make a case against himself with Bush's other top advisers. When they were done, the two men met briefly with the press. Bush said nothing about Cheney's willingness to be considered. In response to a question, Bush set out his criteria. "The main tests are: Can the person be the president? And will the person be loyal to the administration? Can the person bring, you know, some added value to a Bush administration?"[25]

The following weekend, Cheney traveled to Austin to meet with Bush, Karl Rove, and Karen Hughes. Bush's advisers had scrambled to do an independent review of Cheney's background. Bush had asked Rove to make the case against Cheney, a task that Rove performed with vigor. Rove liked Cheney and was one of the few advisers who understood that Bush had not given up on persuading him to take the job. When Rove finished, Cheney filled in the remaining gaps, scrutinizing his own life—professional and personal—and laying bare the vulnerabilities he would bring as a candidate.

He reiterated his negatives: his history of heart trouble, his arrest record, his recent experience in the oil business, his current residence in Texas, the three electoral votes in his native Wyoming. And then there was his congressional record. "I had a very conservative voting record, more conservative than most people realized," he says. "The press never looked at my voting record. They thought I was all warm and fuzzy and they never looked to see."[26]

It was a process that his family was sure would knock him out of consideration. Several days later, Cheney accompanied his wife to Minneapolis, where she attended a meeting of a corporate board on which she served. As they ate dinner with the rest of the board, Cheney stepped out to take a phone call from Bush, who said that he had not seen anything in Cheney's health report serious enough to disqualify him as a running mate. Bush wanted to move forward. The discussions intensified.

Initially, the Cheneys were divided, as they had been when Cheney considered a presidential bid six years earlier. "Lynne and Mary were not enthusiastic. It had nothing to do with the governor. It had everything to do with—we liked our lives. We'd been through this exercise a couple of years before, where I'd thought about running myself and decided not to—partly because I wanted to protect our privacy and didn't want to have to go through all the aggravation that goes with mounting a campaign. Liz was gung ho. She was out in the backyard painting yard signs."[27] As the discussions between Bush and Cheney grew more serious, the opposition from Mary and Lynne softened.

On July 18, Pete Williams, Cheney's former spokesman, now working for NBC News, called contacts back in Wyoming. Williams wanted to know if Cheney was registered to vote in the state. It was a clever newsman's way of finding out whether Cheney was on the ticket. Williams knew that the Twelfth Amendment to the Constitution prohibits electing a president and vice president from the same state. And three years earlier, Cheney had proudly proclaimed himself a Texan. "I am now officially a Texan. We own a home in Dallas, I've got Texas plates on my automobile, paying taxes in Texas," he said. "And I'm thinking about becoming a Dallas Cowboys fan," he told an audience in Washington, D.C.[28] To run on the same ticket as Bush, Cheney would have to register in a different state.

Williams was told that Cheney was not registered in Wyoming. Three days later, the reporter Lisa Myers of NBC followed up Williams's investigation. She called the clerk's office in Teton County, where Cheney would register if he were to use his home in Jackson Hole as his official residence for voting purposes. The person on the other end of the phone said that Cheney had in fact been in that very morning to change his registration from Texas to Wyoming.[29]

Minutes after Myers broke the story on MSNBC, the Democratic strategist Paul Begala confirmed Cheney's warning to Bush that his health would be an issue. "He's a good man. He could serve honorably," said Begala. "But he has a bad heart and I think it's irresponsible for Bush to choose a man as a running mate who has a bad ticker."[30]

The news of Cheney's trip and his likely selection as Bush's running mate came as a surprise to all but the most senior staffers at Bush's campaign offices in Austin. "Here we were in the middle of the headquarters and we didn't know what was going on," says John McConnell, a speechwriter for the campaign who along with Matt Scully had been drafting a vice presidential acceptance speech for an unknown candidate.[31]

On July 25, Bush called Cheney, at 6:22 AM. Lynne answered the phone at their home in Texas and reported that her husband was on the treadmill. Cheney took the call and Bush formally offered the job. He accepted, hung up the phone, and called out to his wife. "Honey, let's sell the house. I quit my job. We're going back into politics."[32]

Cheney told his board at Halliburton that he was being considered as a candidate for vice president and called Tom Cruikshank, the man who had brought him to Halliburton, to share the news.

"Tom, I've decided to run for vice president with George," Cheney said.

"Dick, you've lost your mind," Cruikshank responded.

With that announcement, public life rushed back into the Cheney household. As soon as his candidacy hit the news, the press was literally at his door. One enterprising reporter deposited a disposable camera on his front steps with a note asking the family to record the day's events and return it when all the film had been used. Television crews showed up to look for memorable images. "There was a big window over the front door of the house," he recalls, "and looking out and seeing that antenna—satellite truck and so forth as I trotted down the hallway to the kitchen in my underwear to get a cup of coffee. Suddenly life had changed pretty dramatically."[33]

In one last moment of independence, Cheney drove himself to the Dallas airport in the same 1992 Cadillac STS he had bought as secretary of defense. He boarded the Halliburton plane for

Austin, where he was to join Bush for the announcement at two PM local time.

In front of a raucous crowd at the University of Texas, Cheney stood to the side as Bush spoke, his hands clasped in front of him at his waist. Bush enumerated Cheney's qualifications and reminded the audience that Al Gore had once praised the man whose job it would now be to attack him. "Even my opponent, the vice president, once said, 'Dick Cheney is a good man who is well-liked and respected by his colleagues,' and I agree."

Cheney stepped up to the microphone and matter-of-factly explained his initial hesitance to reenter public life. "I was deeply involved in running a business, enjoying private life, and I certainly wasn't looking to return to public service. But I had an experience that changed my mind this spring. As I worked alongside Governor Bush, I heard him talk about his unique vision for our party and for our nation. I saw his sincerity. I watched him make decisions—always firm and always fair. And in the end, I learned how persuasive he can be."[34]

Looking back, Cheney says he is aware of the conspiracy theories that suggest he and Bush had struck a deal long before he was asked to run the search process. "A lot of people don't believe it," he says, "but I really did not enter into the effort with the notion that somehow I was going to get the job."[35]

Bush and Cheney returned to the governor's mansion, where Cheney and his wife had lunch with George and Laura Bush. After posing for photographs for the covers of the newsweeklies, Cheney was introduced to Scully and McConnell, the two men writing his acceptance speech.

"Nice to meet you," Cheney said. "How fast do you gentlemen work?"

The writers asked Cheney about themes for the speech. Cheney said he wanted his address to emphasize government service and patriotism. He told them about the helicopter flight over Washington, D.C., that he had taken on dozens of occasions as secretary of defense, the short but evocative trip from Andrews Air Force Base to the Pentagon.

Lynne Cheney sat in on the idea session and offered a suggestion that would produce the most memorable line of the speech.

She recalled the refrain from Al Gore's vice presidential acceptance speech in 1992—"It's time for them to go"—and thought it would be particularly effective to use his own words against him. The speechwriters liked her proposal and promised to include it. That Cheney's wife would shape his first words as the Republican nominee was a sign of things to come. He often refers to her as his "closest adviser," and several members of his senior staff confirm that this is not just politesse.

Later that evening, in an interview on CNN's *Larry King Live*, Cheney joked about her role. King had interviewed Cheney one-on-one for several minutes and as he went to a commercial he told viewers to stick around to hear from Lynne.

"This man you're watching could be one heartbeat away from presidency in a few months," King said, excited to have landed Cheney's first television interview following the announcement. "His wife will join us. What do they call the vice president's wife, Dick? Is she second lady?"

Cheney answered without hesitation: "Boss."[36]

The Cheneys spent the night at the governor's mansion and woke early the next morning for a flight to Casper and a rally at Natrona County High School. Cheney had been out of politics for more than five years, and had grown unaccustomed to the smile-for-the-cameras culture of campaigns. When the new candidate walked briskly through the crowd toward the stage, he left his wife several paces behind. Later, an aide gently reminded him that the campaign preferred that he and his wife walk together at all times.

McConnell and Scully delivered a draft of Cheney's speech for the convention at one-thirty PM the following day. The two had set aside large chunks of time before and during the convention to work on the speech. For several days, they waited for Cheney's comments to begin the arduous process of rewriting the text. They heard nothing. It wasn't until after the convention had begun that they were told the Cheneys were happy with their draft.

Reaction to the selection came quickly. The day after the announcement a front-page news analysis in the *New York Times* concluded that the selection was the "least adventurous and least sensational" possibility and "deeply revealing about Mr. Bush's approach to politics and to governing. It underscored that Mr. Bush is generally not a bold risk taker in his public life."[37]

In Washington, the Senate minority leader, Tom Daschle, allowed that Cheney "clearly has a lot of experience" and "is generally well liked," before labeling him "as far right as anybody in the Republican Party today."[38] Martin Frost, a Democrat from Texas, followed the same rhetorical pattern, calling Cheney "a man of integrity," but warning that the selection "raises some troubling questions about what a Bush administration would mean for working families in Texas."[39] Again in line with Cheney's warning to Bush, the Democratic National Committee unveiled a Web site that cataloged his "extreme" voting record.

The Democrats concentrated their attacks on a few votes Cheney had cast in Congress, especially a vote in 1986 on support for the African National Congress (ANC) that included a provision about freeing Nelson Mandela from jail. Chris Matthews, host of MSNBC's *Hardball*, raised the issue the same day Cheney's candidacy was announced. "How do you get African-Americans, black Americans, who are, perhaps, open to the idea of voting for the Republicans, to vote for a ticket which includes a guy, Dick Cheney, who opposed sanctions on South Africa back when it had a white supremacy government?"[40]

President Bill Clinton, immensely popular with black voters, kept up the charge: "All the big publicity is about, in the last few days, an amazing vote cast by their vice presidential nominee when he was in Congress against letting Nelson Mandela out of jail," Clinton said at a fund-raiser for former Representative Bill Nelson of Florida, a candidate for the Senate. "That takes your breath away."[41]

In fact, the vote had been a bit more complicated than Clinton's description suggested. The resolution included two additional provisions, one of which required that the United States recognize Mandela's party, the ANC, which was closely aligned with South Africa's communists. Clinton also left out of his critique the fact that it passed in a Democrat-controlled House, earning the votes of thirty-two members of his party. Most problematic, one of those who cast the same vote as Cheney was Bill Nelson, the man sharing the stage with Clinton when the president made his comments.

Nelson's spokesman explained the candidate's vote. "Bottom line is that Nelson strongly supported two components of the measure, and he considers Mandela one of the century's great leaders. He could not support the third, recognizing ANC because it was dominated by the Communist Party. This vote should be looked at in context." By the start of the Republican National Convention, eight days after Cheney had joined the ticket, most Democrats had dropped this line of attack.[42]

Integrating Cheney with Bush's campaign was a relatively easy process. Typically, a candidate for vice president brings a coterie of staffers who are as enthusiastic about their boss and his new position as they are about the candidate for president. This often produces disunity between the existing campaign and the new arrivals. But Cheney came from five years in the private sector and did not have a large staff to accommodate. Several of his most trusted advisers from years past were already consulting on the campaign. He asked his daughter Mary to serve as his personal aide and hired Kathleen Shanahan, a longtime friend of the Bush family, as his chief of staff. The Bush campaign took the lead in filling the rest of his staff.

As a practical matter, Cheney's willingness to take direction was a small but helpful gesture. But his deference to the campaign sent an early signal to Bush and his advisers that Cheney was willing to subordinate his interests to theirs, an approach he later confirmed by declaring that he would not run for president himself following his tenure as vice president. He had no further political ambitions and no personal agenda. Over time, as Bush would later confirm, this fact more than any other contributed to Cheney's accumulation of power.

On August 2, the First Union Center in Philadelphia was buzzing with excitement in anticipation of Cheney's acceptance speech. After an introduction from his wife, Cheney took the podium and promptly began his remarks.

"Mr. Chairman, delegates, and fellow citizens, I am honored by your nomination, and I accept it."

On NBC, a surprised Tom Brokaw interjected: "Well, that's getting down to business in a hurry."

Having taken care of the task at hand, Cheney paused for reflection. Gerald Ford, who was in town for the convention, had suffered a stroke the previous night. The former president would not be there to see his top aide accept the vice presidential nomination. Cheney asked the audience to keep Ford in their thoughts: "I wouldn't be here tonight if it wasn't for him and the trust and confidence he placed in me twenty-five years ago."

The rest of the speech would be remembered for its combativeness. In his week on the campaign trail, Cheney had shown himself willing to go after Bill Clinton and Al Gore aggressively. The gathering of Republican activists assumed that he would sharpen his critique in this speech. They were right.

The convention up to that point had been carefully staged and cautious. The full text of each speech from the podium was screened for any potentially offensive or inflammatory passages. The censors softened attacks that might sound harsh and sanded down rhetorical edges. It was a convention designed to introduce George W. Bush, compassionate conservative.

But Dick Cheney was not a compassionate conservative. And the campaign was happy to put on display his plainspoken western conservatism.

"For eight years, Clinton and Gore have extended our military commitments while depleting our military power," Cheney said. "Rarely has so much been demanded of our armed forces and so little given to them in return."

He spoke of his own firsthand knowledge of the military, and continued his assault. "Soon our men and women in uniform will once again have a commander in chief they can respect," and "as the man from Hope goes home to New York, Mr. Gore will try to separate himself from his leader's shadow. But somehow we will never see one without thinking of the other. Does anyone, Republican or Democrat, seriously believe that under Mr. Gore, the next four years would be any different from the last eight?"

The crowd roared its approval. "Bill Clinton vowed not long ago to hold on to power until the last hour of the last day,"

Cheney said. "That is his right. But, my friends, that last hour is coming. That last day is near. The wheel has turned. And it is time, it is time for them to go."

As Cheney repeated the refrain his wife had suggested, the crowd joined in: "It is time for them to go!"

It was a dramatic scene, a slice of controlled chaos. And yet when Cheney remembers the moment, the highlight of his long career, a peculiar image fills his head.

As Cheney surveyed the sea of flag-waving partisans, his eyes were drawn to an old man at the head of the California delegation. It was George Shultz, standing directly in front of the stage to Cheney's right. George Shultz, preeminent industrial economist, PhD from MIT, a man who had served in Republican cabinets twice in the previous twenty years, was looking up at Cheney.

"Shultz was just rocking," Cheney marveled. "George Shultz? He was the secretary of the treasury and this guy I'd worked with back in the Nixon administration, and, Christ, he was just like a kid down there—going wild."[43]

Cheney ended his speech with the description he had given the speechwriters of his regular helicopter flights over the nation's capital as secretary of defense:

> *When you make that trip from Andrews to the Pentagon and you look down on the city of Washington, one of the first things you see is the Capitol where all the great debates that have shaped 200 years of American history have taken place. You fly down along the Mall and see the monument to George Washington, a structure as grand as the man himself. To the north is the White House where John Adams once prayed that none but honest and wise men may ever rule under this roof. . . . And then you cross the Potomac on approach to the Pentagon. But just before you settle down on the landing pad, you look out upon Arlington National Cemetery, its gentle slopes and crosses, row on row. I never once made that trip without being reminded how enormously fortunate we all are to be Americans and what a . . . terrible price thousands have paid so*

*that all of us and millions more around the world might
live in freedom.*

No one could have known that these words would acquire
deeper meaning with almost each passing day of his vice presidency.

Cheney and Bush left the convention together for a train trip
through several important swing states. Days later, after they
separated to cover more territory, an event in Florida served as a
reminder that incorporating a running mate is not always easy.

The campaign assigned Cheney to give an address at an elementary school, his first on education. It was Bush's signature
issue as he promoted himself as a "compassionate conservative" devoted to ending "the soft bigotry of low expectations."
Throughout the primaries Bush boasted about the reforms he
had effected in Texas and promised to bring those results to the
rest of the country. On the stump, Bush came to life when he
talked about education.

Cheney was considerably less enthusiastic. He cared about
taxes and national security. As a member of Congress, he had
voted against funding for the Department of Education. And although he had long been concerned about public school curricu-
lums, drawing on his wife's work, in any ranking of Dick Cheney's
public policy priorities education would fall near the bottom.

But it wasn't his campaign, and he had vowed to support Bush
in whatever way he could. So he dutifully agreed to give a speech
touting Bush's proposals for education at an elementary school
in Florida. It did not go well.

"This is my first education speech as candidate," he recalls,
"and I got up in front of the audience, and I looked at the audience and looked at this speech, and the speech was about the
Bush-Cheney sexy new idea for tax-exempt financing of bond
issues to build after-schools. And I'm giving it to a third-grade
audience. These kids are eight and nine years old."

He pauses for an incredulous laugh.

"And some of the parents are there, but it's mostly these kids.
And I'm trying to give this speech, and looking—the eyes are
glazing."

He laughs again, this time with a do-you-believe-that shake of his head.

"The press is hanging around, and they're yakking it up and having a good time. We had a little bit of a mismatch there between the politics and the speechwriting and the advance."[44]

Shortly after the Republican convention, Al Gore had named Senator Joseph Lieberman of Connecticut as his running mate, a choice that many understood as an attempt to distance himself from Bill Clinton's sex scandals, since Lieberman had been an outspoken critic of Clinton. Lieberman also brought some of the qualities Cheney provided for Bush, characteristics the news media summed up as gravitas. A nationwide Gallup poll taken in late August concluded that the race was a dead heat, and surveys in many of the so-called battleground states told a similar story.

On September 4, 2000, just after Labor Day, Bush and Cheney prepared to launch the final two-month push toward the elections in November with a rally in the Chicago suburb of Naperville. Bush would be speaking about tax cuts—a favorite subject of the two Republicans and their audience.

As they waited to begin, Bush noticed Adam Clymer, a reporter from the *New York Times*, standing in the audience. Unaware that his microphone was on and picking up everything he said, Bush leaned in and whispered to Cheney, "There's Adam Clymer. Major-league asshole from the *New York Times*." Cheney agreed. "Oh, yeah. Big-time."

The clip would be replayed hundreds of times on television. And Cheney had a new nickname: Big-Time. During the campaign and even afterward, other campaign officials—those who held senior positions, anyway—called Cheney by his new name.

Some of them thought he earned it again with his performance in the vice-presidential debate. Bush says that Cheney's debate against Joe Lieberman gave the campaign a significant boost. "The pick of the vice president is more about presidential decision making than anything else. . . . The country gets to see the first decision a candidate makes, and the country gets to see what decision making will be like under a president."[45]

For Cheney, who had run the preparation for Ford's debate in 1976 and served as a questioner in a prep session for Ronald Reagan, the process of readying for a debate was intense, but familiar. He gathered a small team of intimates in Jackson, Wyoming, with a few of Bush's campaign strategists. There were policy wonks, debate coaches, media strategists, former colleagues, and family members. The team usually met at Cheney's home, but as the number of advisers grew the campaign sought a larger space. They settled on the Jackson Hole Playhouse and Saddlerock Saloon, an old remodeled theater in downtown Jackson. But what the theater provided in space, it lacked in security. Journalists sneaked into one debate prep session pretending to be tourists interested in the architecture. "They clearly were scoping out the place," Liz recalls. "And we got worried that people would be in there with tape recorders. We needed to keep it private."

Rob Portman, a Republican congressman from Ohio who was close to Bush, would be Cheney's opponent throughout the rehearsals. From the moment Al Gore announced that Joseph Lieberman would be his running mate, Portman spent every spare moment listening to Lieberman on his Walkman. He got tapes of Lieberman's campaign addresses, his floor speeches, his television and radio appearances, and his previous debates. (Cheney himself studied footage of Lieberman's most effective and aggressive debate performance, an evisceration of the Republican Lowell Weicker in 1988.)

By the time Portman faced Cheney in the mock debates in Jackson, he had Lieberman down pat. He could mimic his nasal intonation, his faint New England accent, the leisurely pace of his speech, even his facial tics.

"Rob was very good because he could get under my skin occasionally," Cheney says. In the beginning, Portman could do this by provoking Cheney about Halliburton and questioning Cheney's integrity. "It was good because I had—one of the things you never want to do is lose your temper."[46] In order to prepare Cheney for every eventuality, Portman regularly violated the debate rules agreed upon by the two campaigns, speaking for longer than his allotted time or handing Cheney a campaign pledge.

It was an old debate strategy designed to bait an opponent into something no voter likes—whining about the rules. In the end, Cheney's sessions with Portman would be more contentious than his debate with Lieberman.

Cheney was fly-fishing on the Snake River when he got a call from the campaign workers. They wanted to know whether CNN's anchor Bernard Shaw would be an acceptable moderator. Cheney had watched Shaw's coverage of the Gulf War and liked him. With Lieberman's approval, the debate was set for nine PM on October 5, 2000.

The group sessions were valuable because they put Cheney into a debating mind-set. But most of the work Cheney did on his own. "A lot of the preparation takes place when you're not sitting at the table, maybe nobody else is around," he says. "It's early in the morning and because it's something you're think-ing about doing, you'd think about how to phrase answers and so forth. But it's also a matter of self-confidence and being in control and knowing you're on top of your brief, and then being relaxed."[47]

The debate would be held at Centre College in Danville, Ken-tucky, a small town of 15,000—Cheney's kind of place. Cheney arrived amid the chaos and commotion of a presidential cam-paign: the streets were filled with oversize satellite trucks, and the sidewalks were teeming with reporters from around the world. As nine o'clock loomed, Cheney sought some solitude. Rather than run through his briefing books or rehearse his lines, "I took a nap," he recalls. While his team checked and double-checked its research, he retired to his room. "I literally went and took off my suit, and crawled into bed and took a nap."[48]

The first question, on projected budget surpluses, went to Lieberman. He prefaced his answer with a promise to avoid negative attacks, a perennial line that is often spoken in debates but rarely honored. Cheney reciprocated. "I, too, want to avoid any personal attacks," he said. "I promise not to bring up your singing."

That line had a scripted quality, but a more memorable ex-change in the debate arose from comments by Lieberman that could not have been anticipated. Cheney had counted on harsh attacks about his five years at Halliburton and the generous compensation package he had gotten when he left. Lieberman, though, seemed almost apologetic about having to criticize Cheney.

"I think if you asked most people in America today that famous question that Ronald Reagan asked—'Are you better off today than you were eight years ago?'—most people would say yes. And I'm pleased to see, Dick, from the newspapers that you're better off than you were eight years ago, too."

"And most of it—I can tell you, Joe, that the government had absolutely nothing to do with it."

"Interesting," said Lieberman, searching for a comeback. As Shaw started the next question, Lieberman fired back. "I can see my wife, and I think she's thinking, 'Gee, I wish he would go out into the private sector.'"

Cheney, without pausing, responded: "Well, I'm going to try to help you do that, Joe."

The debate was friendly and substantive, perhaps even a bit wonky. (After one response by Lieberman on taxes, Cheney said to Shaw, "Bernie, you have to be a CPA to understand what he just said.") One observer called the debate a "gentlemanly, civil affair that turned out to be one of the defining moments of Cheney's campaign."[49] Several television commentators remarked that Lieberman and Cheney should have been at the top of their respective tickets.

Mike McConnell, who had worked for Cheney at the Pentagon and would later work for Bush as director of national intelligence, had a similar reaction. "I turned to my wife and said: 'We got the wrong two guys running.' Honest to God! I said, 'There's the two guys we want leading the country.'"[50]

As the two campaigns worked to get their talking points and platform statements out to the public, reality intervened in a terrible way. On October 12, 2000, Al Qaeda terrorists rammed a small boat loaded with explosives into the side of the USS *Cole*, an American destroyer docked in Aden, Yemen. The blast killed the terrorists and seventeen Americans aboard the ship. The candidates responded quickly.

Bush called the attack "cowardly" and talked tough. "There must be a consequence," he told reporters. In Ohio, Cheney used the attack to question once again the wisdom of Clinton's cuts in defense. He left open the possibility that the proper response would be military. "The important thing is to find out who did it and to make absolute certain that they pay a very heavy price for having done it," he said.[51]

Gore, campaigning in Milwaukee, also promised action. "If it is determined to be the result of a terrorist operation, those responsible should know that the United States will not rest until the perpetrators are held accountable," Gore told reporters before returning to the White House to confer with President Clinton. "This is a situation that will bring a response," he promised.[52]

They were wrong. One suspected plotter would be killed by the CIA and two more put to death by a Yemeni court. Many others would escape, and for al Qaeda there was no "heavy price." It would matter.

Over the last week of the campaign, Cheney would fly across the country and back again visiting, for example, Arkansas, Illinois, Florida, Washington, Nevada, Oregon, Tennessee, and Pennsylvania. Two days before the election, Cheney spent the entire day in California, some of it campaigning with Colin Powell. He ended the campaign back in Jackson, Wyoming, with an election eve rally.

When Cheney voted the next morning, at the Wilson Firehouse less than a mile from his house, he was confident that he would end the day as vice president-elect. "I thought we were going to win. I thought, based on what we were hearing from our experts, that we were ahead by at least a couple of points nationwide in the key states that mattered."[53]

So did almost everyone involved in Bush's campaign. Cheney flew to Austin to watch the returns with Bush and the rest of the team. He was exhausted, and although several family friends had joined him on his plane, he excused himself for a nap.

Cheney's campaign chief of staff, Kathleen Shanahan, had gotten the first round of exit polls from the campaign. "Not good," says Cheney. Without going into specifics, Shanahan told Liz Cheney to prepare for a long night.

The Cheneys dug in at their suite at the Four Seasons in Austin. George and Laura Bush had also planned to spend the night watching returns at the hotel, but after some discouraging early returns chose to head back to the governor's mansion with their family. The Cheneys' suite became the gathering spot for top officials of Bush's campaign. Clustered around the TV was a who's who of past and future Republican administrations: Jim Baker, Nick Brady, Don Rumsfeld, Andy Card, Don Evans, Scooter Libby, and David Addington.

The group watched returns all night. From Cheney's perspective, the early results were neither as bad as the exit polls had predicted nor as good as Bush's own polls suggested they would be. Then, at 7:50 PM on the East Coast, CBS called Florida for Al Gore. But many voters in northwest Florida live in the central time zone. Polls there were still open. "The Panhandle was still out," says Cheney, "which really fried us."[54]

The call was premature for other reasons. As everyone would soon understand, the exit polls fed to the networks were badly flawed. Their projections were wrong. Karl Rove understood immediately that calling Florida for Gore this early must have been based on bad numbers. His certainty gave hope to those at the governor's mansion and in the Cheneys' suite. For two hours, Cheney had to reconcile the different conclusions coming from all the major networks and from Bush's top campaign strategist.

The Florida call was reversed at 9:50 PM, putting this decisive state back in the "undecided" column.

The situation was as tense as politics gets. Cheney spent most of the night reviewing election returns with Al Simpson and the others gathered in the suite. Eventually, toward dawn, as he had done "at crucial moments in my career"—on election night in 1976, waiting to meet with King Fahd after the Iraqis' invasion of Kuwait, and on the flight to Austin earlier that day—Cheney went to sleep. "I went in and lay down and took a nap."[55]

But twenty minutes after Cheney turned out his light, there was news. Sitting at a table in the Cheneys' suite, Bush's campaign chairman, Don Evans, took a call from Bill Daley, his counterpart in Gore's campaign. Daley explained that Gore would be calling to concede, but that he needed some time with his family first.

Cheney and his party went to the governor's mansion to join Bush. Everyone was excited, but Cheney thought that something was amiss. "It didn't feel right," he says. Moments later, Gore called Bush to withdraw his concession. The call had been short and tense.

"He was pissed," recalls Liz Cheney. "The governor was pissed."

"Well, he wasn't the only one," adds her father.

At 2:16, when the Bushes' party was gathered at the mansion, Fox reversed its earlier statement and called Florida for Bush.

"It was like a roller-coaster," Cheney says. "'They got Florida. No, they didn't get Florida.' And they call it for us. Daley calls and said, 'Well, he's going to concede.' Then Gore calls and says, 'No, I take it back. I'm not going to concede.' You know, it was one of those nights. How often does that happen?"[56]

The results weren't much clearer the following morning. In time, the difference would be 537 votes in Florida. That's where the recount began, and after thirty-six days of insanity, where it would end.

Cheney, his family, and a small team of his closest advisers remained in Austin for ten days. The group included Paul Wolfowitz, Steve Hadley, Scooter Libby, and David Addington. They were tired and none of them had packed for more than a few days.

In the aftermath of the election, as a legal battle seemed inevitable, Cheney asked Bush how he could be most helpful. The two men decided that Cheney should operate on the assumption that they would ultimately prevail. He quietly began to put together a team to move into the White House in late January.

Two days after the election, Cheney and his campaign advisers gathered in his suite at the Four Seasons. As the group watched televised updates on the recounts in the main room, Cheney held two one-on-one meetings in the bedroom that would have a dramatic influence on the shape of Bush's presidency.

Scooter Libby went first, and reemerged after fifteen minutes. "I talked to him about doing the national security job," Cheney recalls. "Scooter agreed to do that, but he said he also wanted to be my chief of staff. He wanted to be dual-hatted." Cheney had respected Libby's work at the Pentagon, but rarely

had one-on-one interactions with him. It was their time together in the crucible of Cheney's preparation for the debate
which gave him confidence that Libby could handle both jobs.
"I thought about it and decided I'd give it a try," Cheney says.
"He had this sort of two sets of qualities, one in terms of substantive interest, but also, he was a guy that was well organized
in terms of being able to get things done."[57]

David Addington was next. Liz Cheney had taken him aside
during a quiet moment earlier to sound him out about working as Cheney's government lawyer, and Addington said that he
would do any job Cheney asked him to do. When he met with
Cheney, his former boss asked him to join his staff as counsel—
assuming that the recounts went in their favor. Addington, who
had been working at the American Trucking Association, accepted. Libby and Addington had broad experience on national
security matters, and would be working for a vice president with
broad experience on national security matters, so their proximity
to Cheney would give them a strong voice in what would be a
national security presidency.

Cheney and an intimate of Bush's named Clay Johnson would
run the transition from Washington, D.C. Because the election
had not yet been officially decided, Bush's personnel team was not
entitled to use the space designated for the transition. So Cheney
arranged for office space in McLean, Virginia, not far from his
home. While those arrangements were finalized, his kitchen table
served as the worldwide headquarters of the transition.

At first, it was a small group: Cheney, his two daughters,
Scooter Libby, and David Addington. A pile of cell phones sat
in the middle of the table, and each ring brought a mad scramble to determine whose phone was ringing and who was on the
other end.

Cheney shuttled between Washington and Austin, accompanying prospective cabinet officials down to see Bush for an
interview or a press conference. "For a period of time there, the
transition was an occasional press conference," Cheney says. "I
mean, that's all there was."[58]

As the recount continued, news reports changed the story—
and the likely outcome—seemingly every hour. With each new

court filing came an extended press conference, each side accusing the other of dishonesty and willfully excluding the votes of Floridians.

Then, on November 22, came a news story that Cheney had worried about when he joined the ticket. Before dawn that morning, Cheney checked into the George Washington University hospital after experiencing chest pains overnight. Doctors found that he had a slight blockage in one artery, inserted a stent, and reported that the episode would be considered a minor heart attack.

The episode had no bearing on the actual recount, of course, but the campaign press team was concerned about how it might be received in the country at large. If there were to be a Bush White House—still a big if at that point—would Cheney's health further weaken an administration already struggling with issues of legitimacy and, increasingly, the anger of the opposition?

Kathleen Shanahan had an idea. Cheney's heart trouble was the story of the day. She was certain that the cable news shows would talk about it and that writers would write about it. Why not displace some of the inevitable speculation with some reassuring words from Cheney himself, on the phone from his hospital bed? Shanahan thought Cheney's deep, resonant voice would project strength at a time when projecting strength was the top priority. She had a good contact for a producer for Larry King and thought that King, who had spoken about his own history of heart troubles with Cheney at the 1988 Republican convention, would stick to the issue at hand without trying to make news on the recount. Shanahan floated the idea, and Cheney agreed.

A few minutes after nine PM King interrupted a debate between two election lawyers for the brief interview.

"Dick Cheney from his hospital bed at George Washington University is joining us. How are you feeling, Dick?"

"Well, I feel pretty good, Larry."

"Well, you sound great."[59]

Success.

"We're lucky it didn't happen until after the election was over," Cheney would say later. "It worked out OK."

Most of Cheney's time was devoted to helping Bush pick a cabinet. It was a natural role; he had vetted cabinet selections as Ford's chief of staff and had served in George H. W. Bush's cabinet. Cheney's influence would be obvious and would help start the reports about his outsize role as vice president.

While he worked on the transition, he naturally monitored recount developments in frequent conversations with his old friend Jim Baker, who returned to public life to lead the Bush campaign's recount effort. One time when Cheney called for an update, Bush's national campaign chairman, Don Evans, answered the phone for Baker. "It's Big Time," said Evans.[60]

While Cheney concentrated on the cabinet, Scooter Libby began to put together a vice presidential staff. He called friends around Washington soliciting suggestions and conducted preinterviews of candidates to determine who would get an interview with Cheney. Cheney's interviews for his prospective staff were brief and to the point. And each one ended with the same question: "Is there anything in your background that would embarrass me or the president if you were to come to work for us?"

In the end, Cheney's impact on key positions of the Bush administration would be obvious. His good friend and former boss Donald Rumsfeld would run the Pentagon, with Cheney's longtime adviser Paul Wolfowitz as deputy. At Foggy Bottom, Colin Powell, the man Cheney had elevated to chairman of the Joint Chiefs of Staff twelve years earlier, would serve as secretary of state. Cheney's friend Paul O'Neill, from the Ford administration, would become secretary of the treasury.

As a measure of Cheney's influence, the cabinet was revealing.

CHAPTER TEN

The Bush Administration

The Cheneys and Bushes began Inauguration Day, January 20, by worshipping at historic St. John's Church. The weather was dismal, cold and rainy. When the service ended, they climbed into limousines for the one-block trip across Lafayette Park to the White House, where, according to the schedule, they were to have coffee with the Clintons and Gores before the inaugural ceremony.

Clinton wasn't ready.

"We got in the cars and we had to wait," says Cheney. "And then we had to wait some more, and then we had to wait some more, and then we had to wait some more." The famously unpunctual Clinton was finishing his term in signature style. "We must have waited about half an hour before we could go over there."[1]

Once they got to the White House, the couples passed some time in the kind of small talk Cheney typically avoids. "We got over there and made nice-nice and so forth," he says, "waiting to go up to the Hill."[2] Bush and Clinton rode to the ceremony together, while Cheney joined Gore and several congressional leaders for the short drive down Pennsylvania Avenue.

Cheney and Gore had served together in the House for eight years but had never been close. Nonetheless, as the recount began, Cheney had felt some empathy for his opponent. He had been through a very close presidential race with Gerald Ford in 1976, and that experience coursed through his mind during the aftermath of the election. "I always had a certain sympathy for Gore. It had to be tough what they went through," says Cheney. "It had to be tough because the recount went on forever and—when do you end it? On the one hand, you don't want to surrender the presidency if you didn't lose. But on the other hand, if they keep counting and the other guy has still got a bit of an edge . . . When it was clear the court had decided and so forth, I thought he handled it with a certain amount of class."[3]

The conversation on the way to the Capitol was more substantive than the one at the White House. Gore joked about the last-minute presidential maneuvering that required Bush and Cheney to wait for their host. "We were laughing," Cheney recalls, "because Gore was explaining the reason we'd been delayed and they hadn't been ready to receive us on time was Clinton had been upstairs pardoning people."[4]

Cheney felt a sense of familiarity as he arrived at the Capitol. He had taken many oaths of office before—as White House chief of staff, six times as a member of Congress, and later as secretary of defense—and he had been on the platform for the inaugural ceremonies of Ronald Reagan and George H. W. Bush.

"Vice President Cheney, an old, old Washington hand," said ABC News's anchor Peter Jennings, as the cameras captured Cheney first joining the ceremony. "Sometimes said, particularly among the skeptical about George W. Bush here, that he'll be the one with real influence in the city. It's not to say that George W. Bush hasn't had experience here, but Dick Cheney knows this town extremely well. Served in the House of Representatives, served as the secretary in various cabinets, and certainly knows that this town is not—not always paved with goodwill."[5]

Republicans, including large delegations from Texas and Wyoming, were on hand to witness the official end of what in their view had been eight very long years of the Clinton administration. Some Democrats were scattered among the crowd, mainly as participants, witnesses to history, or protesters.

On the west front of the Capitol the bitterness of the previous three months was, if not forgotten, at least mostly set aside for the day. Former presidents Bill Clinton, George H. W. Bush, Jimmy Carter, and Gerald Ford exchanged pleasantries before the formal ceremony began. Supreme Court justices chatted with members of Congress, including the newly elected Senator Hillary Rodham Clinton; and their spouses, including Elizabeth Dole, who would be sworn in as a senator from North Carolina two years later. Current nominees for Bush's cabinet mingled with cabinet secretaries from the presidency of George H. W. Bush—the administration that, with his son's swearing in, would become known as Bush One.

As the ceremony began, Senator Mitch McConnell, a Republican from Kentucky, welcomed the crowd. "In becoming the forty-third president of the United States, George W. Bush will assume the sacred trust as guardian of our Constitution. Dick Cheney will be sworn in as our new vice president, witnessed by the Congress, Supreme Court, governors, and presidents past. The current president will stand by as the new president peacefully takes office. This is a triumph of our democratic republic, a ceremony befitting a great nation."

After an opening prayer by the Reverend Franklin Graham, McConnell turned the ceremony over to Senator Christopher Dodd, a Democrat from Connecticut, who introduced Chief Justice William Rehnquist. Rehnquist turned to Cheney.

"Mr. Cheney, are you ready to take the oath?"

"I am," said Cheney. His wife and two daughters huddled around him, with Liz holding the family Bible.

"Please raise your right hand and repeat after me. I, Richard Bruce Cheney, do solemnly swear . . ."

Cheney vowed to support and defend the Constitution against all enemies foreign and domestic and declared that he took the obligation without any mental reservation or purpose of evasion.

"So help me God."

"Congratulations, Mr. Vice President," Rehnquist said.

The words marked the high point of a long and unlikely career. As Tom Brokaw, anchoring the broadcast for NBC, reminded his viewers, Richard Bruce Cheney had "flunked out of Yale twice, went back, worked for a—as a power lineman for a year, and then ended up at the University of Wyoming, did graduate work at the University of Wisconsin, and came here and became a wonder kid, almost immediately."[6]

When Cheney returned to the White House—twenty-four years to the day after he had left—his role was entirely changed. "It's a very different situation," he says. "I'm not a staffer, I'm the vice president, a constitutional officer, elected same as he is, but my ability to contribute depends on basically my advice to him more than anything else. That has to be confidential. The first time there is a lot of coverage of the vice president urging the president to do X, that starts to put a strain on the relationship. And that linkage to him, that's crucial and needs to be protected. So I don't talk to very many reporters."[7]

The White House press shop did not encourage the new vice president to raise his public profile. Even before the inauguration, several of Bush's top advisers had privately worried that Cheney was overshadowing the president-elect.

Two days after the Supreme Court's decision affirming Bush's victory, *Saturday Night Live* lampooned the situation in a skit, with Will Ferrell playing Bush and Darrell Hammond playing Al Gore.

"Dick Cheney's going to be one tough boss," "Bush" said to "Gore."[8]

"Yeah. Actually, George, you're going to be his boss."

"Don't I wish," Ferrell responded, shaking his head.

Jokes on late-night television were one thing. Questions raised by serious journalists were another, greater concern. Just three days into Bush's presidency, Andrea Mitchell of NBC explored Cheney's power in a story on the *Nightly News*. "Suddenly, the man campaign critics once called a drag on the ticket is everywhere, creating a cabinet that some say looks more like Cheney's than Bush's. Former Cheney mentor Don Rumsfeld for Defense, Cheney ally Paul O'Neill for Treasury, Cheney Gulf War partner Colin Powell together again for State. Even before the inaugural, the question to insiders: 'Is this Bush's White House? Or a copresidency?'"[9]

Bush's advisers, particularly those he brought with him from Texas, wanted to put the issue to rest. The way to do that, they reasoned, was to keep Cheney in the background.

Such an arrangement would be difficult, if not unthinkable, in a typical White House with a typically ambitious vice president. But in this case the inclinations were mutually reinforcing: with few exceptions, the White House communications shop didn't want Cheney engaged with the media; and, with few exceptions, Cheney didn't want to spend time with reporters. He liked the backseat.

"It was a less-is-more strategy," says a senior communications official at the White House. Cheney would be used sparingly, only on big issues and in venues that would generate second-ary press coverage. His personal preference was to do television shows where he could engage in lengthy conversation with the host, and to do them live or "live to tape," so that they couldn't be edited. Because the Sunday talk shows met these criteria, Cheney would be enlisted to appear on shows like *Face the Nation* and *Meet the Press* when the administration had a particular message he could reinforce. But his overall interaction with the press would be kept to a minimum.

"It made sense in the beginning," the communications official said, "because the stories were all that he was going to be the puppeteer. We said, 'Let's have people really listen when you talk.' At some point, though, that needs to change."

It never did. What was conceived as a short-term fix would become the modus operandi for the entire Bush administration.

From the start, Cheney's low profile had the potential to cause problems. Reticent politicians are rare, and the Washington press corps abhors an information vacuum. Journalists assume, often correctly, that politicians who avoid the media are up to no good. Cheney had been the subject of mostly positive media attention for the past quarter century. But as vice president he would receive overwhelmingly negative coverage.

The consequences for the Bush administration and for Cheney would be lasting. The White House would underutilize one of its most effective spokesmen, and Cheney's reputation for competence would be tarnished. In a *Time*/CNN poll taken shortly after the inauguration, 51 percent of the respondents had a favorable impression of Cheney, and only 17 percent viewed him unfavorably—a better rating than his new boss had[10]—yet Cheney rarely appeared in the media. In the end the effort by the White House to enhance Bush's stature by diminishing Cheney's would have the opposite effect.

Several White House officials remember being surprised that Cheney was reserved internally, too. At principals' committee meetings—small gatherings at which Bush's top advisers work through issues before presenting them to the president—Cheney would offer brief introductory remarks at the beginning and, perhaps, a comment or two as the session continued. And when Cheney attended cabinet meetings or another session with Bush, he rarely said a word.

"He's very quiet in his advice," Bush says. "Dick Cheney is not an emotional arguer. He is measured, he's relatively soft-spoken. He doesn't yell; he doesn't gesture a lot, which is probably why when he speaks, people tend to listen. In other words—he's not a butt-in person, doesn't butt in a lot."[11]

When Cheney did speak, though, people listened. "He is the E. F. Hutton of the Bush administration," says one of Bush's confidants. Most important, everyone who worked with Cheney understood that he spoke for Bush.

This understanding flowed from a decision Cheney made even before he was sworn in as vice president: he would not run for president when the Bush administration ended.

"I was persuaded that the only way it would work is if it was clear there was only one agenda in the White House and that was his," says Cheney. "It was essential if I was going to be effective working for him, to be able to have his total trust and confidence, and that of the people around him, that I didn't have my own agenda, that I wasn't there trying to figure out what a particular proposal or speech was going to do for my image or my standing in the Iowa caucuses in January of '08, that I was there specifically as part of a team."[12]

This understanding would come to define the unique relationship between Cheney and Bush. "When you're getting advice from somebody," Bush said, "no matter who he or she is, if you think deep down part of the advice is to advance a personal agenda, if you're an observant person and obviously trying to do what's best for the country, you discount that advice."[13]

Others outside the administration knew that Cheney had the president's trust and that when they spoke to the vice president it was as if they were speaking to Bush himself. This trust allowed Cheney to carve out a role that allowed him remarkable power and independence. "I'm vice president," Cheney said. "I'm not in charge of anything. I don't have any line management responsibilities."[14]

Instead, he built a roster of involvements that grew directly out of his interests and experiences in government. He would work on economic policy, an interest going back to his days implementing Richard Nixon's wage and price controls. He would concentrate on energy policy, using expertise he gained on the Interior Committee and at Halliburton. He would have responsibility for evaluating the U.S. intelligence community, something he had dealt with in the Ford administration, on the House Intelligence Committee, and as secretary of defense. He would involve himself in tax cuts, the Middle East peace process, intelligence, the Anti-Ballistic Missile Treaty, and homeland defense; and he was also an active participant in Bush's initial efforts to promote the education plan known as No Child Left Behind. Education had been an important part of the presidential campaign, and although education reform would not have been an urgent issue in a Cheney administration, it was in the Bush administration.

One of Cheney's defined duties was as the president of the Senate, a role assigned in the U.S. Constitution. The vice president was given the responsibility of casting a tie-breaking vote on those occasions when the Senate was deadlocked. A sign of his formal role is that Cheney's paycheck comes from the Senate.

But if he discharged his duties in the Senate, he still thought of himself as a man of the House. Cheney had intended to spend his career in the House, and he speaks of its traditions and oddities with the peculiar reverence of a member of the club.

So, early in the Bush administration, Cheney's former colleagues in the House approached him with an unprecedented offer. Every vice president has an office in the Senate, but Speaker of the House Dennis Hastert wanted Cheney to have an office on the House side, too.

Cheney and Hastert had gotten to know each other in 1987, when Hastert, a Republican from Illinois, first arrived in Washington. Hastert needed help from the leadership on an issue in his district, and Cheney helped him deliver. Two years later, after Cheney's colleagues had elected him the Republican whip, he began attending meetings of the House Republican Steering Committee. Hastert had been on the committee and sat with Cheney to explain its group dynamics.

When Cheney was sworn in as vice president, Hastert told him that Bill Thomas had offered him premium office space. Thomas, Cheney's classmate from 1978, had risen to become chairman of the powerful Ways and Means Committee. Cheney's new office would be the old Ways and Means conference room, room 210 in the Capitol, right off the House floor.

Thomas got some grief from fellow House members about providing the vice president with the office, so he made the offer conditional. Cheney could have the office if he posed in it for a picture with Thomas and signed the picture: "Thanks for the loan of the office."

"And I want you to underline loan," Thomas told him.

Cheney agreed and the office was his.

Much of the commentary in the first days of the Bush administration concerned whether the White House, newly installed with a sympathetic Republican Congress, would make concessions on its conservative agenda. The country had been bitterly divided by the election and the subsequent recount, and many people argued that pursuing a more centrist, ecumenical agenda could have a healing effect. Cheney summed up this position as, "Gee, the election was so close. The recount went on forever. You're going to have to trim your sails from a policy standpoint. You can't go full force with what you campaigned on."[15]

President Bush rejected this argument. So did Cheney. "That simply wasn't an acceptable proposition," he says. "There wasn't any way we were going to buy that."[16]

It was an early hint of how the two men would regard official Washington and how they would govern. Another followed quickly.

As the Bush administration took office, California was in the throes of an energy crisis. The state had deregulated its whole-sale energy market five years earlier but continued to regulate retail energy. The discrepancy contributed to an imbalance in supply and demand that allowed energy companies, including Enron, to manipulate prices. Costs of essential energy skyrocketed, and shortages of available power created blackouts across the state. Governors representing western states visited Cheney to express their concerns that California's problems might soon be their own.

Within days, Alan Greenspan, chairman of the Federal Reserve, told Cheney that he, too, was worried that the shortages might sweep across the country, leaving a trail of doubts about the strength of the U.S. economy. Everyone listened when Alan Greenspan spoke, and Cheney was no exception. The two men had been close in the Ford administration, discussing policy in the office during the day and over drinks at night. Cheney thought Greenspan had one of the sharpest minds of anyone he'd met and made sure that his ideas made it to the president. He took Greenspan to see Bush in the Oval Office, and the president agreed that the White House had to act.

Cheney and Greenspan had faced such a situation before. In the spring of 1975, when the Ford administration faced an energy crisis of its own, Cheney, then deputy chief of staff, and Greenspan had casually discussed possible solutions. Within days, their brainstorming appeared in a memo to Donald Rumsfeld. The memo was dated May 27, 1975, and its subject was "Decontrol of Oil Prices."

"You should be aware of an option that Alan Greenspan and I have been kicking around on the question of decontrolling oil prices," Cheney wrote.[17] Greenspan and Cheney wanted President Ford to veto a piece of legislation that would extend the Ford administration's authority over oil prices. The lifting of such controls would be an economic shock, Cheney argued, but the long-term efficiency of the free market made removing the government the best option. The memo walked Rumsfeld and Ford through the complicated politics of oil on Capitol Hill and laid out the likely objections. "The main argument against it would be the economic impact of that rapid run up in oil prices, but Alan Greenspan is of the [opinion] that it might not be all that severe."[18]

Cheney had become a skeptic about price controls. Serving on the Cost of Living Council in the Nixon administration, he had seen firsthand the market distortions that result from such deep governmental intervention, and he had emerged convinced that markets function best when left alone. The same free-market thinking would guide Cheney's approach to energy policy in the Bush administration. When others in the administration would tilt toward government-imposed solutions, Cheney would give what one staffer calls his "markets work" lecture, a detailed historical tour of failed attempts by government to regulate the economy. President Bush would later describe Cheney as "hard core free market. Hard core."[19]

President Bush privately asked Cheney to run the administration's energy policy task force. He had his work cut out for him. California's energy problems, routinely described as a "crisis" in the news media, required urgent attention. Andrew Lundquist, a former energy expert in the Senate, was hustled into the White House. Lundquist had worked on the transition, focusing on energy issues, and two weeks before the inauguration had been offered a position on Cheney's staff overseeing energy policy. Lundquist, unsure whether he wanted the job, simply didn't respond. The day of the inauguration, he got another phone call.

"The President gets sworn in on Saturday, and on Sunday morning I'm at the White House because there's blackouts going on in California," Lundquist recalls. "Electricity prices are spiking to $1,000 per megawatt at times in California in certain places. Normally, electricity is $35 a megawatt-hour. Natural gas prices are spiking way beyond."[20]

A renovation of the White House was under way, with painters and carpet layers at work in the hallways and in many West Wing offices. The ad hoc energy policy team, including Bush's economic adviser Larry Lindsey and several others, sat on sofas around a small table in Ari Fleischer's office; having been fixed-up overnight, it was one of the few spaces that were inhabitable. Lundquist was not impressed with the meeting. Nobody in the room knew anything about the policy process at the White

House, and few had expertise in energy. This is it? he thought. These guys have no idea what they're doing.

The next day, Lundquist returned to his office at the Senate, where, after an hour, Scooter Libby called to say that Cheney wanted to see him at one-thirty PM. Lundquist assumed Cheney wanted to talk to him about coming to work in the office of the vice president. Lundquist, who had never before met Cheney, agreed to the meeting.

After a brief exchange of pleasantries, Cheney began to fire questions. Lundquist, he noticed, had worked in the oil-rich North Slope of Alaska, where the Arctic National Wildlife Refuge (ANWR) was located. The refuge had become an issue in the presidential campaign of 2000, when Bush announced that he supported oil exploration there, to the vocal distress of environmental groups.

Lundquist told Cheney that he had at one time in his life worked during summers in the North Slope to make some money. He hesitated. There was more to the story. "To be honest, Mr. Vice President," he confessed, "I got kicked out of school. I sort of took a yearlong weekend and got kicked out of school."[21]

The interview to that point had been businesslike: friendly, but not warm. But at this revelation, Cheney smiled broadly and leaned forward in his chair. I did too, he said. I got kicked out of Yale. To Lundquist, Cheney seemed to be presenting this failure almost as a point of pride.

After a sentence or two about his work laying power lines in Wyoming, it was back to the task at hand. Cheney asked Lundquist to serve as his energy adviser. Lundquist once again demurred, telling the vice president that he wanted to think about the offer. Lundquist wasn't being coy. He liked his job in the Senate—the lifestyle, the hours, the pay. If he accepted a job at the White House, it would mean much more work for much less money. He had not yet given an answer several days later, when Libby called to tell him to report to the White House on Saturday, January 27, to brief the vice president.

The White House needed to put a representative of the administration on the Sunday talk shows who could explain the energy crisis and describe the administration's plans to deal with it. Finding anyone to explain the energy problems in layman's terms was a difficult task. The White House communications team had decided to use the vice president sparingly, in emergencies. This counted as an emergency. Cheney was chosen, and Lundquist was called in to brief him.

After the briefing, Cheney took Lundquist aside. The president has asked me to chair a task force that will direct energy policy from the White House, he explained, and I want you to run it. It was much bigger position than the one they had discussed. Lundquist would be the lead staff member of the White House team assigned to solve the administration's first major crisis. He agreed to take the job.

On January 29, at a meeting about energy policy held in the Cabinet Room, the president publicly named Cheney to lead the development of the new energy plan. Politically, it was not a great place to start. "Today's assignment puts Mr. Cheney at the center of an issue he knows well as the former chief executive of the Halliburton Company, the oil services giant," the *New York Times* reported. "Halliburton and its rivals would all probably benefit from an aggressive government push to promote exploration and to weaken environmental controls."[22]

It was a fair point. After all, both Bush and Cheney had a background in the oil industry, as Cheney had pointed out when he listed his weaknesses as a potential running mate. From the start, critics of the Bush administration hinted that Cheney's project was simply a payback for generous campaign contributions from the energy industry.

Lundquist would be one of the main players in the unexpectedly long life of the Energy Task Force. The other was David Addington, Cheney's general counsel. In the nearly twenty years of their association, Addington had served as a special assistant when Cheney was secretary of defense, had run Cheney's presidential PAC, had worked on the three-man vice presidential search committee, and had been the de facto chief of staff of Bush's transition in 2000. Cheney trusted Addington more than anyone else other than his own immediate family.

In their first meeting after the inauguration, Cheney told Addington he had one job: restore the power of the presidency.

Over the course of a four-year presidency there would be count-less junctures at which the strength of the executive could be in-creased or depleted. He wanted Addington to be aggressive.

On the day Bush announced the Energy Task Force, Adding-ton wrote a memo explaining how to assemble the group so as not to violate an obscure law called the Federal Advisory Committee Act (FACA). A failure do so, Addington understood, could leave the task force open to demands for documents from Congress—the fate that had helped cripple Hillary Clinton's Health Care Task Force. It was a small but important catch.

Clinton's task force had included people who were not gov-ernment employees, and its meetings had thus been subject to strict open-government regulations. Cheney's Energy Task Force was made up entirely of government employees—six cabinet sec-retaries and some staffers. The meetings this group held to solicit the views of outside experts, Cheney and Addington would ar-gue, were not subject to disclosure rules because those outside experts were not members of the task force. The fight about what this distinction meant would end in a bitter dispute about execu-tive power and, eventually, a ruling from the Supreme Court.

The actual work of this task force lasted less than four months, during which it occupied no more than 10 percent of Cheney's time. Lundquist had been working on a broad outline for a na-tional energy policy for eighteen months in the Senate before he came to the White House. It lacked muscle and flesh, but the bones were there: increase energy output, diversify sources, reward in-novation. Not surprisingly, Lundquist's ideas were close to Bush and Cheney's campaign promises. On the stump, Cheney had repeatedly assailed Clinton and Gore's energy policies and had attacked Gore's focus on renewable energy, during the campaign, as unrealistic. The United States, Cheney said, needs to produce more energy. "We have not built a new refinery in this country in more than ten years," he told an audience in Spokane, Washing-ton, two weeks before the election.[23] In a speech at General Mo-tors headquarters in Michigan, Bush savaged Gore for behaving as if "the engines that power your cars are his enemy. . . . Unlike Al Gore, I don't consider the internal combustion engine a threat to the future of mankind."[24]

Lundquist began to meet with a wide variety of outside groups to discuss the work of the task force. He viewed the meetings as necessary but almost never helpful, intended to explain the process to concerned parties rather than to solicit ideas. He and his staff were checking boxes—meet with industry leaders to satisfy the business community, meet with proponents of solar energy to please advocates of alternative energy, meet with environmental groups to please the media.

The environmental groups complained loudly that their views were not being heard and that the task force was spending too much time with representatives of the energy industry, many of them major donors to Bush's campaign. And in reality, there was no question that the task force spent more time in meetings with energy interests than with environmentalists. The question was how much these meetings influenced policy.

Much of the media coverage of the task force seemed to assume that the answer was obvious: quite a bit. But for Lundquist and others on Cheney's staff, the meetings had been largely political, and the content of the plan a foregone conclusion. The meetings might produce an idea or two—or cause the staff to reconsider something already discarded—but nothing that came from them was going to have any real influence in shaping the final policy, much of which had come from the plan Lundquist had drafted more than a year earlier.

This was particularly true of the meetings with environmental groups, many of whom had campaigned actively against Bush and Cheney. "I'm not going to sit through 100 meetings with environmentalists to hear the same thing that just ain't going to work," says Lundquist. Had he proposed what environmentalists had recommended, Lundquist says, "the president and vice president would have fired me."[25]

The report of the task force was released on May 19, 2001. Cheney laid out the goals of the plan in a letter to Bush. "To achieve a 21st century quality of life—enhanced by reliable energy and a clean environment—we must modernize conservation, modernize our infrastructure, increase our energy supplies, including renewables, accelerate the protection and improvement of our environment, and increase our energy security."[26]

"In the end," said Tom Curry of NBC, "many of the task force's recommendations for expanding energy supplies were, not surprisingly, the same ones Bush had made during the 2000 campaign."[27]

Among its many recommendations, the task force called for greater capacity to produce nuclear power, an option that had been virtually off the table for the eight years of the Clinton administration. Cheney had consistently backed nuclear energy as a way to address the growing energy needs of the country. As a member of Congress, he had served on the committee investigating the accident at Three Mile Island in 1979 and had defended the long-term safety record of the nuclear industry. But a front-page story in the *New York Times* suggested that nuclear power might have been left out of the task force's recommendations if it hadn't been for a successful intervention by executives of nuclear power companies. "As the White House was putting together the energy plan that President Bush released last week, there had been almost no talk of nuclear power as a component of the nation's energy strategy," the *Times* reported. But after those executives met with Lundquist and Karl Rove on March 20, "a surprising thing happened." The White House began talking about nuclear power.

The *Times* quoted Christian Poindexter, chairman of the Constellation Energy Group, who was eager to boast about his successful lobbying. "It was shortly after that, as a matter of fact I think the next night, when the vice president was being interviewed on television, he began to talk about nuclear power for the first time."[28]

It was a good story, but it was highly misleading. The task force planned to include nuclear power in its recommendations, and the secretary of energy, Spencer Abraham, had said as much the day before the allegedly decisive meeting between the lobbyists, Lundquist, and Rove: "There hasn't been a new nuclear power plant permit granted since 1979," Abraham complained in a public speech on March 19, 2001, to the U.S. Chamber of Commerce Energy Summit. "Many of the existing 103 nuclear plants are not expected to file for a renewal of their licenses as they expire over the next fifteen years."[29]

Abraham, a member of Cheney's task force, made clear in the same speech that nuclear power would be included in its recommendation. "Our national energy policy will stress the need to diversify America's energy supply. It will be founded on the understanding that diversity of supply means security of supply, and that a broad mix of supply options, from coal to windmills, nuclear to natural gas, will help protect consumers against price spikes and supply disruptions."[30]

Meanwhile, disputes over access to the task force and its records had been playing out in a flurry of letters back and forth between Cheney's office, Democrats on Capitol Hill, and the General Accounting Office (GAO), the investigative arm of the legislative branch. On April 19, two Democrats—Henry Waxman and John Dingell—wrote to Lundquist and to David Walker, comptroller general of the GAO. "We have become concerned about the conduct and composition of the task force," the congressmen wrote in a letter requesting access to the task force's records. "We question the apparent efforts of the task force to shield its membership and deliberations from public scrutiny."[31]

Three weeks later, David Addington wrote back, refusing to provide many of the documents the GAO requested. His argument was based in part on the fact that the task force was not subject to the disclosure requirements covered under FACA, the law he had warned about earlier, in January. It was the first exchange of many.

The Energy Task Force was one of the specific policy assignments Cheney took on at the very beginning of the Bush administration.

In the early months of his vice presidency, Cheney let his staff drive the Energy Task Force, while he focused on other priorities: intelligence, strategic threats, and what would become known as homeland security. When the White House national security team surveyed the state of the world in the days after January 20, 2001, several areas were instantly identified as priorities: China, the Israeli-Palestinian peace talks, and Iraq. Cheney thought the Clinton administration had mishandled Iraq policy, allowing Saddam Hussein to blame the United States and UN sanctions for the suffering of the Iraqi people. The inspection regime had broken down, France and Russia were openly embracing Baghdad, Saddam Hussein's rhetoric had become increasingly belligerent, and Clinton had done nothing but issue empty threats. All this, Cheney believed, made the United States look weak.

But considerations of American prestige paled beside the long-term threat posed by the Iraqi regime. During the campaign, George W. Bush had said that if Saddam Hussein was developing weapons of mass destruction (WMD), the United States would "take him out."[32] When Bernard Shaw asked the Cheney whether he agreed with "such a deadly policy," Cheney responded forcefully.

Uncertainty about Iraq, he had said, is dangerous. "I also think it's unfortunate that we find ourselves in the position where we don't know for sure what might be transpiring inside Iraq. I certainly hope he's not regenerating that kind of capability, but if he were—if, in fact, Saddam Hussein were taking steps to try to rebuild nuclear capability or weapons of mass destruction, you'd have to give very serious consideration to military action to stop that activity. I don't think you can afford to have a man like Saddam Hussein with nuclear weapons, say, in the Middle East."

So reworking the policy toward Iraq—in many ways starting over—was "a top priority," says one of Cheney's advisers on foreign policy. The new secretary of state, Colin Powell, wanted to respond to international concerns that UN restrictions were punishing the Iraqi people rather than their leader. He was advocating "smart sanctions," a new policy that would, in theory, tighten strictures on military equipment but loosen them on humanitarian goods. The changes would help America's image in the region: the United States would appear sensitive to the Iraqi people while getting tough with their leader.

Cheney didn't buy it, and neither did his friends at the Pentagon. Paul Wolfowitz, recently installed as deputy secretary of defense, wanted regime change. He did not support an invasion of Iraq to overthrow Saddam, but he favored doing just about anything short of that. In particular, Wolfowitz wanted to expedite support to Iraqi opposition groups.

This backing had been approved in 1998 with the passage of the Iraq Liberation Act, a law that also made "regime change" the official policy of the United States. In February 1998, in a speech about the dangers of Iraq's weapons of mass destruction, Bill Clinton warned of "an unholy axis of terrorists, drug traffickers, and organized international criminals" and said "there is no more clear example of this threat than Saddam Hussein."[33] Top officials from his administration frequently related their concerns that Iraq could provide WMD to terrorists wanting to strike in the United States. But the Clinton administration had slow-rolled the funding for the Iraqi opposition. The consensus at the State Department was that opposition figures, especially Ahmed Chalabi, the leader of the Iraqi National Congress, were untrustworthy. The CIA agreed. In 1996, Chalabi had sought to foment a coup, with the CIA's involvement, and it had gone bad. "The agency was working it hard, but they were told, I've heard, in '96, the operation they were trying to set up in the northern part of Iraq had been penetrated by Saddam and that they were about to get nabbed—and they did," says an administration official. "The CIA got burned and never recovered."

As the interagency debate continued over the summer, the administration had begun to shape a policy. Through both overt and covert means, the United States would pressure the Iraqi regime by exploiting existing fissures and encouraging new ones, and would fund and train the Iraqi opposition. The policy would be regime change, but it would be gradual; pressure from the outside of the hardened regime would encourage unrest on the inside.

Cheney agreed with Bill Clinton's statements on the evolving threats to the United States—terrorists, rogue states, and WMD—even if he believed the previous administration had done too little to eliminate them. Several members of Cheney's senior staff, including Scooter Libby, had spent time studying how best to defend the U.S. homeland. The vice president shared their interest, and it soon became an official part of his portfolio.

On May 8, Bush publicly announced that Cheney would be leading a comprehensive review of U.S. government homeland security, which the official White House statement described as an effort to ensure "domestic preparedness against weapons of mass destruction." The president noted that an increasing number of terrorist-friendly countries had acquired WMD and "some nonstate terrorist groups have also demonstrated an interest in acquiring weapons of mass destruction."[34] Bush declared that his administration would pursue policies designed to slow the proliferation of WMD and deter their use. Cheney's project, however, would focus on the governmental response after an attack, known in the national security bureaucracy as "consequence management." It was neither offensive nor defensive, but responsive.

A new Office of National Preparedness, housed in the Federal Emergency Management Agency, would coordinate the federal response to any WMD attack. As a practical matter, the announcement meant that Cheney's staff would continue to examine existing plans and offer new proposals. Their report would be due in October.

Cheney discussed this work and the threats that inspired it in May 2001, in an interview with Nicholas Lemann of *The New Yorker*. Lemann asked Cheney whether it was possible to talk about "the threat" in the same way it had been during the cold war.

"No, the threat's much different today, I think," Cheney said. He reaffirmed America's obligation to defend its allies and prevent regions of strategic importance from domination by hostile powers. Then Cheney defined the new threats.

"You've still got to worry a bit about North Korea," he said. "You've got to worry about the Iraqis, what ultimately develops in Iran." It was the "axis of evil" before Bush gave it that name.

He continued: "But beyond that, in terms of a threat to the U.S., and our security, I think we have to be more concerned than we ever have about so-called homeland defense, the vulnerability of our system to different kinds of attacks. Some of it homegrown, like Oklahoma City. Some inspired by terrorists external to the United States—the World Trade towers bombing, in New York. The threat of terrorist attack against the U.S., eventually, potentially, with weapons of mass destruction—bugs or gas, biological, or chemical agents, potentially even, someday, nuclear weapons. The threat of so-called cyberterrorism attacks on our infrastructure, obviously very sophisticated in terms of being based on our intelligence infrastructure."

Asked about efforts to diminish those threats, Cheney spoke of intelligence and preemption. "In terms of the threats to the United States, the terrorism of various kinds, probably intelligence is your first line of defense. . . . You need to have very robust intelligence capability if you're going to uncover threats to the U.S., and hopefully thwart them before they can be launched."[35]

Cheney said that he was particularly worried about the potential that an American city could be subject to an attack with a "dirty bomb," a crude nuclear device with potentially devastating consequences. He had tried to predict the aftermath of such an attack as a participant in the continuity of government exercises in the 1980s. When homeland security experts ran simulations of a dirty-bomb attack, they counted fatalities in the hundreds of thousands. A city victimized by such a detonation would be rendered uninhabitable for years to come, and the few who survived would have debilitating medical problems.

Cheney, normally taciturn, would become animated discussing such doomsday scenarios. One Fourth of July in Jackson, some family members gathered around the big-screen television Lynne had gotten her husband. They started to watch *The Peacemaker,* an action movie from 1997 starring George Clooney and Nicole Kidman. The film follows the two main characters as they track terrorists trying to bring a stolen Russian nuclear weapon across the Atlantic to attack New York City. Cheney watched with rapt attention. When Lynne, who had not been watching with the group, wandered into the room, she casually asked what was happening in the film. Her question was not addressed to anyone in particular, but Cheney excitedly answered before the others could speak up. "He gave a ten-minute, scene-by-scene synopsis of the action," recalls Mark Vincent, Lynne's brother.[36]

His sister interrupted to clarify her question. "What's happening *now*?"

If officials in the Bush administration were worried about responding to future attacks, they were not particularly concerned about retaliating for past ones. Despite all the campaign rhetoric from both parties about the heavy price to be paid for the bombing of the USS *Cole*, the Clinton administration dithered and the Bush administration essentially dropped the issue.

Within a month of the attack, Sandy Berger, the national security adviser, and Richard Clarke, the NSC's director of counterterrorism, reported to Clinton that although they could not show that Osama bin Laden was personally involved, there was no question that al Qaeda was behind the attack.[37] Clinton did nothing.

An editorial in the *Washington Post* ten days before Clinton left office noted: "Yemeni officials say they have developed substantial evidence that the bombing was ordered by the Saudi-born terrorist Osama bin Laden, and financed and coordinated by Muhammad Omar al-Harazi, a bin Laden associate." The editorial concluded that the Bush administration "will also need a coherent strategy for countering Mr. bin Laden; clearly, it will not inherit one."[38]

Richard Clarke, who was held over from the Clinton administration to work on Bush's NSC, wrote a memo on January 25, 2001, to the national security adviser, Condoleezza Rice, calling for the development of a strategy to fight al Qaeda. Among the items Clarke raised was "when and how does the new Administration choose to respond to the attack on the USS Cole?" But Clarke recommended that any decision on a proper response could be delayed. "On the Cole, we should take advantage of the policy that we 'will respond at a time, place and manner of our own choosing' and not be forced into knee-jerk responses."[39]

Cheney doesn't remember participating in any discussions about retaliating for the attack.

"I don't recall spending a lot of time on the USS *Cole* issue," he says. "I mean it was something that had happened previously, that October, I guess, of 2000. I don't recall it cropping up in policy debates or discussions."[40]

The plans for the next attack by al Qaeda were in their final stages by the summer of 2001. The director of the CIA, George Tenet, would later say, "The system was blinking red." But although Cheney says that "al Qaeda was clearly known as a threat," he doesn't remember being particularly alarmed by either the nature or the volume of the intelligence.

Dick Cheney (*left*) with his younger brother, Bob, after a successful fishing expedition in 1952. (*Photograph courtesy of the Cheney family*)

Dick Cheney and Lynne Vincent shortly before they were married.
(*Photograph courtesy of the Cheney family*)

Dick and Lynne Cheney on their wedding day, August 29, 1964.
(Photograph courtesy of the Cheney family)

Cheney and Donald Rumsfeld brief President Gerald R. Ford in the Oval Office.
(Photograph courtesy of the Gerald R. Ford Library)

Cheney argues with Henry
Kissinger, secretary of state
and national security adviser,
as President Ford looks on.
*(Photograph courtesy of the Gerald
R. Ford Library)*

Newly elected
congressman Dick
Cheney takes notes
in his office in 1978.
*(Photograph courtesy
of Jim Steen)*

Cheney, the second-ranking
Republican in the House of
Representatives, listens as
President Ronald Reagan
speaks to members of the
congressional leadership.
*(White House photograph
courtesy of Jim Steen)*

Cheney stands with President George Bush and Vice President Dan Quayle shortly before he is sworn in as secretary of defense in March 1989. Little more than a decade later, Cheney would ascend to the office that both Bush and Quayle had held previously.
(Department of Defense photograph courtesy of Jim Steen)

Secretary of Defense Dick Cheney and Chairman of the Joint Chiefs of Staff Colin Powell speak to U.S. troops in Saudi Arabia during the first Gulf War. *(Department of Defense photograph courtesy of Jim Steen)*

Chief Justice William Rehnquist swears in Dick Cheney as the forty-sixth vice president of the United States. Liz Cheney holds the Bible, as Lynne Cheney and Mary (obscured by Rehnquist) watch. *(Photograph by David Bohrer)*

Dick and Lynne Cheney in the Presidential Emergency Operations Center on September 11, 2001. *(Photograph by David Bohrer)*

Senior White House staff watch television coverage of the South Tower collapsing during the attacks of September 11, 2001, in the Presidential Emergency Operations Center. *From left to right:* Josh Bolten, deputy White House chief of staff; Norman Mineta, secretary of transportation; Cheney; I. Lewis "Scooter" Libby, assistant to the president and Cheney's chief of staff and assistant for national security affairs; Condoleezza Rice, national security adviser; and Mary Matalin, counselor to the vice president. *(Photograph by David Bohrer)*

Cheney meets with senior White House staff and legal advisers on September 12, 2001. *From left:* Andy Card, White House chief of staff; Scooter Libby; Eric Edelman, principal deputy assistant to the vice president for national security affairs; Tim Flanigan, deputy counsel to the president; Alberto Gonzales, counsel to the president; and David Addington, counsel to the vice president. *(Photograph by David Bohrer)*

Cheney and George W. Bush talk during a rare ride together in the presidential limousine in Waco, Texas, on August 13, 2002. *(Photograph by David Bohrer)*

Cheney plays "follow the leader" with two of his granddaughters on Easter Sunday, April 20, 2003, as Lynne Cheney and Liz Cheney look on. *(Photograph by David Bohrer)*

Cheney takes notes aboard Air Force Two in New Windsor, New York, on December 13, 2003, upon learning that American soldiers have captured Saddam Hussein in Iraq. *(Photograph by David Bohrer)*

Cheney speaks to soldiers at al Asad Air Base in Iraq on December 18, 2005. *(Photograph by David Bohrer)*

Aboard Air Force Two on January 15, 2006, Cheney pokes fun of his image by posing with the cover of the *U.S. News & World Report* that labels him "Tough Guy." *(Photograph by David Bohrer)*

"My impression at the time, and as I've looked back on it, was not that an attack was imminent. There was noise in the system. But it's the nature of the business. If somebody comes in and said, 'OK, they're going to strike us at two o'clock tomorrow, and here's where they're going to hit.' OK, it's almost never like that. And it wasn't like that in the summer of 2001."[41]

"We were concerned about terrorism. Al Qaeda was on the horizon out there," he adds.[42]

"In terms of our strategic thinking I don't think there's any question but that we were headed in the right direction. There wasn't any intelligence on this particular attack."[43]

So Cheney went about his day-to-day work, and traveled around the country to promote the administration's energy plan. He made numerous speeches throughout the spring and summer and into the early fall.

David Addington, meanwhile, continued to battle the GAO. The more he refused to provide the task force's records and documents, it seemed, the more interested the GAO became. On September 7, 2001, the GAO's comptroller, David Walker, released a statement that reflected his determination to obtain the paperwork. "This is a very serious matter with significant potential implications for GAO, the Congress and the American people," he said. "We are finalizing our discussions with key Congressional leaders and are preparing for possible litigation."[44]

When it was clear that the GAO was not going to let the issue drop, the vice president began spoiling for a fight: "Where do we sign up, guy?"

He adds: "I looked at it, looked to me like it was a pretty clear-cut case. We had in fact done what we said we were going to do, and we weren't subject to the advisory committee act. And so we fought it."

Cheney's position created problems for the White House communications shop. Every day press officials fielded calls from reporters who wanted to know why the administration was refusing to release the documents. If the task force had truly done nothing wrong, as the White House continued to insist, why not simply disclose the meetings and make the controversy go away?

Andrew Lundquist agreed. He wanted to disclose everything. Make it all available, he argued, and let people use their own judgment. Lundquist understood the principle behind the stand Cheney and Addington were taking, but the controversy was hitting home personally. "I wanted them to put it out there, and I had the most in jeopardy because I ran those meetings," he says.[45] When he broached the subject with the vice president, he says Cheney responded sternly: Don't ever suggest that to me again.

Cheney explains his thinking this way. "One of the things over the years that I've been frustrated by is that government often took the easy way out. Somebody would come along and issue a subpoena or file a lawsuit—they'd cave, give it up. And I didn't like to do that. It struck me as an opportunity to fight all the way, and the president backed me up on it."

So fight it all the way they did, signaling to the world the administration's willingness to fight on behalf of executive power.

"The Bush administration is mounting an aggressive campaign to restore the White House's authority to keep internal information confidential, after the principle of executive privilege took a beating during clashes between President Bill Clinton and Republicans in Congress," said a story in the *Washington Post*.[46]

It wasn't just Cheney's office. Attorney General John Ashcroft was resisting demands from congressional Republicans to produce documents related to campaign finance scandals dating back to the Clinton administration. It was an interesting position for an attorney general whose nomination had been criticized because he was regarded as too partisan. Now, rather than take an opportunity to embarrass the Clinton administration, the Justice Department argued against disclosing the documents on the grounds that the papers were "deliberative documents" and thus not subject to congressional scrutiny.

Abner Mikva, who had been the White House counsel during Clinton's presidency, expressed skepticism that the Bush administration would have much success in reclaiming executive power. "If there's anything that Congress really wants, whatever the status of the law of executive privilege, I don't think a president can withstand the political pressure put on to turn things over," he said.[47]

Mark Rozell, a professor at Catholic University who in 1994 wrote a book called *Executive Privilege*, told the *Washington Post* that defending the Clinton administration made sense politically. But in order to bolster its position on executive power in the law, the Bush administration would be wise to find "something really big, like a national security issue."

His comments were published on September 10, 2001.

CHAPTER ELEVEN

September 11, 2001

O n September 9, 2001, Dick Cheney started his work week at a Sunday barbecue for Prime Minister John Howard of Australia, who was then making his first trip to Washington since Bush took office. It was good turnout for Howard: Secretary of State Colin Powell and Secretary of Defense Donald Rumsfeld joined Cheney for the gathering at the home of Michael Thawley, Australia's ambassador to the United States. Along with a selection of Australian wines, the guests had their choice of shrimp, beef, or barbecued lamb.[1]

On September 10, Cheney addressed the Southern Governors' Association in Lexington, Kentucky. His fourteen-minute speech touched on the struggling economy, because it was in the news, but he focused on the energy policy that he and the Energy Task Force had developed the previous spring.

"I do not accept the false choice between more energy and a cleaner environment," he said. New drilling technology, he said, has made it possible for "oil production to go literally unnoticed and the habitat to be undisturbed."[2]

It was not a hard sell. Most of the governors assembled at the Marriott Griffin Gate Hotel had come to the meeting already supportive of Bush and Cheney's energy plan. And there was strong agreement with Cheney's assertion that "wiser use of energy and more diverse supplies here at home will make us that much less dependent on overseas suppliers."[3]

Instability in the Middle East, Governor Mike Foster of Louisiana told the *New Orleans Times-Picayune,* is "the one thing that would kill this economy deader than a doorknob."[4]

Governor Don Sundquist of Tennessee, also a Republican, echoed that sentiment in a way that made it clear he would not be seeking a diplomatic post after he finished his term. "I don't want to be held hostage by some of those nuts in the Middle East."[5]

President Bush was in Florida that week to promote his proposals to improve elementary and secondary education in a campaign his communications team had designated "Putting Reading First Week." The goal was to shift the national discussion to education and to highlight Bush's plans in a series of speeches and photo opportunities. It wasn't working.

"The president came to Florida today to talk about education," said NBC's David Gregory in a broadcast on September 10. "His advisers know, with the economy in a tailspin, Mr. Bush's initiatives on this and other issues are drowned out—the attention now on what the president can and should be doing to turn the economy around."[6]

And in Washington, the political class was buzzing about revelations in a forthcoming book by the reporter David Kaplan of *Newsweek* called *The Accidental President.* Excerpts from the book, which recounts the battle over the presidential election of 2000, were published in the issue of *Newsweek* on newsstands September 10, 2001. According to Kaplan, Justice David Souter, appointed by President George H. W. Bush, told a group of private school students that he could have flipped Justice Anthony Kennedy if he had only had "one more day" before the Supreme Court made its decision in *Bush v. Gore.* The result might have been President Al Gore. The national postelection hangover continued.

Shortly before seven o'clock the next morning, September 11, Cheney sat for his regular CIA briefing in the small first-floor library of the vice president's residence. The session was unremarkable. On a typical day, Cheney gets into the car waiting to take him to the White House at seven-thirty. In a six-car motorcade that races across downtown Washington, D.C., blowing through stoplights, it takes five minutes to cover the three miles from Cheney's home at One Observatory Circle to the White House. When he arrives, he joins President Bush for his intelligence briefing at seven forty-five.

Cheney's solo briefing is more detailed than Bush's because the vice president asks for more material. Bush is the big-picture guy; Cheney wants details. The vice president will sometimes ask questions in his briefings with Bush to make sure the president is exposed to in-depth treatment of issues Cheney deems important.

On this day, with Bush on the road, there was no intelligence briefing at the White House.

Cheney met briefly in his West Wing office with Scooter Libby. Cheney was wearing a gray pinstripe suit, a crisply pressed white shirt, and a black tie with a silver linked-chain pattern. The vice president has offices in both the White House and the Eisenhower Old Executive Office Building, with most of his staff located in what President Bush calls "The Ike."

When Libby returned to his office in the Old Executive Office Building, Sean O'Keefe, the number two official at the Office of Management and Budget, stopped by Cheney's office for an unscheduled visit. O'Keefe, a tall, slender man whose graying hair and a push-broom mustache give him a striking resemblance to Peter Sellers in *The Pink Panther*, had worked closely with Cheney at the Pentagon as the military's chief financial officer and comptroller. Though he considered Cheney a friend, he knew better than to drop in for an idle chat. Cheney was accessible— he deliberately left gaps in his schedule for "staff time"—but his colleagues at the White House quickly learned to keep their impromptu sessions with him short and to the point.

Nonetheless, O'Keefe spent more than twenty minutes in Cheney's office, discussing a matter that seemed urgent at eight-thirty AM on Tuesday, September 11. In time, neither man would be able to recall what it was that had been so important.

Also waiting to speak to Cheney that morning was John McConnell, the vice president's chief speechwriter. McConnell—an unassuming, erudite man from Bayfield, Wisconsin, a town of 625 at the northern tip of the state—has a square jaw softened by his prominent dimples and easy smile. He had arrived early at his office in the Old Executive Office Building, as usual.

The previous day, McConnell had casually mentioned to Cheney's assistant Debbie Heiden that he wanted to see the vice president. He was preparing for a series of upcoming speeches and needed to discuss the broad themes with his boss. McConnell, who likes to keep a low profile, figured he would just try to catch Cheney in his office for a brief chat as the vice president read the morning papers; instead, he was surprised to find he had been given a one-on-one appointment for eight-thirty that morning. Arriving shortly before his designated time, McConnell found Cheney and O'Keefe deep in discussion, so he waited just outside Cheney's office.

The office—longer than it is wide, with high ceilings—is quite spacious, particularly for the cramped West Wing. Cheney's large mahogany desk sits opposite the entrance to the room, beneath a map depicting the Battle of Chickamauga, one of several battles that his great-great-grandfather survived. In the far left corner of the room is a small television perched on a walnut-stained stand that matches the desk. Four flags sit on either side of a thin table behind Cheney's chair: the American flag and the Wyoming state flag on the left; the secretary of defense's flag and the vice president's flag on the right. The table is filled with photographs of Cheney's wife, his children and their partners, and his grandchildren. The white walls are spare, with only two gold-framed oil paintings—of Thomas Jefferson and John Adams—and a map of the world as decorations. Two windows on the right side of the room provide light. The royal blue sofa in front of Cheney's desk matches the carpet exactly. Two chairs on each side of the sofa form a semicircle facing out from the fireplace by the door. Sewn into the carpeting directly in front of the fireplace is the seal of the vice president of the United States.

As McConnell waited for Cheney to finish his conversation with O'Keefe, he chatted with the Secret Service agent posted at the door and with Debbie Heiden, whose desk is just outside of the office. Their conversation was interrupted by a bizarre news report on the television over Heiden's desk. An airplane had struck the World Trade Center. No one knew what to make of it.

An inexperienced pilot? A wrong turn? Bad weather? The last of these seemed unlikely; it was a beautiful day in New York City, as it was in Washington.

"There wasn't any kind of alarm," says McConnell. "It was just kind of, 'Oh man, look at that.'"[7]

As they watched the television, the Secret Service agent received an urgent call from "the ID"—the intelligence division. He listened for a moment and then hung up.

"He put the phone down and told me: passenger jet. And that's when you go, Geez. And then you start getting a sick feeling. Because a passenger aircraft is *not* going to crash into the World Trade Center."[8]

As O'Keefe came out of Cheney's office, McConnell gestured to the television. O'Keefe nodded; they had been watching the reports inside. McConnell walked in through the door with a stack of papers under his arm and took a seat at the right side of Cheney's desk. The small television on the other side of the desk was tuned to ABC News. The two men watched the fiery scene without saying a word.

In the Old Executive Office Building, Scooter Libby's young assistant Jennifer Mayfield was also monitoring the developments in New York on her television. Libby had just begun a meeting with John Hannah, a national security adviser to Cheney. President Bush had decided to support the creation of a Palestinian state—a major change in U.S. policy. The vexing details of the coming announcement had been discussed for weeks without a resolution. Libby wanted desperately to come up with guidance for the vice president by the end of the day. Before the meeting started, he had given Mayfield strict orders: do not interrupt it.

As soon as she saw that a plane had hit the World Trade Center, Mayfield charged into her boss's office. Libby and Hannah listened to her report. "Unless it's terrorism," Libby responded, "don't interrupt me again."[9]

At the White House, Cheney watched the screen as thick gray smoke poured from the hole in the North Tower. He, too, noticed that it was a clear day in New York. "How in the hell," he asked himself, "could a plane hit the World Trade Center?"

Seconds later, he had his answer. Cheney and McConnell

watched as a second plane appeared on the right-hand side of the screen, banked slightly to the left, and plunged into the South Tower. "Did you see that?" Cheney asked his speechwriter.[10]

In the picture on ABC News, the North Tower largely blocked their view of the plane hitting the South Tower. Still, they were able to see an enormous blast of smoke behind the North Tower and debris from the explosion falling to the ground below.

"We knew then it was terrorism," Cheney recalls.

Jenny Mayfield raced back into Libby's office and told him: "It's terrorism." Within seconds, Libby received a call from Cheney summoning him to the White House.

Moments after the second plane hit, Cheney "popped out of his chair," says McConnell, and walked across the hallway to the office he had occupied as chief of staff in the Ford administration. The current occupant, Andy Card, was traveling with President Bush in Florida. Cheney told Card's secretary that he wanted to speak with Card when Card called back to the White House. Then Cheney returned to his office.

Libby hurried back across West Executive Avenue—the alleyway between the White House and the Old Executive Office Building—to rejoin his boss. He carried an oversize black briefing book under his right arm. Cheney sat behind his desk, leaning back in his large black chair.

Libby's eyes shifted between his boss and the TV to Cheney's right. The bottom of the screen read simply: "ABC News Live Coverage. World Trade Center. New York." The picture—from a different angle now—showed perhaps the top fifty floors of the towers, both engulfed in smoke. The tower on the left-hand side of the screen had a higher point of impact, and there were flames licking skyward from the top of the building.

The morning newspapers, now completely irrelevant, sat next to a thick rubber band on the left side of Cheney's desk. A coffee mug was on the desk in front of Cheney; the television remote control was just to his right.

Mary Matalin, Cheney's top communications adviser, joined the small group around the television. "Is this terrorism?" she asked. Cheney told her that it was.

Liz Cheney, having heard about the first plane hitting the

North Tower on her car radio, called her father on his private line.

"A plane has hit the World Trade Center," she told him.

"Two planes have hit," he responded. "It's a terrorist attack. I've got to go."

Cheney picked up a phone with a direct line to Bush. "I need to talk to the president," he said, and immediately hung up.

As word of the attacks spread throughout the West Wing, many White House officials migrated to Cheney's office: Sean O'Keefe; Condoleezza Rice; Josh Bolten; and the director of counterterrorism, Richard Clarke, among others.

McConnell had never before seen Clarke. "I took it that he was one of these shadowy guys from the sit room"—the White House situation room, the hub of wartime activity—"who you don't really know but who are there doing serious and important work. And I thought this is a guy from the sit room and he's going to go down to the sit room and get people together."[11]

At nine-fifteen AM, President Bush called Cheney. The vice president, sitting at his desk, turned away from the crowd gathered in his office. He spoke to Bush in a voice so soft it was not audible to McConnell, who was leaning on Cheney's desk as he watched the television. Bush and Cheney assessed the situation and discussed what the president would say in his public statement. Better to be cautious, they agreed, and decided that Bush would speak of "an apparent" act of terrorism.

McConnell decided to leave.

Bush spoke to the nation at nine-thirty. He began by saying that he would be flying back to Washington immediately after his remarks.

Today, we've had a national tragedy. Two airplanes have crashed into the World Trade Center in an apparent terrorist attack on our country. I have spoken to the vice president, to the governor of New York, to the director of the FBI and have ordered that the full resources of the federal government go to help the victims and their families, and— and to conduct a full-scale investigation to hunt down and to find those folks who committed this act.

Terrorism against our nation will not stand. And now if
you will join me in a moment of silence. May God bless the
victims, their families and America. Thank you very much.

Moments later, at 9:36 AM, Cheney was standing at the end
of his desk watching television when Jimmy Scott, one of the
agents on his Secret Service detail, hurriedly approached. The Secret Service had gotten word that an unidentified airplane—later
identified as American Airlines Flight 77—was heading toward
Washington. "Sir," Scott said in a firm voice, "we have to leave
now." Before Cheney had a moment to stall, the agent made it
clear that this was not a negotiable matter.

"He put his hand on my shoulder and grabbed me by the back
of my belt," says Cheney, "and propelled me out the door. They
must practice it."[12]

On the way out of his office, Cheney, who never likes to be
caught without reading material, snatched a copy of *The Economist* from a table. The issue, dated September 8, 2001, had a
black cover with small red-and-white lettering: "Why Branding
Is Good for You."

Cheney learned two things in the minute it took the Secret
Service to move him from his office to a secure area underneath
the White House: the Pentagon had been hit and the White House
was next.

McConnell, meanwhile, had returned to his office across the
alleyway. He called his mother in northern Wisconsin. She was
watching news reports about the attacks and told him that the
White House complex had been evacuated. He laughed nervously and told her that he knew this wasn't true because he was
calling her from his office. She was not amused. John, go out
into the hall and see if anyone else is there, she said, exasperated.
Dozens of staffers were hustling down the long hallways to leave
the building. McConnell joined them.

David Addington, counsel to the vice president, had been in
his office, just down the hall from McConnell's, when the second
plane hit. Addington's first move was to place a call Jim Haynes,
general counsel at the Pentagon, who was an old friend, to make
sure that the THREATCON had been raised.

He then began to evacuate the members of Cheney's staff, going from office to office to make sure that those with highest clearances had locked the safe for storing classified material. As he exited the building near the corner of Seventeenth Street and Pennsylvania Avenue, Addington noticed a rapidly growing number of men in black carrying large weapons.

He waited for several minutes and then tried to use his White House ID to reenter the complex. No chance. A Secret Service agent told him that no one was being allowed in. He tried several other entrances without success. With the White House locked down, he decided he would walk the four miles to the Pentagon, and if he couldn't get into the building he would keep walking to his home in Alexandria.

Addington wasn't alone. The buildings in and around the White House complex were emptying, and the streets nearby were filled with smartly dressed young Washingtonians rushing in all directions. Traffic was at a standstill, and many people in the city were hesitant to ride the Metro—Washington's subway system—for fear that it could be the next target. The constant wail of sirens could be heard in the distance. Occasionally the volume would build almost unbearably as a fire engine, a police car, or an ambulance raced past, but then the decibel level of the sirens would diminish. Although most cell phones were not working, rumors of more attacks were spreading quickly by word of mouth: it was said that the State Department had been bombed and that Washington's own twin towers—the buildings that housed *USA Today* and Gannett Newspapers in nearby Rosslyn—had been hit by passenger jets.

Lynne Cheney began the day at the American Enterprise Institute, five blocks from the White House. When the first word of the attacks came, the Secret Service demanded that she move, and a small motorcade began the trip to the vice president's residence, a little more than two miles up Massachusetts Avenue from her office. Moments later, the motorcade reversed course; she would be taken to her husband. A guard at the White House gate wouldn't let Mrs. Cheney's car enter the grounds. Her angry driver gunned the car over the curb to get her to the White House.[13]

McConnell, who was in the alleyway when Mrs. Cheney arrived, spotted Mike Colvin, a shift leader on Cheney's Secret Service detail. John, everybody's got to get out of here, Colvin said. As he walked away from the White House complex, a burly Secret Service agent burst through the doors leading from the Old Executive Office Building to the alleyway. He came bounding down the stairs, shouting, "Everybody run! There's a plane coming at us!"

McConnell did not run. Instead, he walked purposefully to the building Mrs. Cheney had just left. It was a small act of defiance. McConnell had friends at AEI, so he watched television there before heading to the Washington office of Daimler-Chrysler, where the spouse of a White House staffer had invited other displaced staffers to spend the day. In addition to his duties with Cheney, McConnell also served as a speechwriter to President Bush, and he knew better than to leave town.

Less than a minute after he was carried from his office, Cheney had made it to a secure tunnel leading to the Presidential Emergency Operations Center, a shelter beneath the White House. He took a seat on a bench below a secure phone and tried to reach President Bush again. After several minutes of frustration—operators were having difficulty connecting him to Bush, who was now on the move—the vice president and the president spoke again at nine-forty-five AM.

"We're at war, Dick," said Bush. "And we're going to find out who did this and we're going to kick their ass."[14]

Cheney says there was one reason for this call to Bush: to tell the president about reports that the White House had been targeted and to urge Bush not to return to Washington, D.C.

Cheney had practiced for such an attack for two decades. In 1975, when he was chief of staff to President Ford, Cheney had written a memo to the White House counsel ordering the lawyers to study the procedures for presidential succession. As a member of the House Intelligence Committee in the 1980s, and as part of a highly classified continuity-of-government contingency team, Cheney studied the doomsday scenarios designed to decapitate the U.S. government. As secretary of defense, he was sixth in the line of succession.

"What I was immediately thinking about was sort of continuity of government," Cheney recalls. "Once you've got word that a plane is headed for the White House and it looks like Washington may be under attack, this suddenly gets to be a lot more complicated than just an attack. And so I urged him not to return until we could find out what the hell was going on."[15]

Cheney's position had the support of the Secret Service and Andy Card. But Bush had already announced to the nation his intention to return to the seat of power. He was concerned about the message it would send to the world if the terrorists were able to keep the president from the White House. This was a point the two men would debate several times throughout the day.

While Cheney was on the phone, he saw his wife being escorted by the Secret Service past him and into the Presidential Emergency Operations Center (PEOC), the bunker. He spoke to the president for approximately ten minutes, suggesting that Air Force One head to Offutt Air Force Base in Nebraska, and then joined a small group of senior officials in the emergency operations center.

Cheney's experience with issues of the presidential succession forced him to think about the unthinkable: what happens if both the president and the vice president are killed or otherwise unable to serve? Cheney tried to reach Dennis Hastert, speaker of the House, who was third in line for the presidency. But there were problems with the secure telephones—an issue that would frustrate Cheney throughout the day. The difficulties were on both ends of Pennsylvania Avenue.

"We heard Cheney was trying to get a hold of me," says Hastert. "We were having trouble with our secure phone. You had to put the key in and push the button—and it just wasn't working. We decided he'd probably take a call just over the regular lines. He was down in the war room and while I was trying to get that call through we saw the smoke across the mall."[16]

Hastert had no idea what was causing the smoke. He would later learn that a third plane had crashed, this one into the Pentagon.

The speaker called the House chaplain and asked the chaplain to meet him on the House floor. They would close the House in prayer. When Hastert reached the chamber, he saw that Porter Goss was in the chair. He gave Goss, a good friend, a sign indicating he wanted to adjourn the House. As they prepared to pray, Hastert's Secret Service agents rushed to his side and told him that a plane was headed for the Capitol. They hurried him out of the building. "We actually implemented part of the evacuation plan when we moved the congressional leadership to a secure location," Cheney recalls.[17] Still, he had not spoken to Hastert.

David Addington, meanwhile, was part of a large crowd making its way out of Washington, D.C., on foot. He was striding across the Fourteenth Street Bridge and could see the Pentagon in the distance, when his cell phone, which had not been working, rang unexpectedly. It was the Secret Service. The vice president wants you in the shelter, he was told. Addington explained that he had tried unsuccessfully to get back into the White House and received assurances that he would be allowed in this time. He turned around and hurried back, against the current of government workers in loosening business attire. A Secret Service supervisor met him at the East Gate of the White House and escorted him inside.

Much of the response by the U.S. government to the attacks was run from the conference room in the Presidential Emergency Operations Center. In the middle of the room is a rectangular wood table, long enough to seat sixteen people comfortably. At several places around the table, drawers contain a white telephone for secure communications. A second row of chairs along the wall provides room for support staff. Built into the wall closest to the entrance are two large television screens. For most of the day, one was tuned to CNN and the other to the Fox News Channel.

Cheney consulted with Secretary of Transportation Norman Mineta, a Democrat from California who had worked alongside Cheney in Congress. He had served as secretary of commerce in the final months of the Clinton administration. Bush, who came to office on a promise to "change the tone" in Washington and was eager to have a Democrat in his cabinet, asked Mineta to serve as secretary of transportation.

Mineta spent much of the morning at Cheney's side, scrawling notes on a white legal pad with a blue felt-tip pen. Together with Cheney and Rice, he spent the morning trying to track planes by their tail numbers to determine how many might have been a part of the plot. Eventually, Mineta directed that all aircraft be grounded. "Screw pilot discretion," he would later recall saying, "Bring 'em all down."[18] At 9:49 AM, the Federal Aviation Administration carried out Mineta's order.

Cheney took his place at the center of the table, just below the presidential seal. Not long after he sat down, at 9:58 AM, the small group in the PEOC watched as the South Tower crumbled, floor after floor, sending up a cloud of smoke and debris that chased terrified New Yorkers throughout lower Manhattan. One of the landmarks of the New York City skyline was gone, and with it, hundreds or perhaps thousands of lives. In the chaos of the previous hour, it had never crossed Cheney's mind that the towers would crumble. He remembers being surprised. "I think everybody was," he would say later. "I think Osama bin Laden was."[19]

Moments later Cheney spoke to Bush for the third time. The Secret Service had told Cheney that another aircraft was rapidly approaching Washington, D.C. The combat air patrol had been scrambled to monitor the area. We have a decision to make, Cheney told the president: should we give the pilots an order authorizing them to shoot down civilian aircraft that could be used to conduct further attacks in Washington? Cheney told Bush that he supported such a directive. The president agreed.[20]

Within minutes, Cheney was told that an unidentified aircraft was eighty miles outside Washington. "We were all dividing eighty by five hundred miles an hour to see what the windows were," Scooter Libby would later say.[21] A military aide asked Cheney for authorization to take out the aircraft.

Cheney gave it without hesitating.

The military aide seemed surprised that the answer came so quickly. He asked again, and Cheney once again gave the authorization.

The military aide seemed to think that because Cheney had answered so quickly, he must have misunderstood the question. So he asked the vice president a third time.

"I said yes," Cheney said, not angrily but with authority.

"He was very steady, very calm," says Josh Bolten, then deputy White House chief of staff, who was standing at Cheney's side. "He clearly had been through crises before and did not appear to be in shock like many of us."[22]

Cheney says there wasn't time to consider the gravity of the order he had just communicated. It was "just bang, bang, bang," says Cheney, one life-or-death decision after another.[23]

The entire room paused after Cheney had given the final order, as the gravity of this instruction became clear. At 10:18 AM, Bolten suggested that Cheney notify the president that he had communicated the "shoot-down" order. Shortly after Cheney hung up, the officials in the shelter were advised that a plane had crashed in Pennsylvania.

Everyone had the same question, says Rice. "Was it down because it had been shot down or had it crashed?" Rice and Cheney were both filled with "intense emotion," she recalls, because they both made the same assumption. "His first thought, my first thought—we had exactly the same reaction—was it must have been shot down by the fighters. And you know, that's a pretty heady moment, a pretty heavy burden."[24]

Both Rice and Cheney worked the phones in a desperate search for more information. "We couldn't get an answer from the Pentagon," says Rice. They kept trying.

"You must know," Rice insisted in one phone call to the Pentagon. "I mean, you must know!"[25]

Cheney, too, was exasperated. We have to know whether we actually engaged and shot down a civilian aircraft, he said, incredulously. They did not. For several impossible minutes, Cheney believed that a pilot following his orders had brought down a plane full of civilians in rural Pennsylvania. Even then, he had no regrets.

It had to be done. It was a—once you made the decision, once the plane became hijacked, even if it had a load of passengers on board who, obviously, weren't part of any hijacking attempt, once it was hijacked, and having seen what had happened in New York and the Pentagon, you really didn't have any choice. It wasn't a close call. I think a lot of people emotionally look at that and say, my gosh, you just shot down a planeload of Americans. On the other

hand, you maybe saved thousands of lives. And so it was
a matter that required a decision, that required action. It
was the right call.[26]

At 10:28, the North Tower collapsed. The frenzy in the shelter came to a halt, and except for an occasional whisper, the room went silent. On the television, one floor after another gave way, a bit of order amid the catastrophe. The building must have been charged, thought Addington, who was standing against the outer wall of the shelter.

Cheney, seated at the conference table, stared at the screen. Josh Bolten and Norman Mineta stood behind him to his left, Scooter Libby and Condoleezza Rice to his right. All wore virtually the same stunned expressions.

But the group in the shelter had little time to reflect on the tragedy. Two minutes later came yet another warning: an unidentified aircraft was in flight less than ten miles out. Cheney again gave the order to shoot it down.

They waited for news. None came.

At 10:39 AM, Cheney spoke to Secretary of Defense Donald Rumsfeld for the first time. He reviewed the events of the past hour.

"There's been at least three instances here where we've had reports of aircraft approaching Washington," said Cheney. "A couple were confirmed hijack. And, pursuant to the president's instructions I gave authorization for them to be taken out."

There was quiet on the other end of the line.

"Hello?"

"Yes, I understand," Rumsfeld came back. "So we've got a couple aircraft up there that have those instructions at this present time?"

"That is correct," said Cheney. "And it's my understanding they've already taken a couple aircraft out."

"We can't confirm that," Rumsfeld told his former aide. "We're told that one aircraft is down but we do not have a pilot report that did it."[27]

It was mid-morning before Cheney finally spoke to Hastert, who had been moved about twelve miles to Andrews Air Force base despite citywide traffic gridlock. Cheney briefed his friend. Norm Mineta was at his side. "We talked, but it wasn't a long conversation," says Hastert. "You know, Dick never talks for very long about anything. So he gave me the facts and what I needed to know and he gave me a review of what planes they thought might be still in the air that they thought might be dangerous."[28]

Cheney talked to Hastert about presidential succession. "You had the president on the ground in Florida and then in the air in Florida," recalls Hastert, "so the constitutional line had to be kept in order."[29]

Bush had left Florida almost immediately after his brief statement to the press at nine-thirty AM. White House staffers aboard Air Force One were not told where they were going. Reporters traveling with the president calculated that the plane was flying in circles because the televisions on board received a strong enough signal that the passengers could watch the local Fox affiliate for almost an hour with good reception.[30]

In reality, Bush flew west to Barksdale Air Force Base in Shreveport, Louisiana. He spoke to Cheney several times on a secure line, reiterating his desire to return to Washington. Cheney, backed by other senior officials and the Secret Service, continued to advise against it.

The attempt to keep Bush away from Washington would be one of the few decisions that day to draw immediate criticism of the White House. "President Bush made an initial mistake," the presidential historian Robert Dallek told Susan Page of *USA Today*. "The president's place is back in Washington."[31]

In an interview two months later, Cheney dismissed the criticism. "That's crap," said Cheney. "This is not about appeasing the press or being the macho guy who is going to face down danger. You don't think in those personal terms. . . . This is about preserving and protecting the presidency. His importance lies in the office he holds."[32]

The president addressed the nation again at 12:36 PM. His first speech, about three hours earlier, had come off as limp and inadequate. "Terrorism will not stand," he had said, before promising an investigation to find "those folks who committed this act." Much had happened since—the attack on the Pentagon, the plane crash in Pennsylvania, the collapse of both towers at the World Trade Center.

In the White House, the national security team watched from the shelter as Bush spoke. His forceful two-minute speech early that afternoon gave the American public its first hint of the broader war to come. "The United States," he declared, "will hunt down and punish those responsible for these cowardly acts." Several officials continued to monitor the TVs after Bush was finished.

The television reporting throughout the day proved invaluable. For much of the day, the team in the shelter experienced 9/11 as much of America did: through TV. Watching the uninterrupted news coverage not only provided new and timely information; it also allowed officials in the shelter to understand, as they designed their public response, what exactly the American people were seeing. The TVs also were a source of considerable frustration. Although the two televisions on the wall could be tuned to different channels, they could get audio from only one. On several occasions, the officials could see notices of "breaking news" without being able to hear the details of those updates. According to one official in the room, Cheney was "cranked up" about the technical problems and repeatedly demanded that they be fixed. They weren't.

Shortly after Bush's speech, the White House operator sent a call from the Federal Reserve to the communications team in the shelter: Alan Greenspan, the chairman of the Federal Reserve, was stuck in Switzerland. He had been on his way home from a meeting of central bankers in Basel when all aircraft bound for the United States were rerouted. So it was back to Basel. Cheney wanted Greenspan home as soon as possible. The panic following the terrorist attacks could spell disaster for the U.S. economy, and Greenspan was widely considered a calming influence on the markets. Cheney asked Addington to make arrangements with the Pentagon to have Greenspan flown back to the United States the next day.

The conference table in the PEOC became cluttered with pens and paper, platters of sandwiches and cookies, a thermal coffee dispenser and cups, bottles of water, and cans of Diet Coke. As the afternoon wore on, Condoleezza Rice noticed that Cheney hadn't eaten anything. "You haven't had any lunch," she said to the vice president. As soon as she said it, she realized that it probably sounded odd. "I thought, 'Where did that come from? What a strange thing to say in the middle of this crisis.'"[33]

Others were equally self-conscious. "I spent a good part of the time standing up against the wall trying to stay the hell out of the way," says Addington.[34]

Finally, the oxygen in the overcrowded shelter dropped to a level that would not sustain the group very long. Addington quietly asked all but the most senior officials to leave.

At three-thirty PM, Bush called to order a meeting of the National Security Council. Many of his senior advisers were together in the PEOC. For an hour and five minutes a defiant Bush chaired the meeting by teleconference from Offutt Air Force Base. Leaning far forward in his chair as he spoke, Bush told his national security team that the battle against terrorists would be their most important mission—the new purpose of his administration. "We will find these people and they will suffer the consequence of taking on this nation. We will do what it takes."[35]

George Tenet reported that there were strong indications that the attacks were the work of al Qaeda and Osama bin Laden. The National Security Agency, he said, had intercepted conversations of al Qaeda's operatives celebrating their success. The CIA recognized the names of three terrorists from al Qaeda aboard American Airlines Flight 77, including Khalid al Midhar, who had been monitored at a meeting of al Qaeda in Kuala Lumpur, Malaysia, in January 2000.[36] It was good information, but to some Tenet's report provided answers to the wrong question. "I wasn't focused on who as much as how much more," says one of Cheney's advisers.

After the meeting, Cheney and Brian Stafford, the director of the Secret Service, both advised Bush to stay in Nebraska. The vice president believed there was still too much uncertainty to have the president in Washington. Bush was insistent, though. He wanted to speak to the nation from the White House.

"I'm coming back."[37]

With the president returning to Washington, McConnell,

along with his fellow speechwriters Matt Scully and David Frum, walked the few blocks from the Daimler-Chrysler offices to the White House. While they had been drafting the president's address to the nation, the city had emptied out. The streets that only a few hours earlier had been teeming with panicked Washingtonians were now deserted.

When McConnell and his colleagues arrived at the White House, they had to wait for half an hour before they were allowed back into the complex. On the White House lawn, dozens of white tables were sitting empty. They had been set up for an annual picnic the White House hosts for members of Congress and select staffers; this year it had been scheduled for September 11.

Once inside the gates, as the writers walked up the driveway and past the abandoned bank of television cameras that beam images of White House correspondents around the world, they heard a faint but familiar sound growing louder. Just as they reached the front doors of the West Wing, the whirr of the helicopter blades was at its loudest. The president was back.

"That was a very emotional moment," says McConnell. "One of those emotional moments that catches you without any warning. The plan when that day started—he was to be dead and the White House was to be destroyed. And the day is over and he's alive and he's back and the White House is fine and Marine One is delivering him safely. I remember being deeply touched by that."[38]

Bush met with Cheney, Rice, and Card in the shelter shortly after he returned. The four officials compared notes on the day and discussed the objectives of the speech he was to give in ninety minutes.

At eight-thirty PM, the president spoke from the Oval Office. Cheney watched the speech from the shelter. Bush was much tougher than he had been in his two earlier comments. "Today our fellow citizens, our way of life, our very freedom came under attack in a series of deliberate and deadly terrorist acts," Bush said. "The victims were in airplanes, or in their offices; secretaries, businessmen and women, military and federal workers, moms and dads, friends and neighbors. Thousands of lives were suddenly ended by evil, despicable acts of terror."

These were not the words written by McConnell and his colleagues. The White House image maker Karen Hughes had composed a draft of her own, and the president read largely from those remarks. But the most important line that night was written by Matt Scully and came from the speech prepared that afternoon. "We will make no distinction between the terrorists who committed these acts and those who harbor them." It was the first articulation of a policy that would come to be known throughout the world as the Bush doctrine. Bush and Cheney had arrived at this one-sentence policy independent of each other and the speechwriters. It ensured the coming battle would be far more than a law enforcement operation designed to capture individual terrorists.

After the speech, Bush convened a meeting of the NSC in the shelter. Senior officials took their seats around the conference table while staff members crowded along the back wall. The session was unmanageable, so Condoleezza Rice announced that the meeting would continue with only the principals.

When it was over, Cheney stayed to talk with Bush. The two men discussed the events of the day and their plans for a response. The mess on the conference table from earlier in the day had been cleaned up, except for a crumpled white napkin at the far end of the long table. In front of the seat Cheney had occupied for most of the day was his white notepad, and on top of that was *The Economist,* the magazine he had grabbed on the way out of his office some twelve hours earlier. It had never been opened.

Shortly after ten PM, Cheney, along with his wife and his top two aides—Libby and Addington—walked back upstairs from the shelter and out the diplomatic entrance of the White House to the South Lawn, where Marine Two, the vice president's helicopter, was waiting. Joined by three Secret Service officials, a military aide, a communications expert, and Cheney's doctor, they took off under cover of darkness, an unusual precaution. The departure itself was a violation of long-standing protocol: no one takes off from the South Lawn other than the president. Their destination was kept from all but a few of the most senior White House officials.

Cheney had been aboard many such helicopter flights over

the course of his long career, but this was the first time he had flown without a president aboard. With those small changes would come bigger ones.

The helicopter flew over the Pentagon. Dozens of mobile light towers illuminated the deep gash across its facade. Smoke continued to billow out of the hole. "The headquarters of the U.S. military is still smoking, and we're flying over on our way to hide the vice president," Addington thought. "My God, we're evacuating the vice president from Washington, D.C., because we've been attacked."[39]

Addington and Libby, sitting across from one another, exchanged a knowing look. "We'd both had these important government jobs over our career and part of the job was paying attention to doomsday scenarios," says Addington.[40] The unthinkable was suddenly reality.

Less than thirty minutes later, another tradition was discarded when Cheney and his family settled into the Aspen Lodge at Camp David, the facility typically reserved for the president. It would be the first of many nights Cheney would spend in a series of "secure, undisclosed locations." From day to day, only a few senior White House officials knew where Cheney would be hidden.

Liz Cheney, along with her young family, joined her father and mother at Camp David. (Mary, who had been scuba diving in the Caribbean with her partner Heather Poe, was under the protection of her Secret Service detail.)

Liz and her parents gathered in the living room of the Aspen Lodge. For several hours, into the early morning hours of Wednesday, they sat quietly around the television and watched again the images from the day. For Cheney, it was his first opportunity to reflect.

"Having spent all day in the bunker, watching the towers fall and so forth, the president comes back, NSC meeting, president speaks, getting on the helicopter and flying out, looking down on the Pentagon as we left, smoking, lights on it and so forth, and then flying up to Camp David and going to Aspen. And I remember sitting in the living room there, turning on the television, watching the reruns, and I suppose that was the moment, as much as any, that it really hit home what the country had been through that day."[41]

Thousands died on September 11, 2001. In the years that followed, Cheney and his colleagues in the Bush administration said repeatedly that the attacks that day were so significant they changed everything. And yet for all the years that Cheney had thought about an attack on the American homeland, he had imagined something far worse. The attacks of 9/11 produced nowhere near the level of destruction that it would be possible to bring to the United States. The war games Cheney had participated in during the cold war and the worst-case scenarios he role-played as part of the continuity-in-government exercises offered scenarios of nuclear war, with the annihilation of entire cities and millions of deaths.

Still, as the longest day in Dick Cheney's life drew to a close, he knew that there were many more difficult days ahead. He and his colleagues would tell the shaken nation over and over that it would be a long war. And it would. But they were speaking broadly and often figuratively. As one term in office stretched into two, and one war became two in the eyes of many Americans, vast segments of the public questioned the conduct of those wars and lost faith in their leaders. It would indeed be a very long war.

CHAPTER TWELVE

Secure, Undisclosed

Dick Cheney began the morning of September 12 with a question that would define a new era in American life: when is the next attack?

The vice president had spent countless hours during the previous three decades mulling over threats to the country. He had studied various scenarios of an attack. He had contemplated "asymmetric warfare" and had followed the debates about "nonstate actors" and "transregional terrorists."

But those were theories. Cheney had thought that his chance of seeing such a devastating attack in America was remote. We were safe behind our oceans, he had reasoned. The attacks of the previous day made it all real.[1]

"I think probably it is fair to say that 9/11 was a watershed event. And it was the kind of thing that was so significant, such a dramatic change from what had gone before, nineteen guys with boxcutters can kill 3,000 Americans, that it forced all of us to go back and look fresh at what had transpired, at what had led up to that point," Cheney says. "9/11, because of the scale of the enterprise and the number of fatalities, a strike right here in the homeland, in our two great cities, New York and Washington, D.C., it was: 'Whoa, wait a minute. Let's go back and see who are these guys who are able to pull off that kind of an attack. . . . It takes on a different pattern and I think a greater significance for everybody.'"[2]

That morning, Cheney returned to Washington to see Bush. Scooter Libby came with him. When Libby returned to his office in the Old Executive Office Building, he told his assistant, Jennifer Mayfield, that he would be among those staying with Cheney for the foreseeable future at a "secure undisclosed location." In reality, the phrase "secure undisclosed location," used frequently in media reports of the vice president's whereabouts, was the generic description for anywhere Cheney stayed. Most often, this was Camp David, the heavily forested presidential retreat in the mountains of western Maryland. But there would be other locations over the next several months.

Libby asked Mayfield to go to his house and pack a bag for him. Only those considered "essential staff" would move to Camp David. That group included Mayfield and Libby; Debbie Heiden, Cheney's personal secretary; Brian McCormack, Cheney's personal aide; one of Cheney's military aides; and either David Addington, Cheney's counsel, or Neil Patel, his staff secretary.

Libby ordered several changes in the day-to-day operation of Cheney's staff. No one was to mention Cheney's name or his title on the telephone, either cell phones or land lines. Cheney's schedule, which had been distributed by e-mail, would now go out only over secure fax or classified e-mail. And everyone on the staff was required to keep a packed bag at the office at all times. Over the next several months, when Libby was back in town he would perform spot checks to make sure everyone was prepared to bolt at any time.

It didn't take long for Cheney's staff to grow accustomed to life in the undisclosed location. They all worked in the same large, open room in Laurel, the main lodge at Camp David. When the staff members took breaks—this was rare—they hung out together in a seating area in the main room. Their jobs remained largely unchanged, except for the increased face time with Cheney.

The staff ate breakfast and lunch together in Laurel, meals catered by the president's chefs. For dinner, though, they had to eat in the mess hall that serves the soldiers on the small military base at Camp David. The food was standard military issue—a stunning contrast to the four-star catering—and the dining facility was open for only an hour a day. Cheney's staff rarely had time to break away from work for dinner, so the options were limited. With the military on high alert, it wasn't as if they could call the local Domino's for a pizza. Several staffers frequented a small bar on the base called Shangri-La, after the name President Franklin D. Roosevelt had given the retreat. Its specialty was Fritos smothered in chili and cheese.

For one meal, Cheney brought in a pheasant he had shot on a hunting trip that he took after 9/11. The staffers were given a choice between the fresh game and lasagna. While they appreciated the gesture, nearly everyone chose lasagna.

Cheney often worked in an office in the back of the lodge normally used by the president. Each morning, he had his regular CIA briefing and then met alone with Scooter Libby, to compare notes on the new intelligence.

The vice president then typically joined a meeting of the National Security Council by secure videoteleconference (abbreviated SVTC and pronounced "siv-itz"). Cheney did most of his business at Camp David by SVTC, often participating in numerous videoteleconferences each day. In a real-time secure video feed, Cheney could see and be seen by participants from the White House situation room, the State Department, the Justice Department, and the CIA.

For those in the meetings, Cheney was both with them and apart. "We'd see him on the screen," remembers Rice. "He was off being the continuity of government."[3]

Cheney met the same way with those on his personal staff who remained in Washington. It was an odd way of doing business, but the efficiency of the SVTC system minimized the disruption.

The vice president and president agreed that they should avoid being in the same location whenever possible. The White House scheduling office began to produce a color-coded monthly calendar that aggregated the two men's schedules. On those occasions when they were together, it was important not to advertise that fact to the world. But Cheney was back at the White House far more frequently than the White House disclosed.

On Friday, September 14, other senior members of the national security team joined Cheney at Camp David. Colin Powell arrived first, and later Donald Rumsfeld and Condoleezza Rice flew there together. The full war cabinet would be meeting the following morning, but Bush wanted this smaller group to get away from Washington and think big thoughts about the war in front of them.

The small group met in Holly, a cabin across the street from Laurel. As they began to forge a new strategy for the troubled Middle East, a photo display from the negotiations of the Camp David accords during the Carter administration provided a stark reminder of America's difficult history in the region. The four officials dined on buffalo steak, exchanged ideas, and reflected on the past week. They told stories, often about the Gulf War, and there was a sense that this experienced group had seen war and had known war, though only Powell understood it from the battlefield. "They'd seen a lot together before," Rice says. "But they hadn't seen this. This was different, and it was palpable in the room, in the conversation. . . . I could sense there was—I'm trying to find the right word—tension isn't the right word, but anxiety. Anxiety."[4]

When the full war cabinet met the following morning, Cheney was characteristically quiet. George Tenet had made clear several times over the previous days that the attacks were the work of Osama bin Laden and Al Qaeda, his terrorist network. Some people at the meeting thought the attack was too sophisticated to have been executed by terrorists without the assistance of a government hostile to American interests. Thoughts turned to Saddam Hussein and his repeated threats of revenge for his humiliating defeat in the Gulf War.

During a break in the meeting, Cheney chatted with Paul Wolfowitz and Bush. Wolfowitz floated an idea: one way to squeeze Saddam's regime without mounting a full-scale invasion would be to take southern Iraq. As an effect of the cease-fire after the Gulf War, the United States already controlled the heavily Kurdish regions in the north. With American-controlled secure zones in the north and south, Saddam would lose access to important weapons-smuggling routes and oil fields.

All three men suspected that Saddam Hussein may have had a role in the attacks, and both Cheney and Bush were intrigued by Wolfowitz's proposal. But Cheney wanted to defer any discussion of Iraq until a later date.

Back in the large meeting, Cheney spoke up. Iraq needs to be part of our discussion eventually, he said, but right now it is a distraction. Nobody has any doubt that Al Qaeda had planned and executed the attacks, he said, and they should be our focus.

"If we go after Saddam Hussein, we lose our rightful place as good guy," he said.[5]

"He was very firm that we had something to get done, which was Afghanistan," recalls Rice, "and that any move in any other direction was not—you know, we had to get the country ready for a potential of another attack and we had to win in Afghanistan and that was enough."[6]

Cheney thought the United States' response had to be overwhelming, and that the targets must include the Taliban, the radical Islamic regime that had harbored Al Qaeda and Osama bin Laden for more than five years. Any realistic plan had to involve American boots on the ground in Afghanistan.

The next day, Cheney appeared on *Meet the Press*, in a one-hour special broadcast from Camp Greentop, next to Camp David. The vice president explained that "the world shifted in some respects" and predicted, "This is going to be a struggle that the United States is going to be involved in for the foreseeable future."

This administration's approach to terrorism, he said, would represent a significant departure from that of past administrations. "What's different here, what's changed in terms of U.S. policy, is the president's determination to also go after those nations and organizations and people that lend support to these terrorist operators," he said. "If you've got a nation out there now that has provided a base, training facilities, a sanctuary, as has been true, for example, in this case, probably with Afghanistan, then they have to understand, and others like them around the world have to understand, that if you provide sanctuary to terrorists, you face the full wrath of the United States of America. And that we will, in fact, aggressively go after these nations to make certain that they cease and desist from providing support for these kinds of organizations."

"Full wrath," said Tim Russert, the show's host. "That's a very strong statement to the Afghans this morning."

"It is, indeed. It is, indeed."

The host turned to Iraq. "Saddam Hussein, your old friend, his government had this to say: 'The American cowboy is reaping the fruits of crime against humanity.' If we determine that Saddam Hussein is also harboring terrorists, and there's a track record there, would we have any reluctance of going after Saddam Hussein?"

"No."

"Do we have evidence that he's harboring terrorists?"

Cheney reiterated the views he had expressed in the previous morning's meeting.

"There is—in the past, there have been some activities related to terrorism by Saddam Hussein. But at this stage, you know, the focus is over here on Al Qaeda and the most recent events in New York. Saddam Hussein's bottled up, at this point, but clearly, we continue to have a fairly tough policy where the Iraqis are concerned."

"Do we have any evidence linking Saddam Hussein or Iraqis to this operation?"

"No."[7]

At the end of the hour-long interview, the NBC crew that had accompanied Russert to western Maryland applauded Cheney. Eric Schmitt, a reporter on national security for the *New York Times,* wrote that the appearance underscored why Bush wanted Cheney as his vice president.

"Any doubt about Mr. Cheney's star power ended abruptly after his command performance on the NBC News program, 'Meet the Press' on Sept. 16. Not only did he relate vivid details of ordering the evacuation of Congressional leaders and cabinet members, but he also delivered a blunt, clinical assessment of the war ahead."

Not everyone was thrilled by the performance, Schmitt reported. Some Republicans said that the interview "was just a bit too commanding for senior White House aides like Karl Rove and Karen P. Hughes. In allowing Mr. Cheney to describe his role, the aides walked a fine line between reaffirming the vice president's stature and having him overshadow the president."[8]

Not coincidentally, Cheney would make only a few major media appearances over the next several months. In each one, he would reiterate a point he had made just five days after the attack. "There may well be other operations that have been planned and are, in fact, in the works."

Later the same day, at a CIA briefing on Capitol Hill, members of Congress were told that the attacks of 9/11 were meant to be the first part of a longer campaign of terror. Senator Bob Graham, chairman of the Senate Intelligence Committee, reported that "very credible" intelligence showed that terrorists might have been plotting to blow up bridges or unleash a chemical attack on a city water system.[9] Americans were paying attention. Issues that had been the province of a small subset of the national security bureaucracy had been thrust on a dazed public.

For the next month, Cheney remained mostly behind the scenes, but his presence was felt. On September 20, President Bush announced that Governor Tom Ridge of Pennsylvania would be the new director of the White House Office of Homeland Security—an office that grew out of the homeland security study Cheney's staff had been conducting since the spring. Its deputy director would be Admiral Steve Abbot, who led that effort.

Cheney was also at work erecting a new national security framework for a nation at war. He had favored a strong executive function during his twenty-five years of public service in peacetime America and he had regularly endorsed the notion that a president should have "extraordinary authority" in times of war. There was little doubt he would seek to expand executive power now that the man he served was a war president.

These early efforts would spawn some of the Bush administration's most controversial policies: the Terrorist Surveillance Program, military tribunals, the guidelines for interrogation of detainees, programs for tracking the finances of terrorists. This was a time when Cheney's visibility was at its lowest and his influence at its greatest. Ideas sprang up from the intelligence world, one after another. In some cases, these ideas had been considered previously but abandoned for lack of urgency.

Those decisions did not seem like close calls at the time. "Anything that anybody told you you could do legally, you do it, because the thought that something else was coming was just ever-present," says Rice.[10]

On *Meet the Press*, Cheney had been asked about whether there was any international law that would forbid the United States from killing Osama bin Laden. His answer was revealing: "Not in my estimation, Tim. But I'd have to check with the lawyers on that, obviously. Lawyers always have a role to play, but one of the interesting things here is the way in which people have rallied around, other governments have rallied around this notion that, in fact, this is a war."[11]

That new reality required attention to the economy, too. If the U.S. economy had been in a "tailspin" before 9/11, as NBC's David Gregory had reported, the potential impact of the attacks was catastrophic. Could the terrorists win by causing so much distress that an already weak economy would effectively collapse?

"The losses are mounting," reported an article in *The Economist* eleven days after the attacks. "By the end of trading in New York on September 20th, the Dow Jones Industrial Average had shed nearly 13% of its value since Wall Street reopened on September 17th."[12] The airline industry, already teetering, was devastated. Hotels and restaurants laid off tens of thousands of employees. Initial estimates showed more than $100 billion in economic losses to New York City alone.

The Federal Reserve quickly cut interest rates, and Alan Greenspan reported to Congress that the long-term prospects of the U.S. economy were good. But most analysts were not optimistic. "There is little doubt that the scale of the destruction of economic value, combined with the loss of confidence, will precipitate one of the most severe global recessions in recent memory," said David Folkerts-Landau, an economist with Deutsche Bank.[13] In June 2001, Bush signed a $1.35 billion tax cut passed by Congress. After 9/11, Cheney pushed for additional tax cuts to stimulate the economy; as always, bigger was better.

But Cheney spent most of this time on the new war on terror. He was consumed—some would say obsessed—with preventing the next attack. "He read every intelligence report," says Rice. "I mean every intelligence report no matter how minor. And I was feeling kind of driven crazy by not knowing how to—at that time we didn't really have a good system for sorting what was reliable and what wasn't. I mean, we were just getting basically the raw reporting of everything that was coming in, you know. So I remember thinking that he had an extraordinary memory . . . for all of these things."[14]

As fall went on, the country's anxiety about potential new attacks took a frightening new shape. Early in October, a sixty-three-year-old photo editor named Robert Stevens entered a hospital in Florida with mysterious symptoms: a fever, an odd sore throat, labored breathing. On October 4, hospital officials confirmed that he had been exposed to anthrax. Two days later, he was dead. Tommy Thompson, secretary of health and human services, at first downplayed the infection as an isolated incident, and federal officials insisted that there were no indications of a terrorist attack. But two days later, another man in Florida tested positive for anthrax, and three days after that a third person was found to have the deadly virus. All three victims worked at American Media, a tabloid newspaper publisher in Boca Raton, Florida; the anthrax was determined to have come in with the mail. The FBI was called in to investigate and promptly put out a statement saying it had no evidence connecting the anthrax to terrorists. Within days, an employee of NBC News in New York tested positive. The FBI claimed there was "no connection whatsoever" with the 9/11 attacks.[15] In time, similar letters would go to the offices of Senator Tom Daschle and Senator Patrick Leahy; congressional office buildings were evacuated and some staffers were given antibiotics to fight possible infection.

The FBI's statements were intended to keep people from panicking, but much of the nation was nonetheless seized with fear of additional attacks. The attackers of 9/11 had focused on Washington and New York City. These were natural targets: the seat of the U.S. government and the capital of the world economy.

And although the events of that day were intended to affect all Americans, the reality was that a factory worker in Sioux Falls, South Dakota, was less concerned with his day-to-day safety than a broker in Manhattan or a legislative aide on Capitol Hill. The anthrax attacks in Florida started to change that.

Suddenly, national and local leaders were using an entirely new vocabulary: inhalation, cutaneous, weaponization, the Ames strain, Cipro, spores, aerosolize, silica. Postal facilities and businesses were shut down as investigators in white biohazard suits conducted tests for deadly substances. Americans everywhere were warned against opening their own mail unless they knew the sender. Anthrax was reported in such unlikely targets as Reno, Nevada.

As this new wave of fear consumed Americans at home, overseas the U.S. military was in the first stages of destroying those who had brought terror to the United States.

At one PM eastern daylight time on October 7, President Bush announced that a war in Afghanistan was under way. "On my orders, the United States military has begun strikes against Al Qaeda terrorist training camps and military installations of the Taliban regime in Afghanistan. These carefully targeted actions are designed to disrupt the use of Afghanistan as a terrorist base of operations, and to attack the military capability of the Taliban regime."[16]

On October 12, Cheney agreed to a rare public appearance, an interview with Jim Lehrer on PBS. One week earlier, Cheney had emerged from the undisclosed location to attend a two-minute statement by President Bush in the Rose Garden, encouraging Congress to support the tax cuts the White House had proposed. Then Cheney was gone again, and he had remained out of sight.

Lehrer immediately asked about the anthrax. Without challenging the FBI directly, Cheney questioned the logic of its claims that the attacks had not been made by terrorists. He methodically listed things he said "we know" about Osama bin Laden and Al Qaeda: they attacked America, they have attacked American interests before, they have tried to acquire weapons of mass destruction, and they have trained people in Afghanistan to use those materials. "So, you start to piece it all together. Again, we have not completed the investigation and maybe it's coincidence, but I must say I'm a skeptic."[17]

"What suddenly became really obvious was that we weren't prepared for a major biological event of any kind," says Condoleezza Rice. Cheney was the driving force on those issues. His staff had been studying how the federal government would respond to an attack involving WMD, and he peppered his advisers with questions during his regular meetings. "He became really one of the leading—probably the leading—the kind of tip of the spear in terms of homeland protection," says Rice, "especially after the anthrax attack."[18]

The country was in a panic, and most Americans were seeing only a fraction of the frightening reports coming in to the U.S. government. Among the dozens of false alarms never disclosed to the public in the months after 9/11 was a positive test for botulinum toxin, a substance so deadly that one gram could kill 1 million people.[19]

For Cheney, the national mood literally came home, in several frightening moments involving his children and grandchildren. Liz Cheney, the vice president's elder daughter, who had taken a job at the State Department, received a call from the Secret Service at her office: there had been a positive anthrax test taken at her home. Her three children were still in the house with their nanny, unaware that there was anything wrong.

The Secret Service told her that they would evacuate the children to a nearby facility and test them for exposure. Liz wanted to accompany them, but she was told not to come to the house. She called the nanny and, without explaining exactly what was happening for fear of upsetting her, strongly suggested that the kids might like to take a trip to the mall.

Cheney's immediate family would be subject to emergency drop-everything-and-leave-now evacuations several times in the two months after September 11.

The vice president hinted at his personal concerns in the interview with Lehrer. "I know it's difficult—I've talked to my own family—what should they be worried about? How should they operate? We find ourselves under a much higher level of security now than ever before. It's necessary, and we have to adapt to that."[20]

Cheney emerged once again on October 18 to deliver remarks at the annual Alfred E. Smith dinner in New York City, a politically important fund-raiser. He spent several hours earlier in the day meeting with city officials and others involved in the massive effort to get New York functioning again. The police blocked the streets as Cheney's motorcade made its way down to ground zero and normally impatient drivers in the heavily Democratic town stood outside their cars and cheered.

Cheney received a similar reception that evening. His brief speech was interrupted by applause eighteen times as he paid tribute to the city and promised a relentless—and often unseen—war against terrorists.

"These operations will require a broader, more realistic approach to gathering intelligence. We are dealing here with evil people, who dwell in the shadows, planning unimaginable violence and destruction," he said. "We must and we will use every means at our disposal to ensure the freedom and security of the American people."[21]

In Afghanistan, the fighting continued. The American planners believed that the difficult terrain required a new kind of war, one that harnessed the technological superiority of the U.S. military and intelligence services and deployed it in lethal cooperation with Afghan anti-Taliban fighters. The first three weeks of the war had featured heavy aerial bombardments, attacks that made it difficult to gauge success. The lack of visible progress stirred worries at home. Many commentators, particularly conservatives, were concerned about the lack of ground troops in Afghanistan. Senator John McCain, Bush's chief rival for the nomination in 2000, began to openly question the Pentagon's tactics. The conservative columnist Charles Krauthammer gave a blunt assessment in the *Washington Post*. "The war is not going well and it is time to say why," Krauthammer wrote. "It is being fought with half-measures."[22]

Donald Rumsfeld had come to the Pentagon with grand plans for transforming the U.S. military. Among other changes, he wanted to reduce the size of the Army. Many hawks opposed this plan, and now they worried that Rumsfeld was more concerned with field-testing his ideas than with winning the war.

Others joined in, with many veteran reporters returning to the paradigm they knew. R. W. Apple of the *New York Times* wondered whether the United States was destined to lose. In an article headlined, "A Military Quagmire Remembered: Afghanistan as Vietnam," he wrote: "For all the differences between the two conflicts, and there are many, echoes of Vietnam are unavoidable."

"Despite the insistence of President Bush and members of his cabinet that all is well, the war in Afghanistan has gone less smoothly than many had hoped," Apple contended, later adding, "Signs of progress are sparse."[23]

Although Bush had never made a habit of paying much attention to his critics, especially those who aired their views in the media, he was concerned. Are we making the kind of progress we should be? he wondered. Is our strategy working?

Bush raised the questions at a meeting of the National Security Council. No one had any doubts, or at least no one was willing to share any doubts in front of the group. But Bush wanted to be sure, so he went to Cheney privately.

They had been in office nearly one year, and Bush had come to trust that Cheney would be straight with him. Is this the right strategy? Cheney told the president that he had no reservations about the war plan and counseled patience.

A month later, things would look quite different.

Cheney spent October shuttling mainly between the nation's capital and Camp David. He had virtually no public presence. This prompted questions from journalists to the White House: Where is the vice president? Are the threats against Washington so great that it isn't safe for him to stay in town? Doesn't his absence send mixed messages to Americans who are hearing from their president that they should resume their normal lives?

There were persistent rumors about Cheney's health, and there was one report that journalists were getting calls to see if Cheney had died.[24] Cheney's fight with lawmakers over the records of the Energy Task Force had already earned him a reputation for secretiveness. His absence added to this image. Within a month of 9/11, it had become part of pop culture. On *Saturday Night Live*, the master impersonator Darrell Hammond, as Dick Cheney, announced that the undisclosed location was, in fact, Kandahar, Afghanistan, and that he would be acting as a "one-man Afghani wrecking crew."

Even in the sober mood of wartime, the enforced camaraderie of Camp David allowed for some moments of fun. One morning, Bryan McCormack, Cheney's personal aide, was summoned to Laurel lodge at the vice president's request. Cheney had a long list of relatively minor tasks for him, and the vice president walked through his instructions for each task with great deliberation. The meeting stretched on for several minutes.

When McCormack returned to his temporary home, a well-appointed cabin called Rosebud, he found a piece of paper attached to the front door. The note said he had been relocated in order to accommodate Joe Hagin, the White House deputy chief of staff, whose arrival was expected that day.

No big deal, McCormack thought. I know how this place works. There's a pecking order, and I'm at the bottom.

The note further explained that his belongings had already been moved to his new temporary home, Sweet Gum cabin. McCormack thought he had heard of all the living units on the premises, but Sweet Gum was unfamiliar. He filed the name away in the back of his mind and got on with his day.

As Cheney's personal aide, McCormack had grown accustomed to last-minute changes in accommodations and being bumped by more senior staffers. Flexibility was the first requirement of the job. In fact, McCormack had been assigned to Rosebud in the first place because a problem with the septic system had made it unfit for Liz Cheney and her family. So McCormack did not spend much time thinking about the switch. He would look for the cabin as he continued down his list of jobs from Cheney. He was in no hurry to find it.

For the pranksters who had moved McCormack's belongings, watching in the distance as he read the note, this was a problem. For the joke to have its full impact, they needed to be able to watch his reaction on arriving at Sweet Gum.

McCormack jumped into the golf cart he used to get around

the grounds. As he resumed his errands for the vice president, he ran into Phil Perry, Liz Cheney's husband. Perry was very friendly and full of questions. Where are you staying? he asked. How are the living arrangements?

The interest struck McCormack as odd. Phil had always been cordial, but they had never discussed the quality of his accommodations. What's more, no one at Camp David cares where anyone else is staying. In order to reach one another, the staffers don't dial cabin-to-cabin but put calls through the operator. McCormack explained that he had been transferred to Sweet Gum and Phil helpfully pointed him in the right direction.

A short time later, McCormack ran into Liz Cheney. She also showed unusual interest in his lodging. When he told her that he had been moved to Sweet Gum, Liz Cheney gave him directions and encouraged him to find his new cabin.

McCormack knew something was not right. He looked for several minutes. When he finally located the cabin, he understood why the others had wanted him to find it. Sweet Gum was a kids' playhouse—maybe nine feet square—located in the front yard of the house provided for the Camp David base commander. McCormack's American flag boxers were tacked to the front door. He entered to find the rest of his clothes draped over the rafters.

McCormack suspected that Liz Cheney was behind the prank. But when he blamed her, she insisted that the entire joke was her father's idea.

Later in the day, McCormack saw Cheney and his wife. The vice president, unaware that McCormack had already been to Sweet Gum and discovered the prank, asked him where he was staying. McCormack said he was in Sweet Gum. Cheney feigned ignorance and asked where it was located. McCormack played dumb, too, and said that he hadn't yet been there. I've been coming up here for years, said Cheney, and I've never heard of Sweet Gum. You'd better find it before it gets dark.

Mrs. Cheney, perhaps in the spirit of post-9/11 empathy, intervened and urged her husband to let McCormack in on the joke. But McCormack preempted his confession.

"Sir, I know everything. I know who was involved and who did what."

Cheney began to protest, but McCormack politely cut him off.

"It was all Liz," he said. "I know it was all Liz."

Cheney nodded and said nothing.

The next morning, Laura Bush buttonholed McCormack at the main lodge. "So," she said, "how was Sweet Gum?"

On November 5, Cheney's undisclosed location moved to the Paul Nelson Farms in Gettysburg, South Dakota. He goes to this private retreat each year with friends to hunt ring-necked pheasant. His daughter Mary, who often joins him on his hunting and fishing trips, came along on this one.

The commercial hunting resort advertises itself as "solace from the world's daily assault on your senses." As the owner, Paul Nelson, says, "We're for people who love a great hunt, but don't really want to rough it."[25]

The well-appointed lodge sits on a tranquil pond. The Cheney party, organized by Cheney's good friend from Jackson, Dick Scarlett, had the entire facility to itself. It would be a welcome break from the intensity of Camp David and Washington. The weekend started on a light note when Mary showed up for the first hunt wearing a Dick Cheney mask.

That mood was shattered when a National Guard helicopter mistakenly entered the airspace above the farm, which was restricted because of Cheney's presence. Cheney saw the helicopter and turned to Mary and the Secret Service agent Pat Smith.

Is that supposed to be there? he asked.

Before Smith could answer, a fighter jet appeared from nowhere, ripped across the sky, looped around the nose of the chopper, and took off in the other direction. The helicopter froze in place. If the pilots hadn't understood the urgency of clearing the restricted airspace from their radio communications, they got it now. For Cheney and his friends, it was a stark reminder of how much things had changed.

On November 26, 2001, Cheney traveled to the Secret Service headquarters in Beltsville, Maryland. Shortly after the inauguration, President Bush asked Cheney to review and assess the components of the U.S. intelligence community. Cheney was eager to do it. In his eight years out of government, there had been an influx of new leadership and new technology. In many respects the U.S. intelligence community was not the same one he had known at the Pentagon.

So he spent a day at the FBI, getting acquainted with its senior officials and seeking a better understanding of its mission. He did the same at the CIA, the National Security Agency, the Defense Intelligence Agency, and the National Reconnaissance Office.

His schedule for the visit to the Secret Service training facilities called for a tour, a briefing, and some hands-on training. Cheney would bring a small group of his staff, as well as the White House photographer assigned to him.

Cheney had handpicked David Bohrer from a small group of potential photographers the previous spring. Bohrer, a Pulitzer Prize–winner, was working as a photojournalist for the *Los Angeles Times* when he heard that the White House Photographer's Office had an opening. He applied, and among the materials in the portfolio forwarded to Cheney were photographs Bohrer had taken for a book he had produced three years earlier. It was a subject likely to appeal to the former secretary of defense—*America's Special Forces: Weapons, Mission, Training.*

Bohrer took a trip to the training facility in Beltsville the day before Cheney's visit, in order to prepare. In addition to a stop at the gun range, the vice president was scheduled to do some high-speed driving on an open course. Bohrer had been photographing Cheney for eight months and knew that demonstrative facial expressions were uncommon.

The lesson in evasive driving had the potential for some of these rare shots. So Bohrer mounted a camera on the windshield of the Chevrolet Camaro Cheney would be driving the next day. The instructor would be in the passenger seat, and Bohrer would hide in the backseat.

The training went as planned. The Secret Service instructor taught Cheney how to execute a "J-turn." Cheney had not driven at all since he drove his Cadillac STS to Love Field in Dallas to join Bush's campaign as the candidate for vice president. Cheney put the car in reverse, accelerated to forty miles per hour, and jerked the wheel violently in order to turn the car a full 180 degrees.

Bohrer couldn't see Cheney's face, but he clicked away by remote control at those times he was certain that the thrill of speed or the adrenaline rush of abrupt spinning would be greatest.

The photographs were taken on film, and Bohrer was eager to have it developed. He was stunned when he saw the pictures. Even as he whipped around in a Hollywood car-chase turn, Cheney was expressionless.

"It was as if he was taking a Sunday drive," says Bohrer.[26]

As fall turned into winter, Cheney gave a series of interviews looking back at 9/11 and explaining the administration's response to the attacks. In each interview, Cheney was asked for his personal feelings about that memorable day and its aftermath. And each time, he avoided an answer.

"How has this crisis changed you?" asked Gloria Borger on *60 Minutes II*. Cheney said that living through something like 9/11 makes people think of their country and their family. Borger persisted. "What about you, though?"

Cheney told her that he enjoyed his job and couldn't imagine more important work than what he was doing.

Evan Thomas from *Newsweek* also asked him about 9/11. "At any point do you ever feel in physical danger?" Cheney said that he hadn't thought about it. "You've got a job to do," he said.

Thomas asked: "Were you ever worried about your family?" Cheney, without answering directly, noted that his wife had been in the bunker and explained that the Secret Service had relocated his elder daughter and his grandchildren "So everybody did their job."

The only glimpse the vice president gave into his emotions came in response to a question from Thomas. Cheney recalled hearing that his friend Barbara Olsen, the wife of Bush's solicitor general, Ted Olsen, had been aboard the plane that struck the Pentagon.

In an interview held just before the five-year anniversary of 9/11, Cheney was still reluctant to talk about how the attacks had affected him personally.

The question was simple: "If the September 11 attacks changed everything, what did they change in you?"

"Well," he started, "it moved a whole set of concerns from the hypothetical realm to real-world everyday problems. It became—has been throughout our administration now, the Bush-Cheney administration, the dominant concern. And it—I try to think how to say it succinctly. You can wander all over the lot on this."

With that, Cheney returned to policy.

"It affected clearly how we've allocated our time, how we spend a good part of every day, the content of our intel briefs, the organization of the federal government, the people that we've hired in key positions, the reorganization of the intelligence community, the creation of the Department of Homeland Security, our spending priorities, our foreign policy, military policy, the use of military force, our legal system. Another way to put it is: What hasn't it touched? It clearly affected our political campaigns. It was front and center in both the 2002 and 2004 elections. I think it'll probably be important again this year. It also took us to absolute peaks in terms of public approval and also the pits. The president is 80 percent in the polls after we do Afghanistan and after three years in Iraq we're in the thirties. It's shaped how the world looks at the United States and looks at this administration."[27]

The interviewer tried again, reminding Cheney that President Bush had gotten emotional on national television in the Oval Office, and Condoleezza Rice remembered crying when she heard "The Star-Spangled Banner" played by the Queen's Marching Band in England. It was an emotional time.

"Did you ever have a moment where everything sort of caught up to you and you had a good cry?"

"No. Not a cry." Cheney found the question odd. He pressed on. "I can remember going to ground zero, in October, for the Al Smith dinner. And this was several weeks after 9/11. It was stark and moving, but by then we were dealing extensively with it."[28]

There is little question that the attacks influenced virtually every aspect of the Bush administration from September 11 forward. Still, Bush says that he doesn't talk about the events of the day with Cheney. "You don't need to," he says. It's always there. "Look, we don't sit around psychoanalyzing each other."

Bush finds Cheney's stoicism reassuring. He says it provides a helpful counterbalance to his own demeanor. "He is measured and he's from that western stock that's kind of the quiet—there's strength in quietude, in a way. That's the opposite of me. I'm an emotional guy. I cry, I go to see the fallen and spend hours with them weeping, laughing, hugging," says Bush.

"Cheney is not a hugger," he adds. "But he loves deeply."[29]

By early December, one month after Bush and Cheney spoke privately to evaluate progress in Afghanistan, the United States had destroyed Osama bin Laden's training camps and eliminated many Taliban refuges. The military had captured hundreds of terrorists associated with Al Qaeda and the Taliban, and interrogations were leading to additional captures. Discussion turned to postwar Afghanistan.

Cheney felt strongly that the United States should avoid becoming an occupying force. One of the reasons the war had gone as well as it had, in his view, was that so many Afghans had fought alongside the coalition's soldiers. Cheney often spoke of the need to "get locals into the fight." The immediate benefits were obvious: they knew the terrain, and more Afghans fighting meant fewer Americans dying. In many ways, Cheney thought the long-term consequences for postwar Afghanistan could be equally significant. After having been subjugated by the Taliban for nearly five years, the Afghan people could rightly claim that they played an important role in overthrowing a regime that had oppressed them.

It would be one of many important differences between Afghanistan and Iraq.

CHAPTER THIRTEEN

Back to Baghdad

In the early morning of February 12, 2002, David Terry, a senior official from the CIA, arrived at One Observatory Circle to give Cheney his daily intelligence briefing. Among the materials Terry presented was a classified report from the Defense Intelligence Agency (DIA) published internally the previous day. The report carried a startling title: "Niamey Signed an Agreement to Sell 500 Tons of Uranium a Year to Baghdad." Niamey is the capital of Niger, an impoverished African country whose struggling economy, according to the CIA, is built on "some of the world's largest uranium deposits." The Iraqi regime was prohibited by United Nations sanctions from pursuing the technology for weapons of mass destruction, to say nothing of purchasing large quantities of uranium.[1]

The vice president wanted to know more. A voracious and discriminating consumer of intelligence since his days in the Ford administration, Cheney is rarely satisfied with the formal intelligence products he's given. Richard Kerr, who was the CIA's deputy director for intelligence when Cheney was in Congress, remembers the questions Cheney would ask. "How do you know this? How confident are you? What about this source? Why did you choose to think about this particular information in one way and this in another? How did you make that choice, and how did you integrate those two ideas?" It wasn't hostile questioning, Kerr remembers. "From an intelligence officer's point of view, the most impressive thing you can have is consumers that are interested in what you do, question it, and test the system."[2]

The time Cheney spends as vice president with intelligence officials "involves getting the analysts to say more than they've written," says one of Cheney's aides, who sometimes participated in those briefings. So when Cheney asked Terry for additional information about the DIA's report on uranium, Niger, and Iraq, it was hardly out of the ordinary.

Terry later reported Cheney's interest back to the CIA in a memo titled "Briefer's Tasking for Richard Cheney on 02/13/2002."

"The VP was shown an assessment (he thought from [the Defense Intelligence Agency]) that Iraq is purchasing uranium from Africa. He would like our assessment of that transaction and its implications for Iraq's nuclear program. A memo for tomorrow's brief would be great."[3]

That afternoon, officials at the CIA's Counterproliferation Division discussed how they might investigate further. An employee of the division, Valerie Plame Wilson, suggested that the agency send her husband, Joseph Wilson, to Niger to make some inquiries. Joe Wilson was a former U.S. ambassador to Gabon with experience in Niger. In a memo to the deputy director of the Counterproliferation Division, Valerie Wilson wrote: "My husband has good relations with the PM [prime minister of Niger] and the former Minister of Mines (not to mention lots of French contacts), both of whom could possibly shed light on this sort of activity." Mrs. Wilson would later say she asked her husband, on behalf of the CIA, if he would investigate "this crazy report" on a uranium deal between Iraq and Niger.[4]

Wilson agreed to go. It was a trip that would have great and unforeseen consequences.

Cheney would soon set out on a momentous trip of his own—a tour of the Middle East, then roiling with speculation that the United States would invade Iraq. The trip began on March 10, 2002; it was Cheney's first journey overseas—and one of only a few he would make—as vice president. Over the course of the Bush administration, the White House would use Cheney in foreign diplomacy the same way it used him to communicate with the American public: not often, but always in times of real importance.

Cheney's trip would take him to twelve countries in ten days. In designing it, the White House had two specific goals: to listen to allies in the region about the next phase of the war on terror and to ask them about Iraq.

One of the most important stops would be Egypt. Cheney met with Egyptian President Hosni Mubarak in Sharm al-Sheikh on March 13. Mubarak noted the effects of UN sanctions on Iraq and promised to press the Iraqis to permit a resumption of weapons inspections by the UN.

"We'll try hard with Saddam Hussein to accept the UN inspectors to go there," Mubarak said, standing next to Cheney at a press conference after their meeting. "We are going to meet with some of his special envoys and tell them that this is a must."[5]

Iraq's special envoy, vice president Izzat Ibrahim al Duri, arrived to see Mubarak the following day. It was a replay of the run-up to the Gulf War. More than a decade earlier, Izzat Ibrahim had shadowed Cheney as he shuttled throughout the region to build support for the deployment of U.S. troops. On August 7, 1990, both men had visited Alexandria to see Mubarak, in meetings separated by only a matter of hours.

After Izzat Ibrahim met with Mubarak this time, the Iraqi News Agency quoted one of Mubarak's political advisers, Usamah al-Baz, who emphasized another aspect of Egypt's position. He said that Mubarak had encouraged Iraq to comply with UN resolutions, but added: "It is unacceptable for the people of Iraq to continue to suffer" from international sanctions. "It is also unacceptable for the resolutions to be open-ended, as if Iraq is destined to continue to suffer in this way." Asked about the possibility of a war in Iraq, Usamah said, "Our position in this regard is very clear; it is against striking Iraq and against any attempt to change the ruling system in any sisterly Arab country through foreign forces."[6]

Cheney's next stop was Yemen, a small tribal nation just to the south of Saudi Arabia on the Gulf of Aden. The president of Yemen, Ali Abdallah Salih, had never been a reliable ally. Yemen is one of the poorest countries in the Arab world, and Salih has often shrugged off the requests of American diplomats by pleading poverty. American officials considered this diplomatic blackmail, but for Salih, it was not without political sense. Senior Yemeni clerics have close ties to Islamists, including Osama bin Laden, whose family had Yemeni roots, and openly threatened to bring down the Yemeni regime whenever Salih tilted toward Washington.

"He's not one of those guys who makes long-term alliances," says one senior administration official. "So periodically you have to renew the lease."

In Egypt, Cheney had gotten disturbing intelligence about his upcoming trip to Yemen. According to reports, terrorists affiliated with Al Qaeda were plotting to bring down Air Force Two with shoulder-fired antiaircraft rockets. (This sort of weaponry would be easy enough to come by there; one official in the Bush administration who worked in the region referred to Yemen as "the ultimate arms bazaar.") The reports were specific and highly credible.

The trip was scheduled to be short, and Cheney did not want to cancel it. So his security detail devised an alternative plan. The press had not been scheduled to go along, a fact that made the changes easier. When Cheney boarded Air Force Two in Sharm al-Sheikh, he climbed up the front stairs as he always did. But instead of settling in for the flight, he walked briskly to the back of the airplane, down a set of stairs on the other side, and into a waiting vehicle that surreptitiously transferred him to a hulking gray military transport plane, one of several on the tarmac.

Everything about the departure of Air Force Two was routine, except that the vice president was not on the plane. The C-17 carrying Cheney and a few aides trailed behind. Not only was the military aircraft a less obvious target than Air Force Two—especially to prospective attackers who would have no idea that Cheney was on board—it was also more capable of taking evasive measures. On approach, Air Force Two came in low and very fast; and it didn't land on its initial approach. The trailing C-17 remained at a higher elevation and only at the last moment began a rapid, nose-down descent that included several didoes, or evasive loops, before landing safely.

Cheney met Salih at the airport. The vice president brought only Scooter Libby and a military aide; Salih brought a top intelligence official. The two men discussed the Middle East peace process, Yemen's role in the war on terror, and Iraq. Salih, who had good relations with Saddam Hussein, warned Cheney that any invasion of Iraq by the United States would be certain to provoke the Iraqis to use their weapons of mass destruction.[7]

Cheney continued on his trip. In public, virtually everyone he met warned against removing Saddam Hussein. In private, however, people were telling him something different. Asked in a later interview whether the public statements matched the views he was given in one-on-one meetings, Cheney, predictably, declined to answer: "I don't want to talk about it."[8]

But others who accompanied him say that the United States' allies in the region wanted to have it both ways. In public, they would oppose the use of military force in Iraq. There were legitimate reasons for them do so, in both the long term and the short term. In some ways, the prospect of a democratic Iraq posed as many complications as living with a neighbor who had a history of hostile excursions into sovereign nations. The vision of an Iraq after Saddam Hussein as an island of democracy in a sea of despotism that made George W. Bush hopeful made Arab leaders nervous. If the Iraqis were free to choose their own leaders, wouldn't the Saudis want to do the same? And the Egyptians? And the Jordanians?

Adding to those complications, the anti-American elements in the region—the so-called Arab street—were strongly opposed to an invasion of Iraq by the United States. Government leaders who voiced support for such an invasion risked domestic unrest that could threaten their own grip on power.

In private, however, several of the Arab leaders Cheney met with were much more supportive. Twice in the past twenty-five years, Saddam had invaded his neighbors unprovoked. His belligerence was legendary, and his recent history meant that the United States had to keep at least some military forces in the region. This American presence caused its own problems. Islamists who opposed their own countries' governments often cited alliances with the United States as their reason. Osama bin Laden had targeted the Saudi regime for allowing "crusader armies" to occupy the "lands of Islam in the holy places." Ayman al Zawahiri, the Egyptian who became Osama bin Laden's deputy, had sought to overthrow Hosni Mubarak. And Abu Musab al Zarqawi, a young jihadist not yet known to the world, was well-known to the Jordanian government for his opposition to the relatively moderate King Abdullah II.

To help mollify their own populations, several Arab allies asked Cheney to say something—anything—that could be used to demonstrate the United States' sympathy with the Palestinians. Cheney said he was prepared to meet with the Palestinian leader Yasir Arafat on the trip, but only if Arafat denounced anti-Israel terror and took measurable steps to stop it. That did not happen.

The vice president had long been skeptical about Arafat. The United States had been trying to broker peace in the Middle East when Cheney first came to the White House in 1974. There had been countless agreements and numerous "breakthroughs," but the problems—and the fighting—persisted. Cheney also recalled that in those few moments when he had discussed issues with Bill Clinton on Inauguration Day, the former president had disparaged Arafat harshly, saying that the Palestinian leader had extracted numerous compromises from the Israelis at Camp David in 2000 and then walked away.

But on his return to the United States, Cheney discussed his trip on the Sunday talk shows, sounding more conciliatory toward the Palestinians. Prime Minister Ariel Sharon of Israel had imposed a travel ban on Arafat, and Cheney publicly asked that it be lifted in order to allow Arafat to attend an Arab summit in Lebanon. Cheney also promised the Palestinians a greater role in the Middle East peace process. The message to the United States' allies in the region was unmistakable: we're listening.

The months after 9/11 were marked by an unusual spirit in Washington, a rare feeling of bipartisan unity against a hated enemy. That spring, though, politics had returned to Washington.

In May, the Republican National Committee sent contributors a picture of President Bush taken aboard Air Force One on September 11. Democrats and even some Republicans were outraged at the use of the attacks for political purposes. The politicization of the war on terror had begun in earnest.

Then on May 15, 2002, a story on *CBS Evening News* provided an opportunity for the Democrats to go after Bush on his strongest issue. They took it. "The president's daily intelligence brief delivered each morning by the director of Central Intelligence warned in the weeks before 9/11 that an attack by Osama bin Laden could involve the hijacking of U.S. aircraft," reported David Martin.[9]

Cheney had long regarded Martin as one of the best reporters on national security working in Washington; many others shared his opinion. He thought this story was misleading. But the damage had been done.

Martin's report was followed by hundreds more, and by the following day the revelation had triggered a full-scale political war in Washington. Democrats accused the president and his top advisers of failing to "connect the dots" before September 11. The Bush administration, they said, had not done enough to prevent the attacks. Hillary Clinton took the Senate floor, demanding answers. She waved a copy of the *New York Post* from that morning. The headline made an incredible charge: "9/11 Bombshell: Bush Knew!"[10]

"I am simply here today on the floor of this hallowed chamber to seek answers to the questions being asked by my constituents, questions raised by one of our newspapers in New York, with the headline: Bush Knew," she said. "The president knew what?" She insisted that she was not trying to place blame; her constituents simply deserved answers to their many questions.[11]

At a press conference the same day, the House minority leader, Dick Gephardt, invoked the Watergate-era formulation: "What we have to do now is find out what the president—what the White House—knew about the events leading up to 9/11, when they knew it, and most importantly, what was done about it."[12]

Cheney and his colleagues at the White House were angry.

The president's daily briefing (PDB) typically includes intelligence on a wide variety of subjects in numerous individual reports. In the briefing of August 6, most of the information on Al Qaeda's potential attacks was from 1999, and all of it lacked specificity. Cheney thought the Democrats were using the report in a dishonest—and dishonorable—way.

In a speech in New York that same evening, he fired back: "What I want to say to my Democratic friends in the Congress is that they need to be very cautious not to seek political advantage by making incendiary suggestions, as were made by some today, that the White House had advance information that would have prevented the tragic attacks of 9/11. Such commentary is thoroughly irresponsible and totally unworthy of national leaders in a time of war."[13]

Members of Congress called for the White House to release the entire PDB, titled "Bin Ladin Determined to Strike in U.S." Although doing so would have likely quelled the controversy, Cheney counseled against it. That release wouldn't happen for another two years. And when it did, the warning did not seem as stark as the Democrats had made it sound.

Much of the information was old. "A clandestine source said in 1998 that a Bin Ladin cell in New York was recruiting Muslim-American youth for attacks."

And the segment of the briefing that generated so many headlines and so much speculation came with a seldom-quoted disclaimer: "We have not been able to corroborate some of the more sensational threat reporting, such as that from a [redacted] service in 1998 saying that Bin Ladin wanted to hijack a U.S. aircraft to gain the release of 'Blind Shaykh' Umar Abd al-Rahman and other U.S.-held extremists.

"Nevertheless, FBI information since that time indicates patterns of suspicious activity in this country consistent with preparations for hijackings or other types of attacks, including recent surveillance of federal buildings in New York."

It was hardly the kind of intelligence that could have been used to prevent an attack. And it certainly did not indicate that the White House had been warned about 9/11. The episode provided an important lesson. The fellow feeling that came after 9/11 was trickling away. The inhabitants of Washington, D.C., were reverting to form. The months ahead would be contentious.

There was another lesson, directly related to the drive toward war in Iraq. Saddam Hussein had been openly defiant of the UN's weapons inspectors for more than a decade, and had finally kicked them out in 1998. He had attempted to assassinate President George H. W. Bush in April 1993. He was harboring several well-known terrorists, including Abu Nidal and an Iraqi fugitive from the attack of 1993 on the World Trade Center in New York. He was known to be financing suicide attacks in the Middle East.

There was more. The Clinton administration had cited Iraqi support for the production of chemical weapons at the al Shifa chemical plant in Sudan to justify bombing the facility in retaliation for the al Qaeda bombings of the embassies in East Africa in 1998. The American media had reported that the Iraqi regime offered safe haven to Osama bin Laden in late 1999.[14] And after 9/11 the Iraqi regime had gloated over the attacks, declaring them a "lesson" in the perils of aggression. And all this time, Saddam Hussein had consistently claimed that the "mother of all battles," begun in 1991, was not yet finished.

If the Democrats were willing to accuse the White House of failing to connect the few dots available to policy makers before 9/11, what would they do if Iraq once again turned to aggression?

The White House put Cheney on *Meet the Press* to answer the criticism and defend the president. He used the appearance to highlight the nexus between terrorists and rogue states, particularly Iraq. "It takes us back into the 'axis of evil' speech the president made at the State of the Union, our concerns about Iraq, our concerns about the possible marriage, if you will, between the terrorist organization on the one hand and a state that has or is developing weapons of mass destruction on the other. And if you ever get them married up—that is, if somebody who has nukes decides to share one with a terrorist organization, with the expectation they'll use it against us—obviously we've got another problem."[15]

Over the course of the spring and summer, Cheney and his

top aides took several day trips to Langley, Virginia, a leafy suburb of Washington, D.C., that is home to the CIA. As he had
during his days in Congress, Cheney wanted to go beyond the
analyses delivered to him by his briefers.

"If you really want to dig into an issue, then there are lots
of other things you can do—ask for special studies, you can
have briefers come to the office, or to my house. We've done
all of that—or go out to the agency itself, NSA, CIA, whatever
it might be, and spend time going over, getting briefed, hearing
from a whole group of people on particular subjects. So I did that
on a number of occasions."[16]

Jami Miscik, deputy director for intelligence, and John
McLaughlin, deputy director of the CIA, attended virtually all
the briefings; the CIA's director, George Tenet, sat in on some.
Supplementing this core group were analysts and regional experts who were best prepared to answer Cheney's specific inquiries. Cheney usually brought Scooter Libby; John Hannah,
his deputy national security adviser; and Neil Patel, his staff
secretary.

Cheney asked most of the questions. If the briefings slid into
issues that involved operational details, Cheney would ask his
staff and the CIA personnel to step out of the room so that he
could speak one-on-one with the highest-ranking CIA official
present.

In some cases, the vice president and his staff came away impressed at the depth of knowledge demonstrated by the analysts,
who sometimes gave thoughtful and comprehensive responses to
detailed queries. The meetings were always professional. "I don't
recall that these sessions were ever particularly contentious,"[17]
McLaughlin would say.

Although Iraq's WMD programs would become the center-
piece of the Bush administration's case for war, most of Cheney's
sessions at the CIA—one participant remembers that there were
ten in all—focused on the nature of the Iraqi regime and its history of support for terrorism. "He didn't ask much on the WMD
side, because the agency's position was clear," says one official.
And on those other issues, Cheney and his staff were not impressed with the level of interest or the quality of analysis.

The CIA's experts couldn't answer the most basic questions about the Iraqi regime or its relationships. Several of them seemed uninterested in the issue of Iraq's support for terrorism, even resentful that they were being asked about it.

One point of disagreement that came up repeatedly was the assumption shared by many at the CIA—particularly those on its Near East–South Asia desk—that religious differences would preclude cooperation between the secular Iraqi regime and Islamic fundamentalists like Osama bin Laden. Cheney and his staff pressed the analysts on the issue. Why had Saddam Hussein put *Allahu akbar* on the Iraqi flag after the Gulf War? Why did he finance religious extremists? And what about the many reports of a nonaggression pact between Iraq and al Qaeda?

In several instances, the team from Cheney's office seemed to know the CIA's own material better than the analysts who had prepared it. Jami Miscik was embarrassed. On more than one occasion, she apologized to Cheney after an analyst left the room. "I don't think we were really focused on the [counterterrorism] side," one analyst in the intelligence community would later admit to Senate investigators, "because we weren't concerned about the [Iraqi intelligence service] going out and proactively conducting terrorist attacks. It wasn't until we realized that there was the possibility of going to war that we had to get a handle on that."[18]

As the administration privately weighed the costs and consequences of a preemptive war in Iraq, a similar argument was taking place in newspapers and on television screens across the country. In late August, the public debate exploded in op-eds from Democratic and Republican heavyweights in foreign policy, including one column each from two of Cheney's good friends.

In his time at the Pentagon, Cheney had met almost every Wednesday morning for breakfast with Brent Scowcroft and James Baker III. The three men had discussed the prospect of a war with Iraq and their respective roles in planning for such a possibility. The meetings had been friendly. There was almost always more agreement than disagreement, and whatever disputes they had stayed among the three of them.

Things were different now. Within eleven days, Scowcroft,

Baker, and Cheney would lay out sharply divergent views of the threat from Iraq and how to eliminate it. Scowcroft went first, in an op-ed in the *Wall Street Journal*, "Don't Attack Saddam." The former national security adviser wrote that "scant evidence" connects Saddam Hussein to terrorism and that Saddam was "unlikely to risk his investment in weapons of mass destruction, much less his country, by handing such weapons to terrorists who would use them for their own purposes and leave Baghdad as the return address." He concluded: "An attack on Iraq at this time would seriously jeopardize, if not destroy, the global counterterrorist campaign we have undertaken."[19]

Baker disagreed. "Regime change in Iraq is the policy of the current administration, just as it was the policy of its predecessor," he wrote in the *New York Times.* "That being the case, the issue for policymakers to resolve is not whether to use military force to achieve this, but how to go about it. The only realistic way to effect regime change in Iraq is through the application of military force." But before taking unilateral action, Baker wrote, the United States should take its case to the UN Security Council for a vote on the many resolutions Saddam had violated.[20]

Cheney followed Baker by one day, when he spoke in Nashville to the 103rd Convention of the Veterans of Foreign Wars. His speech would be the last word from the "Gulf War threesome" and the first words of the Bush administration's concerted campaign for war in Iraq. Promising a "candid appraisal of the facts," Cheney started at the beginning.

"After his defeat in the Gulf War in 1991, Saddam agreed under UN Security Council Resolution 687 to cease all development of weapons of mass destruction. He agreed to end his nuclear weapons program. He agreed to destroy his chemical and his biological weapons. He further agreed to admit UN inspection teams into his country to ensure that he was in fact complying with these terms."[21]

Saddam did not abide by this agreement, Cheney continued. Not only had he enhanced his capabilities in the field of biological and chemical weapons, Cheney declared; he had resumed his nuclear program. "Many of us are convinced that Saddam will acquire nuclear weapons fairly soon," he said. "Just how soon, we cannot really gauge. Intelligence is an uncertain business, even in the best of circumstances."[22]

Cheney offered an example. "Prior to the Gulf War, America's top intelligence analysts would come to my office in the Defense Department and tell me that Saddam Hussein was at least five or perhaps even ten years away from having a nuclear weapon. After the war we learned that he had been much closer than that, perhaps within a year of acquiring such a weapon."[23]

Some people in the intelligence community bristled at Cheney's public accusation. But, if anything, Cheney softened the judgments others had made at the time. Eleven years before Cheney's speech to the VFW convention, Les Aspin, a Democrat, who was chairman of the House Armed Services Committee, appeared on ABC's *This Week* and disparaged the intelligence on Iraq that preceded the Gulf War. Asked about the quality of intelligence on Saddam Hussein and Iraq, Aspin described it as "not very good, not very good. We missed a whole different program he had there. He had a program that was developing a much more crude, more primitive way of developing nuclear weapons that we missed entirely. This was a much bigger program than we thought at the time, and in fact, as the polls showed at the time, the one really rallying call to the American public for the war was the nuclear threat. If the public had really known that, there probably would have been a lot more support for the war, if they'd really known what was really going on a year ago, in August, if they'd have known the extent of the Iraqi program, I think there would have been a lot more support for the use of force."[24]

If the flawed intelligence had led to flawed policy, Aspin said, an even more dangerous Saddam Hussein would be in power. "If we'd have adopted a policy of letting sanctions work to get him out of Kuwait, we clearly would have faced a nuclear threat well before those sanctions had any chance of biting to the degree that was necessary to get him out."[25]

Cheney's speech came less than three weeks before President Bush was scheduled to appear before the UN General Assembly. Cheney was not sanguine about Bush's prospects there; he believed that the UN had done too little to thwart Saddam Hussein's ambitions regarding WMD over the past decade and he had little confidence it would be more successful if given another chance. Although the UN's inspectors had found and destroyed stockpiles of weapons, the record showed that at other times Saddam's regime had duped them. Cheney noted the defection of Hussein Kamel, Saddam's son-in-law, who, after leaving Iraq in 1995, pointed inspectors to a chicken farm where they found evidence of the continuing programs. The lesson, said Cheney, is "that we often learned more as the result of defections than we learned from the inspection regime itself."[26]

Then he went farther.

"Against that background, a person would be right to question any suggestion that we should just get inspectors back into Iraq, and then our worries will be over. Saddam has perfected the game of cheat and retreat, and is very skilled in the art of denial and deception. A return of inspectors would provide no assurance whatsoever of his compliance with UN resolutions. On the contrary, there is a great danger that it would provide false comfort that Saddam was somehow 'back in his box.'"[27]

The cable networks had taken some of Cheney's speech live. Reporters took Cheney's comments about inspections to mean that Bush had decided to forgo the United Nations. They wanted to know whether the president stood behind Cheney's remarks. The White House press secretary, Ari Fleischer, said he did.

Cheney had been skeptical of going to the UN. Over the previous few weeks, he had had several sharp exchanges with Colin Powell about the wisdom of seeking a new resolution from the UN Security Council.[28] In Cheney's view, the Security Council had already passed seventeen resolutions that Saddam Hussein had violated. Why get another one? If we try and fail, he argued, it will be much harder to do what we may need to do.

The national security adviser, Condoleezza Rice, had seen the speech in draft form. She thought it was an aggressive speech, but didn't see it as anything that would cause problems. But predicting what would make news was never easy. "I didn't think much of it, really. It wasn't the first time that I missed and others missed how it would be read," she says.

Rice called Cheney to voice her concerns. "I just told the vice president, I said I think it's been really misread and I think it's cut off the president's options." Cheney was dubious of the UN, but he hadn't intended to foreclose any alternatives with his speech. He simply thought it was useful to lower expectations about what an inspection regime might accomplish. Or, more important, might not accomplish. He agreed to clarify his remarks in his next speech. Although he remained skeptical about inspections, he did not repeat his argument that they provided a false sense of security to the world.

On September 12, 2002, after close consultation with Powell and British prime minister Tony Blair, Bush made clear that he would seek another UN Security Council resolution. It would be an important decision that would shape the case for war. There was a broad consensus in the international community that Iraq had weapons of mass destruction and, separately, there was no question he had failed to comply with UN resolutions requiring Iraq to demonstrate that his weapons had been destroyed.

"It ended up that the WMD thing got emphasized as opposed to, say, Saddam's ties to terror—I expect in part because it was easier, because there were a lot of resolutions already on the books that dealt with that," says one senior administration official. "It was the easy subject to go to during the course of the debate and trying to pull together an international consensus to take action against Saddam Hussein. So I think there's probably some truth to the notion that the focus on the UN made WMD more important in terms of the case that was being argued."

The Bush administration took its case to Congress. Cheney did not think the White House needed congressional approval to send troops to the Persian Gulf, and as he had before the Gulf War, Cheney warned that seeking approval and failing to obtain it would be a political setback. Still, he was confident they would prevail.

Even the most vocal war opponents conceded that Saddam Hussein posed a threat. Senator Ted Kennedy articulated this view in a speech opposing military action in Iraq shortly before Congress would vote to authorize war. "There is no doubt that Saddam Hussein's regime is a serious danger, that he is a tyrant, and that his pursuit of lethal weapons of mass destruction cannot be tolerated. He must be disarmed."[29]

Former president Bill Clinton weighed in on the debate about Iraq from a trip to South Africa. He rejected suggestions that efforts to disarm Iraq would be a distraction from the broader war on terror. "I think we can walk and chew gum at the same time. That is, I think we can turn up the heat on Iraq and maintain an intense focus on terror," he said in an interview with ABC News. Clinton encouraged the Bush administration to return to the United Nations, but left little doubt that he remained concerned about Iraq's WMD programs.

"We should go back to the United Nations, get a tough resolution that says, okay, Saddam Hussein says he'll allow unfettered inspections, he will go along with the will of the international community to totally disarm his chemical and biological and nuclear capacity. . . . He's got a dangerous program. We need to eliminate it."[30]

Many Democrats had supported regime change in Iraq when Bill Clinton was president, and Cheney believed it would be hard for them to argue that Saddam Hussein, after four years without UN weapons inspections, was less dangerous than he had been under their president.

John Murtha, a man Cheney had regarded as his greatest ally on Capitol Hill during his time at the Pentagon, was struggling with how to vote on the resolution. He had been a leader of the House Democrats who voted in favor of authorizing the first Gulf War in 1991. But this time he wasn't sure. Brent Scowcroft and his associates had been calling Murtha to urge him to oppose the resolution. But while Murtha respected Scowcroft, he trusted Cheney. So he asked to see the vice president.

Cheney and Murtha met in Cheney's House office in the Capitol. Murtha explained his concerns. He believed Saddam Hussein had weapons of mass destruction and was worried that the Iraqi leader would use them on American soldiers in the event of an invasion by the United States. But he did not believe that Iraq had a significant relationship with al Qaeda and did not share the Bush administration's sense of urgency regarding regime change. Cheney promised to take Murtha's concerns back to the president. Then he made a strong case for war.

Despite his reservations, Murtha voted to authorize the war.

And after a trip to the region in late November, he was glad that he had. "I came back convinced that I'd made the right vote because the military said to me: 'We've got a red line drawn around Baghdad. And we think that they're going to use weapons of mass destruction—matter of fact, we verified that, monitoring their cell phones.'"[31]

Eighty of Murtha's Democratic colleagues in the House of Representatives voted with him to support the resolution. The final vote total in the House was 296 to 133. The Senate, with a one-seat majority for the Democrats, voted 77 to 23 to authorize war. The *New York Times* called the vote "overwhelming."[32]

Shortly before the vote, the U.S. intelligence community produced an authoritative document for policy makers. The National Intelligence Estimate (NIE) was written to reflect the consensus of the agencies that make up the intelligence community. Its title reflected their collective certainty: "Iraq's Continuing Programs for Weapons of Mass Destruction."

The "key judgments" in the document were equally categorical:

- We judge that Iraq has continued its weapons of mass destruction (WMD) programs in defiance of UN resolutions and restrictions. Baghdad has chemical and biological weapons as well as missiles with ranges in excess of UN restrictions; if left unchecked, it probably will have a nuclear weapon during this decade. (See INR alternative view at the end of these Key Judgments.)
- We judge that we are seeing only a portion of Iraq's WMD efforts, owing to Baghdad's vigorous denial and deception efforts. Revelations after the Gulf war starkly demonstrate the extensive efforts undertaken by Iraq to deny information. We lack specific information on many key aspects of Iraq's WMD programs.
- Since inspections ended in 1998, Iraq has maintained its chemical weapons effort, energized its missile program, and invested more heavily in biological weapons; in the view of most agencies, Baghdad is reconstituting its nuclear weapons program.

- We assess that Baghdad has begun renewed production of mustard, sarin, GF (cyclosarin), and VX; its capability probably is more limited now than it was at the time of the Gulf War, although VX production and agent storage life probably have been improved.
- We judge that all key aspects—R&D, production, and weaponization—of Iraq's offensive BW program are active and that most elements are larger and more advanced than they were before the Gulf war.

The NIE specified that the intelligence community had "high confidence" in several of its most pointed conclusions.

- Iraq is continuing, and in some areas expanding, its chemical, biological, nuclear and missile programs contrary to UN resolutions.
- We are not detecting portions of these weapons programs.
- Iraq possesses proscribed chemical and biological weapons and missiles.
- Iraq could make a nuclear weapon in months to a year once it acquires sufficient weapons-grade fissile material.

The intelligence Cheney and President Bush were seeing on Iraq's WMD programs was even more unequivocal.[33] Most intelligence products, Cheney knew, are full of "on-the-one-hand," "on-the-other-hand" analysis. He found the reporting on Iraq's WMDs different.

There were caveats—the State Department would dissent on one item; the DIA would register a different view on another. But the cumulative views of the intelligence community on Iraq's WMD programs struck Cheney as remarkably unified and uncharacteristically emphatic. "They were not presenting qualifiers," he says.[34]

At the same time, Cheney wasn't exactly looking for nuance. Twelve years earlier, days after Iraq invaded Kuwait, he had dropped the CIA briefer from his presentation to King Fahd. The analyst had been too wishy-washy. He was out.

If the intelligence community was in agreement on Iraq's WMD programs, the same could not be said for its assessments of Iraqi support for terrorism. Many analysts remained skeptical about a connection between Iraq and al Qaeda, but there were pockets of dissent. And one of those who increasingly believed that there was a relationship was the director of Central Intelligence, George Tenet.

Tenet had assigned analysts from the Counterterrorist Center (CTC) to review existing reporting on Iraq and al Qaeda. He told them to be dogged and to look for worst-case scenarios. When the CTC gave its assessments to Tenet, the analyses lent credence to a relationship between Iraq and al Qaeda, and the CIA Director found them persuasive.[35]

On October 7, 2002, as the Senate was preparing to vote on a resolution authorizing war in Iraq, Tenet sent a letter to Senator Bob Graham, who was then chairman of the Senate Select Committee on Intelligence. Tenet laid out his evidence for a relationship between Iraq and al Qaeda:

- Our understanding of the relationship between Iraq and al Qaeda is evolving and is based on sources of varying reliability. Some of the information we have received comes from detainees, including some of high rank.
- We have solid reporting of senior level contacts between Iraq and al Qaeda going back a decade.
- Credible information indicates that Iraq and al Qaeda have discussed safe haven and reciprocal nonaggression.
- Since Operation Enduring Freedom, we have solid evidence of the presence in Iraq of al Qaeda members, including some that have been in Baghdad.
- We have credible reporting that al Qaeda leaders sought contacts in Iraq who could help them acquire WMD capabilities. The reporting also stated that Iraq has provided training to al Qaeda members in the areas of poisons and gases and making conventional bombs.
- Iraq's increasing support to extremist Palestinians coupled with growing indications of relationship with al Qaeda sug-

gest that Baghdad's links to terrorists will increase, even absent U.S. military action.

The phrasing of the fourth item on that list may have been a result of questions Cheney asked about reporting he had seen. In 2002, the vice president had been briefed on fresh intelligence that members of the Egyptian Islamic Jihad had made their way to Iraq and had begun setting up safe houses in Baghdad. Cheney found the report interesting, but odd. He had understood that Egyptian Islamic Jihad had merged with al Qaeda several years earlier. Ayman al Zawahiri, the group's longtime leader, was now Osama bin Laden's chief deputy. Cheney wanted to know why the report did not simply conclude that al Qaeda was setting up safe houses in Baghdad.

He returned the report to the CIA with a question: Would it be accurate to substitute "al Qaeda" for every mention of "Egyptian Islamic Jihad?" The answer did not come immediately, but when it did, the CIA finally acknowledged that members of al Qaeda were operating in Baghdad.

To Cheney, the episode was one example of many that demonstrated the unwillingness of some CIA analysts to take an objective look at Iraq and its support for radical Islamic terrorists, al Qaeda in particular. In this case, analysts were so determined to avoid reporting the presence of al Qaeda members in Iraq that they presented Cheney with a less-than-accurate description of the situation in Baghdad.

On November 8, 2002, the UN Security Council passed Resolution 1441 on a vote of fifteen to none. In direct language, the resolution warned of the "threat Iraq's noncompliance with Council resolutions and proliferation of weapons of mass destruction and long-range missiles poses to international peace and security" and declared Iraq in "material breach" of its previous obligations. The Security Council would provide a "final opportunity" for Iraq to disarm, and failure to do so would result in "serious consequences."

Four days later, the French foreign minister Dominique de Villepin explained France's vote to his countrymen in a radio broadcast. For years France had carried Baghdad's water on the Security Council. And though Villepin and President Jacques Chirac of France said they wanted to avoid a war in Iraq, their willingness to sign the resolution seemed to indicate that they would not try to block any attempt by the United States to use force. "If Saddam Hussein does not comply," said Villepin, "if he does not satisfy his obligations, there will obviously be a use of force. . . . The security of the Americans is under threat from people like Saddam Hussein who are capable of using chemical and biological weapons."[36]

As the Bush administration moved closer to war, the message that policy makers were getting from the nation's top intelligence officials was unambivalent. "It's a slam-dunk case," Tenet told Bush at a meeting in the Oval Office.[37]

Bush, Cheney, and others would later come under intense criticism for seizing on—and publicizing—the most alarming intelligence as they contemplated a war in Iraq. Cheney says that their case reflected the intelligence they were given.

"The doubts were not what was being reported," he says. "What I remember when you talk about doubt was sitting in the Oval Office with the president and he asked Director Tenet, how good is the case against the Iraqis on WMD? And George responded, 'It's a slam dunk, Mr. President. It's a slam dunk.' We took that as pretty solid. No qualifiers."[38]

On January 28, 2003, Cheney sat directly behind the lectern, next to his friend Denny Hastert, as President Bush used his State of the Union address to make the case against Iraq to the American people. "Some have said we must not act until the threat is imminent," he said. "Since when have terrorists and tyrants announced their intentions, politely putting us all on notice before they strike? If this threat is permitted to fully and suddenly emerge, all actions, all words, all recriminations would come too late. Trusting in the sanity and restraint of Saddam Hussein is not a strategy, and it is not an option."

Bush then got specific. He listed the vast amounts of banned weapons that Iraq had failed to account for when the UN inspectors left at the end of 1998. After each one, Bush said: "He hasn't accounted for that material. He's given no evidence that he has destroyed it."

He continued. "The International Atomic Energy Agency confirmed in the 1990s that Saddam Hussein had an advanced nuclear weapons development program, had a design for a nuclear weapon, and was working on five different methods of enriching uranium for a bomb. The British government has learned that Saddam Hussein recently sought significant quantities of uranium from Africa."[39]

The last sentence would become the most heavily scrutinized claim made by the Bush administration. It would become so familiar that it developed its own shorthand: "The Sixteen Words." It set off a battle over intelligence that would stretch into the final years of the Bush administration and would result in the prosecution of Scooter Libby, Cheney's chief of staff.

Weeks before the State of the Union address, top officials of the administration understood that they would have to make their case to the world in a comprehensive presentation. The UN Security Council, having passed Resolution 1441 threatening "serious consequences" for noncompliance by Iraq, was thought to be an appropriate venue.

Cheney's office would prepare the materials to be used in making the case. Scooter Libby handled the WMD component; Neil Patel compiled the brief on Iraq's support for terrorism; and John Hannah organized the case on Saddam Hussein's long record of abusing human rights. When they started their work, President Bush had not yet decided who would present the United States' case to the Security Council.

Libby told his colleagues to prepare for a lengthy presentation, one that might last two days. His idea, supported by his boss, was that the most persuasive case would be the most exhaustive. It would be an indictment of the regime in Iraq, to be sure, but also of the members of the Security Council who sat by passively as the Iraqis violated tough-sounding agreements. The United States had plenty of strong intelligence, they believed, though probably not something that most people would regard as a "showstopper," in the words of one official. Libby and his colleagues spent the last half of January compiling three six-inch binders for the presentation.

President Bush decided that his chief diplomat, Secretary of State Colin Powell, would represent the United States at the Security Council. Almost immediately after he was chosen, Powell rejected the proposal for a two-day presentation. He wanted something short and hard-hitting. The address would focus on Iraq's proscribed weapons, he decided, because most of the UN resolutions on Iraq related to its WMD programs. It was a logical decision, given his audience, but a second departure from the presentation envisioned by Cheney's staff, which would have devoted considerably more time to Iraq's support for terrorism and Saddam Hussein's history of brutality.

In the end, Powell's presentation took less than three hours. He spent 80 percent of his time discussing Iraq's weapons of mass destruction.

On March 2, the International Atomic Energy Association (IAEA) reported that documents it had been given by the U.S. government purporting to show a uranium deal between Iraq and Niger were forgeries. The UN agency disputed the claims of the Bush administration that Iraq had an active nuclear program. Cheney was asked generally about the IAEA's conclusions regarding Iraq's nuclear program during an interview on *Meet the Press* on March 16, 2003.

Cheney did not address the forged documents directly, but he did make a startling claim. "We believe he has, in fact, reconstituted nuclear weapons."[40] It was a claim not supported by the U.S. intelligence community's assessments. Cheney would later say that he simply misspoke, and the transcript seems to support him: three other times on the same show he spoke only of the possibility that Iraq was reconstituting its nuclear program. Still, the claim—reconstituted—would be used dozens of times to illustrate Cheney's eagerness to overstate the threat from Iraq.

On the afternoon of March 19, 2003, George Tenet and Donald Rumsfeld came to the Oval Office with exciting news. The CIA had gotten a solid tip that in a matter of hours, Saddam Hussein was expected to be at a complex on the Tigris River called Dora Farms. The CIA's sources on the ground in Iraq reported that they had seen his sons, Uday and Qusay Hussein, at the complex and understood that Saddam, who had been there the night before, would return after midnight.

If we want to be sure to get him, Tenet told Bush, this is our chance. "Tenet was fairly firm in his opinions," says a senior administration official involved in the debate. "He thought the sourcing was good."

Others on Bush's national security team considered the many questions such a targeted attack raises. There were legal questions in addition to concerns about strategy and public relations. Was Saddam Hussein a legitimate target? As a commander of enemy forces, they determined, he was. What does this do to our war plan? What if we open the war by accidentally killing women and children?

Cheney remained quiet throughout most of the discussion. He was skeptical about the intelligence. He understood from his meetings with analysts that the CIA's sourcing in Iraq was weak. He worried that they were being set up.

But Tenet was persistent. "It's as good as it gets," he said. "I can't give you 100 percent assurance, but this is as good as it gets."[41] He had said the same thing in meetings at the Pentagon earlier in the day.

Bush was particularly concerned about collateral damage. "I needed to make sure that when we took the attack that we weren't being set up and that the first rocket attack into Baghdad—that had advanced, by the way, the strategy that we had already agreed on wasn't going to end up killing a busload of children, for example. I needed more information."[42]

As Cheney considered the pros and cons, he thought that the potential benefits of a successful attack outweighed the harm that could be done if they didn't get Saddam. "It was a good signal to send to go after Saddam Hussein at the outset," Cheney says. "If he were there and we did get him, it could, in fact, save lives and shorten the conflict by a significant margin. And if he wasn't there and we didn't get him, it still wasn't a bad way to start the enterprise."[43]

Cheney shared his view with Bush. "His attitude was, decapitate and we get this thing over quicker," Bush recalls. The president was surprised by the advice. "I thought he was at times cautious," Bush says. "But not in this case."[44]

The others in the Oval Office—including Tenet, Powell, Rice, and Rumsfeld—all supported bombing Dora Farms. Cheney's advice was decisive.

"If you trust it, go."

At 8:35 PM Cheney began calling congressional leaders. He telephoned the Senate majority leader, Bill Frist; the Speaker of the House, Dennis Hastert; and the House Majority leader, Tom DeLay. Cheney briefly told them about Bush's decision and his imminent speech to the nation.

President Bush had also assigned Cheney the job of reaching out to several world leaders. One of those on the vice president's list was Bashar al-Assad, president of Syria, which has a long border with Iraq.

"He wouldn't take my call," says Cheney.

When Cheney finished his work, he spoke briefly with Bush in the Oval Office and then returned to his office down the hall to watch the president's speech. With the channel tuned to Fox News, Cheney listened to Bush make the case for war one last time. The war would open with an attempt to decapitate the Iraqi regime by targeting Saddam at Dora Farms.

"My fellow citizens, at this hour, American and coalition forces are in the early stages of military operations to disarm Iraq, to free its people and to defend the world from grave danger. On my orders, coalition forces have begun striking selected targets of military importance to undermine Saddam Hussein's ability to wage war," Bush said.

He added, "A campaign on the harsh terrain of a nation as large as California could be longer and more difficult than some predict. And helping Iraqis achieve a united, stable, and free country will require our sustained commitment."[45]

Mike Gerson, who wrote the speech, was prescient when he included those words. But few at the White House that night expected a long war.

For a short time, Cheney was alone with his photographer, David Bohrer, who quietly snapped pictures of the vice president. Cheney had been pushing for this war as hard as anyone at the White House. He had known war as secretary of defense and vice president. As he sat at his desk, the gravity of the undertaking, and the seriousness of its purpose, was written on his face. Bohrer's photographs from that night reveal precisely the same expression Cheney wore in the bunker on September 11.

After the Gulf War, Cheney had been an outspoken defender of leaving Saddam Hussein in place. Removing Saddam, he often said, would not be worth the cost. Fighting in Iraqi cities could be messy, and creating a government among Sunnis, Shiites, and Kurds would be difficult. Arab countries wouldn't support an occupation. Americans would die, perhaps in significant numbers. "And the question in my mind," Cheney had said, "is how many additional American casualties is Saddam worth? And the answer is not very damned many."[46]

The risks of removing Saddam Hussein remained. So what had changed?

"I think after 9/11 when you move to a situation where your biggest threat is the possibility of terrorists, state-sponsored terrorists, or a terrorist with a relationship with a rogue government able to get their hands on deadly technologies, Saddam Hussein is a hell of a problem. And he was a problem before 9/11, but he became a bigger problem after 9/11 in light of that threat that we're living with still to this day, the possibility of an al Qaeda cell in the middle of one of our cities with a deadly biological agent or a nuclear weapon."[47]

Moments after the bombing started, Scooter Libby joined him. Libby stood about ten feet behind his boss. Two oversize cardboard maps—one of the greater Middle East and the other of Iraq—were propped up against the wall at his feet. Libby watched the TV over Cheney's shoulder. For more than an hour, the two men watched the first news footage of the bombing of Baghdad. Another long war had begun.

CHAPTER FOURTEEN

The War over the War

Baghdad fell on April 9, 2003.

An enduring image of the war took place that day when jubilant Iraqis teamed with U.S. Marines to topple a statue of Saddam Hussein in al Firdos Square, in the heart of the Iraqi capital. American troops had stormed into Baghdad, meeting unexpectedly little resistance. In the square, they stopped to wait for reinforcements, and when the Iraqis there could not bring down the statue on their own, the Americans used their heavy equipment to lend a hand. A rope was looped around the neck of the statue, which was hauled to the ground before a cheering crowd. It was a moment of triumph, and of hope. The Iraqi people—at least the ones captured on camera that day—were happy. And although pockets of intense fighting remained throughout the country, only 102 American soldiers had been killed in three weeks of combat.[1]

As during the first Gulf War and on 9/11, Cheney got an important part of his information about the war from the news media. He followed the work of journalists embedded with American units and regularly tuned in for televised updates. Sometimes he woke in the middle of the night to watch reports coming in from the desert.

As more Americans reached the Iraqi capital, it seemed more and more apparent that the end of that war was near. For the most part, U.S. troops were told to look the other way while Iraqis looted government buildings and Saddam Hussein's palaces. Some Baghdadis walked the streets toting lamps and office equipment. Others gathered old files of the Baath Party looking for information on relatives who had gone missing under Saddam Hussein's regime. There was gunfire from spontaneous celebrations throughout the country, as well as gunfire from sporadic battles between coalition troops and Iraqi diehards. That fighting seemed haphazard and indiscriminate, and it may have been. But the seeds of the coming insurgency had been sown. And though some American officials knew this even before "major combat" was over, many senior policy makers and intelligence officials did not.

In late April, Ambassador David Dunford, escorted by U.S. troops, led a small team of American officials to the building of the Ministry of Foreign Affairs in Baghdad. Others soon joined them, including Arabic-speaking Iraqi-Americans working with the Pentagon. The ministry building had been partially burned, and many of its offices had been ransacked. But some documents survived. Among them was a memo from the director of Iraqi intelligence, the Mukhabarat, dated February 2003. The document contained instructions to Iraqi intelligence officials and agents on how to create unrest in postwar Iraq—a blueprint for the insurgency. A second, similar document was "a list of jihadists, for want of a better word, coming into Iraq from Saudi Arabia before the war," said Dunford. "That suggested to me that Saddam was planning the insurgency before the war."[2]

Although the documents were provided to a "fusion cell" run by the CIA in Baghdad, senior policy makers, including Cheney, would not learn of their existence until 2006. The strength of the insurgency would be apparent by then.

Another enduring image of the war came on May 1, 2003, when President Bush spoke from aboard the USS *Abraham Lincoln,* off the coast of southern California. "Major combat operations in Iraq have ended," he said. In overthrowing the Iraqi regime, he said, the United States had won "a crucial advance in the campaign against terror. We've removed an ally of al Qaeda, and cut off a source of terrorist funding. And this much is certain:

No terrorist network will gain weapons of mass destruction from the Iraqi regime, because the regime is no more."[3]

Television cameras there to record the speech captured a large banner behind Bush that read, "Mission Accomplished." The president's speech told a slightly different story: "The transition from dictatorship to democracy will take time, but it is worth every effort. Our coalition will stay until our work is done. Then we will leave, and we will leave behind a free Iraq."[4]

One week later, on May 7, Cheney traveled to Dallas, Texas, for an event at Southern Methodist University. To end the spring semester, he had agreed to be interviewed by Hugh Sidey, a former Washington bureau chief for *Time* magazine. Sidey was something of a journalistic icon, having covered nine presidencies over the course of his distinguished career. Cheney had been in government for five of them. The two men had a lot they could discuss. The session was to be off the record, meaning that Cheney's words could not be quoted or characterized by journalists attending the event.

So he was surprised to learn the following day that several of his remarks had been published in the *Dallas Morning-News,* which had cosponsored the event. The newspaper carried Cheney's comments about the opening night of the war and the attack on Dora Farms. "I think we did get Saddam Hussein," Cheney had said, adding that the information was unconfirmed. "He was seen being dug out of the rubble and wasn't able to breathe."[5]

Ironically, the article also quoted Cheney on the subject of his low public profile. "From time to time, they trot me out when it makes sense to do so," he said.[6]

Cheney had been burned by reporters many times before, and his opinion of journalists had steadily declined during his three years in office. But this was a first. In the very story in which Cheney's remarks appeared, the newspaper admitted violating the ground rules of the event.

"Before Mr. Cheney's remarks, university officials announced late Tuesday afternoon that the session would be considered off the record," wrote the reporter Gromer Jeffers Jr. "News cameras were allowed to film the event without sound for the first 10 minutes."

The article quoted Robert W. Mong Jr., the president and editor of the paper. "As one of the event's sponsors, we were not consulted in advance about the ground rules for Vice President Cheney's lecture. Had we been asked, we would not have found the rules acceptable. The event should have been open."[7]

The story also quoted Cheney on what would become one of the most intensely discussed issues of Bush's presidency: Iraq's missing weapons of mass destruction. It had been six weeks since the start of the war and six days since President Bush declared that major combat operations had ended. And although there had been dozens of reports from the military about potential caches of WMD, none of them had checked out.

Where were they? Three months earlier, Colin Powell had focused his presentation to the UN Security Council on the evidence obtained by the United States of Iraq's WMD programs. "Every statement I make today is backed up by sources, solid sources," he had said. "These are not assertions. What we're giving you are facts and conclusions based on solid intelligence." And yet Saddam Hussein had not used these proscribed weapons during the war, as everyone including many war opponents had feared.

"We had a lot of evidence and intelligence reporting numerous sources before the war indicating that he had continued his chemical and biological programs," Cheney told his audience at SMU. "If he had none, it would have been a simple matter for him to comply with the UN Security Council resolutions." Hussein, he said, "had years to hide it. . . . We have to look for it and take time to find it. But I think they will be found."[8]

One day before Cheney's speech, the columnist Nicholas Kristof of the *New York Times* wrote an article called "Mystery of the Missing WMD." It focused on the sixteen words President Bush had spoken in his State of the Union address as evidence for Iraq's ongoing nuclear-arms program: "The British government has learned that Saddam Hussein recently sought significant quantities of uranium from Africa."

Much of the reporting on the vanishing Iraqi weapons raised the possibility of an intelligence failure—a mistake—but Kristof's column suggested something far more sinister: that the Bush administration had knowingly used false and discredited information to take the nation to war. It was an extraordinary charge, but it was backed up by a source who claimed inside knowledge of the deception. Kristof's article included this passage:

> *I'm told by a person involved in the Niger caper that more than a year ago the vice president's office asked for an investigation of the uranium deal, so a former U.S. ambassador to Africa was dispatched to Niger. In February 2002, according to someone present at the meetings, that envoy reported to the CIA and State Department that the information was unequivocally wrong and that the documents had been forged.*
>
> *The envoy reported, for example, that a Niger minister whose signature was on one of the documents had in fact been out of office for more than a decade. In addition, the Niger mining program was structured so that the uranium diversion had been impossible. The envoy's debunking of the forgery was passed around the administration and seemed to be accepted—except that President Bush and the State Department kept citing it anyway.[9]*

The envoy was Joseph Wilson, whose wife had recommended him to her bosses at the CIA more than a year earlier. Kristof had met Wilson shortly before the column ran at a meeting of the Senate Democratic Policy Committee, and at a subsequent meeting Wilson gave the columnist permission to use the story without attributing it him.

Cheney's staff was perplexed by the allegation. One line in particular caught Scooter Libby's eye: "The vice president's office asked for an investigation of the uranium deal."

Libby was Cheney's national security adviser in addition to his chief of staff, and he would have known had his boss been involved in such an undertaking. But he had never heard of it. So Libby began asking questions. Who is this former ambassador? Who sent him to Niger? Why him? And why does he think the request originated with the vice president?

Libby instructed Cathie Martin, Cheney's communications

director, to tell reporters that Cheney didn't know about the trip or about an investigation of the uranium deal. Martin did this, and after a quick burst of interest, the issue seemed to fade away. It would be a temporary reprieve. Kristof's article created an easily understood story line: there are no weapons; the Bush administration lied. But the truth was considerably more complicated and would take years to unravel entirely.

Cheney, meanwhile, turned his attention to domestic policy and tax cuts. In 2001, after President Bush signed a $1.35 trillion tax cut, the administration pushed for and won passage of several additional tax cuts to stimulate the economy after 9/11. The country had avoided the economic disaster some had predicted in the wake of the attacks, but the economy had remained sluggish at the end of 2002. In January, President Bush proposed a new tax cut: $726 billion over ten years.

Earlier in his career, such a measure might have given Cheney pause. Although he had supported Ronald Reagan's tax cuts, his motivation was more institutional and partisan than economic or ideological. Reagan's fiscal plan was "the new game in town and we were there to support the president," he recalls.[10]

Reagan believed that cutting taxes would expand the economy and at least partially offset the revenues lost to the tax base with additional taxes paid at the lower rates. Cheney was familiar with the theory. He had been at the Two Continents Bar near the White House in 1974 when the economist Arthur Laffer first drew what would become known as the Laffer curve. Cheney remembers that Laffer roughed out the curve on a linen napkin.

Cheney embraced Laffer's theory, but he had his reservations. "I felt there was merit in the argument that Laffer made, that tax rates were important not just in terms of how much money you could collect from the society to finance government, but because it was a key incentive or disincentive to unleashing the creativity and entrepreneurial spirit of the American people, that you had to think about how tax rates affected individuals, and individual behavior, and individual decisions." But throughout the 1980s he remained concerned about deficits and was not convinced that the tax cuts would "pay for themselves."[11]

The situation in early 2003 was different. The country, already in a recession, had experienced a traumatic shock and remained at war. And although the deficit was growing, Cheney thought it was still manageable; in early 2003 it stood at a little more than 4 percent of the gross domestic product, compared with nearly 6 percent in the 1980s.[12]

The Senate Republicans, with a slim majority, found the price of Bush's proposal too high. Senator Charles Grassley, a Republican from Iowa who chaired the Senate Finance Committee, had committed to a tax cut no greater than $350 billion. Moderates in his caucus like Senator George Voinovich of Ohio and deficit hawks like John McCain refused to go higher.

The House Republicans, however, always more aggressive and usually more conservative, wanted a tax cut of $550 billion; and Bill Thomas, chairman of the House Ways and Means Committee, had substituted a cut in capital gains tax rates for the dividend tax cut preferred by the Senate and the White House.

Two things were clear. President Bush, whose ratings in the polls were still high after the toppling of the Iraqi regime, would get a significant tax cut. And Republicans in the House and Senate would battle one another fiercely before they reached a compromise.

For Cheney, the negotiations over the tax cuts in 2003 recalled the differences between the House and the Senate. They also reminded him why he always considered himself a man of the House.

The members from each chamber played to type. Depending on who was doing the characterizing, the House Republicans were either doughty and indomitable or impetuous and reckless. The senators were either stubborn and paternalistic or cautious and responsible.

Some of the differences reflected political reality. In the Senate, Republicans had a slim two-seat majority. The margin was greater in the House, 229 to 206. But Cheney thought the dispute between the Republican leaders had as much to do with institutional temperament as with numbers. The Republican leaders in the House were so skeptical about their counterparts in the Senate that they sometimes placed bets—sitting in the Oval Office—on how many times the Senate majority leader, Bill Frist, would come up with an excuse as to why he couldn't accomplish one thing or another that the House wanted passed.

From early in the Bush administration, Cheney had attended the Senate Republicans' policy luncheons on Capitol Hill. These weekly lunches are essentially brainstorming sessions, where the senators review pending legislation and discuss items likely to be on the agenda soon. Cheney usually comes without a specific goal and says little.

"He's very quiet," says Senator John McCain. "If it's an issue that he feels strongly about, then he stands up and speaks. But generally he doesn't speak up. I would say, three out of four lunches that he comes to he doesn't say anything. He just has lunch, sits at a big table with guys, converses, and after lunch people talk to him."[13]

The interactions give the senators an opportunity to report to their colleagues and constituents that they spoke to the vice president about one issue or another. Says McCain: "It's very good because given the egos of the senators, they can say, 'Well, I told Cheney . . .'"[14]

"I've seen him listen to some tirades from senators that would try anyone's patience," says McCain. "He stands there, smiles. Polite."

McCain sits up; straightens his face; and, speaking in an exaggerated monotone, does his best impersonation of Dick Cheney: "Thanks very much. Thank you. Yes."

"I've seen a guy come up to him and say, 'We've got to reauthorize the ag bill. Understand? My farmers, they've got to have this emergency funding. You've got to get the president to say, 'We need this ag bill.' He smiles," says McCain, continuing as Cheney. "'Thank you very much. Yes. Yes, Pat.' In fact, now that I think about it, I've never seen him fire back at any one of these guys when they do that. I just never have."[15]

Cheney, of course, has a reason for subjecting himself to this gantlet. "What I try to do is maintain those relationships when you don't need them so that they're there when you do need them," he says.[16]

This was one of those times. The discussions over a compromise tax bill between the Republican leaders in the House and

their counterparts in the Senate had grown acrimonious. Bill
Thomas and Charles Grassley, the chairmen of the tax-writing
committees, were not speaking to each other.

Frist called Cheney and asked him to help broker a deal. It
was a familiar role.

"Especially in the first term, he had a sporadic role as an emis-
sary of the president in a particularly tough spot or where there
were particularly difficult personalities to deal with," says one
senior administration official.

Cheney's colleagues were motivated less by his skills as a ne-
gotiator than by the mere weight of his presence. The vice president is coming, they thought; we'd better get it done.

"When I'd get a phone call from the vice president, that was
bad news, because he wanted me to do something," says former
speaker of the House Dennis Hastert. "When I'd get a call from
the president, it was an attaboy or a slap on the back for getting
something done. [If] Dick was going to call, there was a tough
deed to follow."[17]

As the negotiations proceeded, Cheney served as a mediator
between the various parties involved—Thomas, Grassley, Hastert, and Frist. The issue came to a head at a meeting President
Bush hosted at the White House late in the afternoon on May 19,
shortly before a state dinner for President Gloria Macapagal Ar-
royo of the Philippines. The president and vice president talked
with the members of Congress outside on the Truman Balcony,
overlooking the South Lawn of the White House, with the Washington Monument in the distance.

Grassley had made clear before the meeting that he would
not back a tax cut larger than $350 billion, the amount the Senate
had agreed on in negotiations the previous month. That created
an impasse: what Grassley saw as honoring a commitment to his
colleagues, Hastert and Thomas viewed as intransigence. So they
came up with a plan.

Hastert would provoke Grassley in the meeting by trying to
cut him out of the negotiations. "I suggested to Senator Frist that
probably, in all reality, Chairman Grassley ought to step down
and not be in the conference because he'd already had a plan that

he was locked into. And why go into negotiation with somebody who was locked in?"[18]

It was a mischievous proposition—to have the chairman of the Senate's tax-writing committee cut out of negotiations on the tax bill—made even more outrageous by the fact that Grassley was in the room. Hastert looked at the vice president.

"Cheney"—Hastert interrupts his story with loud guffaws—"had to swallow his tongue to keep from kind of bursting out laughing. To say that to the senator—well, he took great umbrage."[19]

But as the bickering dragged on, Cheney grew less amused. "He'd look at you with those cold, steely blue eyes and kind of level a look at you—never really change expressions that much—until you kind of knew he'd had enough," says Hastert.[20]

At the end of the meeting, Bush spoke up, telling the group that he wanted a bill to sign by Memorial Day, just one week away. He left to the others the thorny problem of working out the details. For Cheney, that meant spending much of the week on Capitol Hill. And it would mean lots of quality time with Bill Thomas and Charles Grassley.

Despite their differences, Cheney personally liked both men. He regarded Grassley as an authentic Iowa farmer, hardworking, tough, genuine, dependable; several times when Cheney called Grassley's cell phone, he reached the senator on his tractor. Anytime Cheney traveled to Iowa, Grassley could be counted on to be at his side. And Cheney had old ties to Thomas, a longtime ally whom Cheney regarded as one of the sharpest thinkers in the Republican caucus.

Cheney spent the next three days shuttling between meetings with House and Senate leaders, the tax writers, and moderate Republicans. The entire time he consulted with Thomas, his old friend. On several occasions, Cheney met with Thomas and a Republican senator to hammer out agreements on specific provisions of the complicated legislation. When disagreements arose in meetings between the parties, the vice president would keep silent, then confer with Thomas later.

"Cheney would call me and ask for the rundown," says

Thomas. "I'd say: first one's not true, second one is partially true, third one is true. Then we'd go in there and lay it out."[21]

On virtually all the smaller agreements, Cheney had the last word. Everyone understood that he was speaking for the president.

Says Thomas: "If he said, 'I'm with Thomas,' it was over. If it's a member of the House who says that to them, they'd say, 'Go to hell.' But they know better than to push him. He does that Western stare: Don't pull out your gun."[22]

On May 21, the talks stalled again. The White House dispatched Cheney to Capitol Hill, where he met first with Thomas in the Ways and Means Committee office at the Capitol. Although the president's deadline was only five days off and relations across the Capitol were strained, the two men took their time before getting to the issues that had brought them together. They chatted about their shared interest in political science, reminisced about their time in Congress, and marveled at the contours of their respective careers since they had first taken the oath of office in January 1979. "Nobody would have ever bet then that he would have ended up chairman of Ways and Means and I would have ended up vice president," says Cheney.[23]

Then, Thomas briefed Cheney on the latest developments in the negotiations. The vice president was sympathetic to his former colleagues in the House and strongly favored their plan to stimulate the economy by cutting capital gains tax rates. (Cheney had embraced cuts in capital gains taxes since his very first days in politics; his first campaign brochure touted similar cuts, originally proposed by his old boss Bill Steiger.)[24]

Communication between Thomas and Grassley had broken down completely. Grassley would later say that he wanted to teach Thomas about "the respect he should have for someone of equal rank."[25] Cheney stayed in touch with Grassley, but asked senator George Voinovich of Ohio to meet with Thomas. Cheney and Thomas knew that Voinovich, who had been among the moderates insisting on a $350 billion cap on the tax cuts, was very likely the vote that would clinch the tax cut, if they could get him to agree to a slightly amended proposal from Thomas.

"Cheney sits there and doesn't say a word," says Thomas.

"He put me out there and I would perform, so that he wasn't making the arguments, I was. It was a classic good cop, bad cop routine."[26]

Two days later, after Voinovich cast his vote in favor of the package to make the count fifty-fifty, Cheney cast the tie-breaking vote for the $350 billion tax cut in the Senate.

"He's kind of like a pastor coming in and settling a family dispute within the congregation," said Grassley. "The Republicans are a unified congregation because of Vice President Cheney."[27]

Bush's signature the following week made this the third-largest tax cut in American history. A month earlier, Bush had dismissed a proposal of the same size as a "little bitty" tax cut, but now the White House claimed victory.

The vote had split closely along party lines. Republicans thought the cuts would stimulate economic growth and offset the lost revenue by collecting a relatively smaller part of a bigger pie. Democrats thought the tax cuts favored the rich and would deepen the deficit by taking revenues away from the government. "This tax bill is one of the most dangerous, destructive, and dishonest acts by the federal government I have ever seen," said Mark Dayton, a Democratic senator from Minnesota. "The tax base of the American government is being destroyed."[28] Liberal economists questioned the administration's claims that the cuts would spur economic growth and scoffed at the notion that the cuts would raise revenues. One of these economists said that any such suggestion was "undoubtedly wrong."[29]

Cheney wasn't concerned. Although he'd been a deficit hawk for much of his congressional career—and had never been considered one of the true supply-siders—he had come to believe that economic growth would outweigh the effects of an increased deficit. When the former secretary of the treasury Paul O'Neill told a reporter that Cheney had said, "Deficits don't matter," Cheney did not deny the comment, though those who know his views say it was an oversimplification. Often, when someone on his staff would express concerns about deficits, Cheney would jokingly repeat this line before taking up the issue seriously.

In a speech to the U.S. Chamber of Commerce earlier that year, Cheney had challenged the view that bigger deficits lead to increased interest rates. There was, he said, "one slight flaw" in that argument: "The evidence of recent years simply doesn't support it."[30] As the surpluses of 2001 turned to deficits in 2003, he argued, interest rates remained low, in some cases lower than when the budget was expected to maintain a surplus.

Cheney hadn't always thought that way. In Congress, he had cosponsored legislation mandating a balanced budget; and a letter to the *New York Times* that he signed along with several colleagues advanced the very argument he would later dismiss. "In the real world beyond the editorial offices of *The Times*, it is clear that our high real interest rates and high dollar are the result of excessively high deficits and an inability to control Federal spending," the Republican congressmen wrote.[31] When the staff flashed the letter before him as they prepared for the speech, Cheney seemed surprised that he had once agreed with it.

On February 4, three weeks after Cheney's speech, the chairman of the Federal Reserve, Alan Greenspan, sent him a Fed study: "Budget, Deficits, Debt, and Interest Rates." The handwritten note from Greenspan said simply: "As requested." The study challenged the economic theory behind Cheney's claims. So Cheney asked Cesar Conda, his top domestic policy adviser, to write up a report so that he might respond to Greenspan.

Cheney was willing to challenge Greenspan, but he held Greenspan's views in very high regard. The two stayed in touch throughout the first five years of the Bush administration until Greenspan's retirement in early 2006. They were close, but Cheney kept his friendship with Greenspan to himself. "You knew it," says the former director of the Office of Management and Budget (OMB), Mitch Daniels, "but as with so many things about the vice president, he goes about his business quietly."[32]

Cheney tracked the Federal Reserve closely and regularly asked for briefings from his economic staff on monetary policy. One time a staffer expressed concern that the Fed's policy was too tight and, noting the Fed's independence, lamented the inability of the administration to do anything about it. Cheney smiled. There are ways to put our oars in the water over there, he said.

Of course, one certain way to lower deficits is to decrease spending, something the Bush administration had been disinclined to do through much of the first term. Defense spending could have been expected to grow; the country was at war. But free-market critics of the administration pointed out that nondefense discretionary spending grew by 28 percent by the summer of 2003. A study by the libertarian Cato Institute criticized Bush as "the most gratuitous big spender to occupy the White House since Jimmy Carter."[33]

When asked, in an interview in June 2004, about the administration's spending, Cheney didn't challenge the criticism but sought to put it in context by pointing out the size of the deficit in relative terms — as a percentage of the gross domestic product — and the demands of the war on terror. Would the Bush administration seek to shut down government agencies, as the House Republicans had sought to do just ten years earlier? "I can't make any predictions," he said. "I think we will be relatively tough in a second term with respect to spending limits because it's important to do so and the president's clearly indicated that it's a concern."[34]

One way Cheney could exercise some restraint over spending was through his seat on the Budget Review Board. The White House chief of staff, Andy Card, and Josh Bolten, his deputy for policy, created the five-person panel at the beginning of the Bush administration. The idea had been Card's. As secretary of transportation under George H. W. Bush, Card had spent what he felt was an inordinate amount of time fighting over budget issues. It's a ritual of life in Washington: cabinet secretaries want as much money for their departments as they can squeeze out of the White House allotment; the OMB, which oversees the federal government's annual budget request, often tries to limit those expenditures.

Cheney had endured this process from the other side, as cabinet secretaries filed into the Oval Office one after another to persuade President Ford that their agencies needed more money. Ford, polite to a fault, had a difficult time saying no, sometimes leaving his staff in the uncomfortable position of negotiating between an OMB eager to hold the line on budget requests and a president willing to try to accommodate his cabinet secretaries. In other recent administrations, the president had devoted a significant amount of his time to mediating these funding disputes.

Card's idea was simple: create an advisory board that can settle these issues and save the president valuable time. With Cheney's encouragement, Card and Bolten brought the group into existence. It included the director of the OMB, the secretary of the treasury, the White House chief of staff, the director of the National Economic Council, and the vice president. Cabinet secretaries who were unhappy with the OMB's decisions about spending for their departments would first submit their appeals in writing. Then, three or four days later, they would come to the White House to meet with the board, gathered around a long table in the chief of staff's office, and make their case in person.

Cheney's presence on the board was critical. All the cabinet secretaries understood that he was a full partner in the president's decisions. He was naturally skeptical about any additional funding, though he was much more open to additional money for homeland security and defense. "If the vice president thought someone's appeal had merit, it was probably going to happen," says Daniels, the former OMB director.[35]

If the board turned down requests for additional funding, the cabinet members could make a final appeal to the president. But over the first three years of the Bush administration, that never happened.

Daniels offers two reasons. "Dick Cheney was the chairman and people knew they were getting a fair hearing, and sometimes there were adjustments, and I would say one out of four or five, anyway. And secondly . . . I think no question the team felt that if the vice president didn't agree on an increase then it wasn't very likely that the president would either."[36]

While Cheney managed intramural skirmishes inside the White House and on Capitol Hill, the administration was embroiled in a much bigger fight. Although most of Iraq was relatively calm, sporadic combat continued there. More than two months after the beginning of the war, U.S. military and intelligence agents and inspectors had failed to find stockpiles of Iraqi WMD. In public, senior Bush administration and intelligence officials expressed confidence that the weapons would be discovered. But privately, they were concerned.

Elected officials, including some who had supported the

war, soon began to launch attacks. Two prominent Democratic senators—Joe Biden, the ranking member on the Senate Foreign Relations Committee; and Bob Graham, former chairman of the Senate Select Committee on Intelligence—publicly accused the Bush administration of manipulating the intelligence on Iraq.

Cheney had seen this before, and he understood that it would soon get worse. At a panel discussion on April 23, 1977, just four months after he left the Ford administration, Cheney had described congressional accountability on issues of foreign policy and national security. "Well, what they do, though, they complain if you don't consult; if you do consult in advance and get agreement . . . frequently when the time comes to stand up and be counted, when the decision is criticized, none of them are there. It won't be sort of outright denial that they were involved in the process at all, but they all head for the cloak room when it comes time to stand up and be counted."[37]

Day after day, newspaper and television reports contained the same accusation: the administration exaggerated the intelligence of Iraq's WMD programs. An assessment made in September 2002 of Iraqi WMD and issued by the Defense Intelligence Agency was leaked to *U.S. News & World Report.* "There is no reliable information on whether Iraq is producing and stockpiling chemical weapons."[38]

Cheney's trips to the CIA were the subject of a front-page article in the *Washington Post.* "Vice President Cheney and his most senior aide made multiple trips to the CIA over the past year to question analysts studying Iraq's weapons programs and alleged links to al Qaeda, creating an environment in which some analysts felt they were being pressured to make their assessments fit with the Bush administration's policy objectives, according to senior intelligence officials," wrote Walter Pincus and Dana Priest. The article quoted a "senior agency official" who claimed that Cheney's visits "sent signals, intended or otherwise, that a certain output was desired from here."[39]

Bush administration officials defended their statements on WMD by pointing out that they were backed by the CIA's assessments. One week after the article in the *Washington Post* on Cheney's visits to CIA headquarters, Pincus published another story, this one following up on the columns by Nicholas Kristof. The story claimed that Joseph Wilson—who was still unnamed—had visited Niger and returned to tell the CIA that reports of attempts by Iraq to buy uranium there were false, and that documents related to the purchase were probably forged. The "dates were wrong and the names were wrong," Wilson told Pincus.[40]

It was an extraordinary accusation. The unnamed ambassador, portraying himself as a high-level insider, was accusing the Bush administration and the vice president of knowingly using discredited information to take the nation to war. Cheney had spent the better part of his career ignoring criticism of him in the press. This was different. The issues were too serious. So Cheney's office quietly worked to knock down Wilson's claims.

Cheney reportedly spoke to George Tenet about the Wilson trip. Various news accounts suggest that Tenet told Cheney that Wilson's wife, Valerie Plame, worked at the CIA and recommended her husband for the trip. It was information Cheney passed on to Scooter Libby.[41]

Libby, meanwhile, was making his own efforts to learn more about the Wilson trip and its origins. He had asked Marc Grossman, a senior State Department official, for information on the former diplomat and his trip, and raised the issue with Craig Schmall, his CIA briefer.

But they could give reporters few concrete reasons to be skeptical about Wilson's allegations; the details of his trip were still classified.

In the June 30, 2003, issue of *The New Republic* Wilson appeared again, with an even more audacious set of claims. The article, "The First Casualty," was heavily sourced to current and former intelligence officials. Its bottom line: the Bush administration had lied the country into war. A chief exhibit was the story of Joseph Wilson—still unnamed.

Wilson told reporters from the magazine that Cheney's office had received the forged documents from the British a year before Bush spoke the "sixteen words" in the State of the Union address in January 2003:

Cheney's office had received from the British, via the Italians, documents purporting to show Iraq's purchase of uranium from Niger. Cheney had given the information to the CIA, which in turn asked a prominent diplomat, who had served as ambassador to three African countries, to investigate. He returned after a visit to Niger in February 2002 and reported to the State Department and the CIA that the documents were forgeries. The CIA circulated the ambassador's report to the vice president's office, the ambassador confirms to TNR. But, after a British dossier was released in September detailing the purported uranium purchase, administration officials began citing it anyway, culminating in its inclusion in the State of the Union. "They knew the Niger story was a flat-out lie," the former ambassador tells TNR.[42]

The claim seemed odd to a staff member on the Senate Select Committee on Intelligence who had been following the story of Iraq and Niger. It was odd, she thought, because it could not be true. Wilson had traveled to Niger in late February 2002. The U.S. intelligence community had received the forged documents on October 9, 2002, more than eight months after Wilson's trip, when they were delivered to the U.S. embassy in Rome.

The committee asked Wilson to submit to an interview. When confronted with evidence that contradicted his public allegations, Wilson claimed that he must have "misspoken." But if it was a simple mistake, as Wilson suggested, he had made it on three separate occasions in interviews with three separate publications. Not surprisingly, Wilson changed his story after being confronted by the Senate's investigators.

On July 6, 2003, he told his story in his own name for the first time in an op-ed published in the *New York Times*. He acknowledged that he "never saw" the forgeries, and he later conceded the same point in an appearance on television.[43] It should have been a crucial admission, giving pause to the editorialists and politicians who had relied on Wilson to support their claims about the administration's mendacity. Wilson had been a compelling source precisely because he presented himself as a fact witness regarding the forged documents.

His evolving narrative apparently had escaped the notice of his editors at the *New York Times*. The same editorial pages that had published Kristof's columns, including the former ambassador's claim that he had debunked the forged documents, now carried his admission that he had never seen them.

The *Times* published Wilson's op-ed under the headline, "What I Didn't Find in Africa":

> *Based on my experience with the administration in the months leading up to the war, I have little choice but to conclude that some of the intelligence related to Iraq's nuclear weapons program was twisted to exaggerate the Iraqi threat.*

Wilson gave readers of the *Times* a rundown of his activities in Niger:

> *Those news stories about that unnamed former envoy who went to Niger? That's me. . . . In late February 2002, I arrived in Niger's capital, Niamey, where I had been a diplomat in the mid-70s and visited as a National Security Council official in the late 90s. . . .*
>
> *I spent the next eight days drinking sweet mint tea and meeting with dozens of people: current government officials, former government officials, people associated with the country's uranium business. It did not take long to conclude that it was highly doubtful that any such transaction had ever taken place.*[44]

White House officials were stunned. They had obtained from the CIA the Agency's one-and-a-half-page report on Wilson's trip.

"We were given the contents of what the report had said," says one White House official. "The guy goes over there and comes back and says Iraq was looking for uranium. We thought, 'Shit, we should declassify that and put it out.'"

Wilson had not submitted anything in writing upon returning from Niger. Instead, two CIA officials chatted with him at his house about his trip shortly after he returned. For the most part, they found the details of his time in Niger unsurprising and not particularly significant. There was one exception: Wilson's meeting with the former Nigerien prime minister, Ibrahim Mayaki.

Not surprisingly, Mayaki told Wilson that Niger had signed no contracts with rogue states while he served in the government, first as foreign minister and then as prime minister, from 1997 to 2000. But he provided one tantalizing detail. Although Mayaki denied that Niger agreed to provide Iraq with yellowcake, he told Wilson that the Iraqis had come looking. He had spoken with a top Iraqi official in 1999 who told Mayaki he wanted to explore "expanding commercial relations" between the two countries. As Niger has virtually nothing else the Iraqis might have desired, Mayaki concluded that this request for enhanced trade meant one thing: the Iraqis wanted uranium.

The CIA's analysts thought Wilson's report corroborated earlier intelligence that Iraq had sought uranium from Niger. Not only did Wilson's debriefing fail to knock down what his wife had derided as a "crazy report" when she recommended him for the mission; his conversation with Mayaki had bolstered the earlier reporting on the Iraqis' interest in uranium.

But journalists covering the story had no way to know this. So the White House considered declassifying the report and releasing it.

It would be a pivotal moment. Several of Bush's advisers—a group that included such normally cautious officials as the White House Communications Director Dan Bartlett and Anna Perez of the National Security Council—wanted to declassify and release Wilson's report. But there were risks. Confronting Wilson on his fabrications might further antagonize the CIA.

Several White House officials thought senior CIA officials were embarrassed about the Wilson trip. Sending a former ambassador to drink tea with Nigerien officials—officials who would certainly deny claims that they even considered illicit trade with the Iraqis—hardly counts as a rigorous investigation. And the mere fact that the CIA had to outsource its inquiry to an ex-diplomat suggests that the CIA had very limited sources of its own.

Worse, the CIA's brief report on Wilson's trip was written in such a way that it made his expedition sound more serious than it had been. Wilson was not named in the report, but instead was described as a clandestine source, allowing readers of the report to assume that he might have had independent knowledge of any Iraq-Niger talks.

The deputy national security adviser, Stephen Hadley, was on the phone several times a day with George Tenet, handling the sensitive diplomacy between the White House and the CIA. Hadley did not want to do anything to further antagonize the CIA leadership. So despite the fact that Joe Wilson was free to discuss his trip and mischaracterize his report—the CIA never made him sign a nondisclosure agreement—Wilson's report would remain classified.

The Wilson story line quickly became conventional wisdom. A front-page story in the *New York Times*, published two days after Wilson's op-ed piece, simply declared that Wilson "reported back that the intelligence was likely fraudulent."[45]

Wilson's op-ed began a week—White House staffers came to call it "hell week"—in which the administration struggled, sometimes in public, to respond to his charges. Their effort was complicated by the fact that President Bush, Condoleezza Rice, and several other senior officials left for a trip to Africa on July 7. Confusion reigned, but one thing was clear: the CIA no longer stood by the sixteen words in Bush's State of the Union address. Tenet called Rice and told her that both his agency and the White House would have to accept partial blame for including the sentence. Rice and her colleagues at the White House wanted to prevent a public fight between President Bush and the CIA, so officials from the NSC and the CIA worked to come up with language for a "press guidance" that would share responsibility for the sixteen words.

As they debated how to handle the worsening public relations disaster, White House officials made two crucial errors. First, they ignored what was arguably the most important question: were the sixteen words accurate? Had Iraq in fact sought uranium in Africa? "I don't remember us ever having a debate about the truth of the allegation," says one White House official who was involved in the discussions. "It was all about managing the PR fallout."

"It was a process issue," says Rice. "It wasn't the issue of whether or not what he said was true. It was an issue of whether or not, given that apparently there had been all of these concerns expressed about it, whether or not . . . we should have put it in a speech for him. That was the issue."[46]

Second, the deputy national security adviser, Stephen Hadley, misunderstood Cheney's position on the sixteen words. As White House officials debated whether they should retract the sixteen words—or at least admit that the words should not have been in Bush's speech—he asked for Cheney's thoughts. Working through an intermediary, Hadley came to believe that Cheney had offered his reluctant approval of a public concession that the language should not have appeared in the State of the Union address. Cheney had done no such thing.

Those familiar with Cheney's thinking say that he was strongly opposed to any concession. He was joined in this view by Scooter Libby and Bob Joseph, an expert on nonproliferation at the NSC. Any perception that the White House was disowning its claim could have damaging long-term effects. And, not incidentally, Cheney, Libby, and Joseph believed Bush's statement was true. The British had, in fact, reported that Iraq was seeking uranium from Africa, and U.S. intelligence agencies had previous reports of Iraq's nuclear activities on the continent. In any case, they argued, leaks from the CIA did not begin with the issue of Iraq and Niger and wouldn't end with it.

They were right. For much of the week leaks and counter-leaks appeared on the front pages of the nation's newspapers. CIA officials told reporters that the agency had long been suspicious of the underlying intelligence on the deal between Iraq and Niger. The administration responded that the CIA had continued to approve language on the Iraqi efforts to obtain uranium from Africa through the State of the Union speech. An article in the *New York Times* called the situation "an unusual exercise in finger-pointing" within the Bush administration.[47]

Then, on July 10, Libby spoke to NBC's Washington bureau chief, Tim Russert. Libby called Russert to complain about Chris Matthews, host of MSNBC's *Hardball*, who had become increasingly critical of the role played by Cheney and his "neocon" allies in planning the Iraq War. Libby believed that Matthews's frequent use of the word "neocon" was anti-Semitic and sought to put a stop to it. It was in this conversation, Libby would later claim, that he first learned that Wilson's wife worked at the CIA. Russert disputed Libby's version of their chat, telling colleagues, and later a grand jury, that he had not known about Valerie Plame's employment at the time.

On July 11, 2003, Tenet released a statement in which he sought to explain how the "sixteen words" had ended up in the State of the Union speech. "First, CIA approved the President's State of the Union address before it was delivered. Second, I am responsible for the approval process in my Agency. And third, the President had every reason to believe that the text presented to him was sound. These sixteen words should never have been included in the text written for the President."

Tenet said the reporting from Wilson was mixed, but he called into question the truth of at least part of Wilson's story. "There was no mention in the report of forged documents—or any suggestion of the existence of documents at all," he said.[48]

That same day, President Bush and the national security adviser, Condoleezza Rice, told reporters traveling with the president in Africa that the CIA had approved the language in the State of the Union address. "I gave a speech to the nation that was cleared by the intelligence services," said Bush. Rice added, "The CIA cleared the speech in its entirety."[49]

Tenet was livid. He thought he had worked out a deal with the White House in which the CIA and the Bush administration would share the blame. Now the White House wasn't holding up its end.

The Democrats, meanwhile, seized on Wilson's claims for political leverage. Howard Dean, at the time a leading contender for the Democratic presidential nomination, called for the resignation of top officials in the Bush administration. The Democratic

National Committee (DNC) produced a television ad accusing Bush of knowingly including false information in his State of the Union speech; and the chairman of the DNC, Terry McAuliffe, called for an independent probe: "This investigation is needed to uncover what President Bush has refused to explain."[50]

On July 12, Libby accompanied Cheney to Norfolk and discussed inquiries from the media about Wilson's trip. Cheney's staff was frustrated. Tenet had directly contradicted Wilson's claims that he had debunked the forged documents, but the reporters seemed to be unmoved. "It was a disaster trying to deal with reporters on this," says one Cheney adviser. Eric Edelman, deputy national security adviser to the vice president, wanted Cheney to personally correct the record by saying publicly that he had not known Wilson and knew nothing of Wilson's trip until it had been reported in the press.

Two days later, Robert Novak wrote the column that would eventually result in a criminal probe. Novak identified Wilson's wife, Valerie Plame, as "an agency operative on weapons of mass destruction," and sourced this information to "two senior administration officials."

In the last paragraph of the column, Novak raised the issue of Wilson's oral report to the CIA. To determine whether the Bush administration ignored Wilson's warnings about the alleged uranium deal, Novak wrote, "requires scrutinizing the CIA summary of what their envoy reported. The agency never before has declassified that kind of information, but the White House would like it to do just that now—in its and in the public's interest."[51]

On July 16, 2003, David Corn wrote an article for *The Nation*'s Web site, with the headline "A White House Smear."

"Did senior Bush officials blow the cover of a U.S. intelligence officer working covertly in a field of vital importance to national security—and break the law—in order to strike at a Bush administration critic and intimidate others?" The answer, Corn's article implied, was yes.

Wilson spoke to Corn for the article and refused, at least on the record, to discuss whether his wife worked for the CIA. "Let's assume she does," wrote Corn. "That would seem to mean that the Bush administration has screwed one of its own top-secret operatives in order to punish Wilson or to send a message to others who might challenge it."

Corn ended the piece with a direct accusation. "The Wilson smear was a thuggish act. Bush and his crew abused and misused intelligence to make their case for war. Now there is evidence Bushies used classified information and put the nation's counter-proliferation efforts at risk merely to settle a score. It is a sign that with this gang politics trumps national security."[52]

The flawed story line was hardening. Although George Tenet, director of the CIA, had personally and publicly dismissed Wilson's claim of having debunked the forged documents, and although Wilson himself had been forced to admit that he could not have done so, politicians who should have known better—and in some cases did know better—repeated Wilson's story.

"The vice president is the one who went to the CIA on several occasions," explained Senator Bob Graham of Florida, who had been the ranking Democrat on the Intelligence Committee. "He asked specifically for additional information on the Niger-Iraq connection. The United States sent an experienced ambassador, who came back after a full review with a report that these were fabricated documents. You cannot tell me that the vice president didn't receive the same report that the CIA received, and that the vice president didn't communicate that report to the president or national security advisers to the president. So I have to believe that the president knew or should have known that this information had been classified as unreliable by the CIA."[53]

On July 23, 2003, Senator Ted Kennedy and Senator Carl Levin held a press conference to keep alive the controversy over the "sixteen words." Levin said:

> *There's been a great deal of discussion as to how it was possible that a statement which the CIA did not believe itself ended up in a State of the Union message, but attributed to the British instead, because the British apparently did continue to believe it. Of course, the purpose of the State of the Union message was to create the impression that the administration believed what the British believed, since it*

was a credible source. But the facts of the matter were, and
George Tenet has now acknowledged it, . . . that for many,
many months the CIA did not believe that Iraq had sought
to obtain uranium from Africa.[54]

It was a command performance and, as press conferences are designed to do, it generated coverage in the media. But Levin failed to share an important detail with reporters.

On January 29, 2003, the day after the State of the Union address, Levin had personally written to George Tenet demanding more information on Bush's claims concerning Iraq, uranium, and Africa. Levin instructed the CIA to provide "what the U.S. [intelligence community] knows about Saddam Hussein seeking significant quantities of uranium from Africa."[55]

The CIA responded on February 27, 2003—one month after the State of the Union address—by defending the intelligence included in Bush's speech. Although the CIA found the Nigeriens' denials of a deal credible, it wrote to Levin that "reporting suggest[s] Iraq had attempted to acquire uranium from Niger." The CIA continued to question "whether Baghdad may have been probing Niger for access to yellowcake in the 1999 time frame."[56] And not only did the CIA defend Bush's speech; it did so by relying for support on a document based on the "intelligence report on the former Ambassador's trip to Niger."[57]

Levin chose to ignore the CIA's letter and over time stepped up his accusations. "In October, the president gave a speech in Ohio. That speech contained the same misstatement, but it was struck," he told the editorial board of a small newspaper in Michigan. "Then, it shows up in the State of the Union address. That makes it more than a mistake."[58]

On July 30, the CIA wrote to the Justice Department to request an investigation of the leak. The CIA referral was classified and its existence was known only to a small group of officials from the CIA and at Justice. In a typical year, the CIA files fifty such referrals. Like the others, the referral in the Plame case went to John Dion, a career official at the Justice Department who handles investigations of classified leaks. To Dion, the Plame case was one of many that required his attention; nothing suggested to him that it had a higher priority than any of the others that had ended up on his desk.

For two months very little happened. Wilson gave fiery speeches in which he promised to have Karl Rove "frog-marched" out of the White House in handcuffs. Writers for *The Nation* and indignant bloggers published editorials. Far-left congressional Democrats made occasional calls for an investigation. But the story seemed destined to disappear.

Then, on September 26, 2003, Andrea Mitchell of NBC News and Alex Johnson of MSNBC broke a big story on the MSNBC Web site. "The CIA has asked the Justice Department to investigate allegations that the White House broke federal laws by revealing the identity of one of its undercover employees in retaliation against the woman's husband, a former ambassador who publicly criticized President Bush's since-discredited claim that Iraq had sought weapons-grade uranium from Africa, NBC News has learned."[59]

The same day, the Justice Department ordered the FBI to begin an investigation into the leak.[60]

Over the next two weeks the *New York Times* would run nearly three dozen stories on the case, and the *Washington Post* more than forty. News reports noted the close relationship between Attorney General John Ashcroft and the White House. Editorials called for Ashcroft to recuse himself. Prominent Democrats stepped up their calls for a special prosecutor.

White House officials and those close to Cheney believe that the referral was made public as a result of a deliberate leak from the CIA, part of the broader war between the CIA and the Bush administration. It embarrassed the White House and put pressure on the Justice Department to appoint a special prosecutor. On September 29, 2003, a news story in the *New York Times* said, "The very fact that Mr. Tenet referred the matter to the Justice Department comes as a major political embarrassment to a White House that is famously tight-lipped, and a president who has repeatedly vowed that his administration would never leak classified information."[61]

The second consequence of the leak was a stream of news stories in which journalists reported uncritically the claims of the

CIA and Joseph Wilson regarding Iraq and Niger, and stated unequivocally that the White House had simply ignored their strong warnings about the intelligence.

The story of September 29, 2003, in the *New York Times* also said, "The agent is the wife of Joseph C. Wilson 4th, a former ambassador to Gabon. It was Mr. Wilson who, more than a year and a half ago, concluded in a report to the CIA that there was no evidence that Saddam Hussein tried to buy uranium ore in Niger in an effort to build nuclear arms. But his report was ignored, and Ambassador Wilson has been highly critical of how the administration handled intelligence claims regarding Iraq's nuclear weapons programs, suggesting that Mr. Bush's aides and Vice President Dick Cheney's office tried to inflate the threat."[62]

In any case, the disclosure of the CIA's original referral in the Plame case was a pivotal moment. Without this leak, it is unlikely that any serious investigation—by the Justice Department or a special prosecutor—would have taken place.

In a column of October 1, Novak sought to clarify his role in the story. The disclosure of Plame's name and place of employment, he wrote, came as "an offhand revelation" at the end of a "long conversation" with a senior administration official who is "no partisan gunslinger."[63]

On October 3, the White House counsel Alberto Gonzales conveyed to the staff an order from the Justice Department requiring White House officials to turn over anything related to the leak, including phone logs, e-mails, letters, and personal notes. Throughout October and November the FBI interviewed a number of senior officials from the White House and Cheney's office, including Libby.

In early November, Cheney hired Kevin Kellems as his new chief spokesman. Kellems had worked on strategic communications at the Pentagon for three years, serving for most of that time as the senior communications adviser to Deputy Secretary of Defense Paul Wolfowitz, Scooter Libby's mentor.

Kellem immediately noticed that Libby had an unusual interest in Joseph Wilson. "He was nervous from the very first day about Wilson," says Kellems. "There's only one word that captures his feelings about Wilson: he was obsessed."[64]

Kellems said Libby gave the small communications staff "weird research assignments" related to the Wilson affair, with some requests for information coming late at night. "He was nervous as hell—freakish about it. He'd go through an article about Wilson and freak out about specific sentences," says Kellems. "Then he'd look at you and wonder why you didn't see it the same way."[65]

As the investigation continued, the White House in general, and Cheney's office in particular, became a seedbed of paranoia. Staffers never knew if a person they passed in the halls had been interviewed by the FBI, much less what anyone else had said. And no matter how much time White House officials spent thinking about the investigation, any discussion of it with their colleagues was out of the question.

On December 30, 2003, Attorney General John Ashcroft formally recused himself from the case and Deputy Attorney General James Comey announced the appointment of Patrick Fitzgerald, a U.S. attorney from Chicago, as special prosecutor. In his announcement, Comey noted that he had given Fitzgerald a broad mandate. "I have today delegated to Mr. Fitzgerald all the approval authorities that will be necessary to ensure that he has the tools to conduct a completely independent investigation, that is that he has the power and authority to make whatever prosecutive judgments he believes are appropriate without having to come back to me or anybody else at the Justice Department for approvals."[66] It would be another important moment.

CHAPTER FIFTEEN

War and Politics, Politics and War

T he vice president was late, and for good reason.
Air Force Two had landed at Stewart International Airport in New Windsor, New York, shortly after three PM on December 13, 2003. Cheney was in town for a quick fund-raiser at the Pawling Mountain Club, an exclusive hunting preserve about ninety minutes north of New York City. He had attended many of these events in recent weeks, making trips to Missouri, Texas, Oklahoma, Ohio, and New York. This particular event was expected to raise $150,000 for Bush and Cheney's reelection campaign.

A crowd would be waiting for him at the club, and Cheney made a habit of arriving on time. But before he deplaned he was stopped by an urgent phone call from the White House. It was the president, and he had some extraordinary news. Cheney, wearing a light blue dress shirt, took notes at the desk in his cabin as Bush spoke. American soldiers had captured Saddam Hussein at a farm in Ad Dawr, Iraq, just south of his hometown, Tikrit. No one had been hurt. The identification was preliminary, Bush cautioned. Cheney spoke to Rumsfeld briefly before departing for the fund-raiser.

Cheney's staff noticed with some surprise that he seemed to be in a good mood when he arrived, despite being forty-five minutes late. But he gave no hint of Saddam's capture, and his high spirits didn't spill over into the speech; instead, he used his brief remarks to counsel patience in the war on terror. "He said that it is not going to be over in a hurry," one guest told a local newspaper.[1]

Also at the event was Mary Cheney, who would serve as her father's personal aide for the reelection, just as she did for the campaign of 2000. As the two flew back to Washington, after the fund-raiser, Cheney suggested stopping off at Don Rumsfeld's Christmas party. Mary was surprised; she figured her father had had enough socializing for one day. Cheney couldn't share the news with her in earshot of anyone else, so he waited until Marine Two deposited them back at the vice president's residence in Washington, D.C. With the rotor blades still whirring loudly, he told Mary that he was going to give her a note in the car but warned her that she wasn't to react on reading it. The note indicated that U.S. soldiers had captured a "high-value target" in Iraq and that they were awaiting DNA confirmation of the person's identity. Mary understood that he meant Saddam Hussein.

The following morning, L. Paul Bremer, head of the Coalition Provisional Authority (CPA) in Iraq, mounted the stage at a press conference in Baghdad. "We got him," he exclaimed to sustained applause from the Iraqi journalists gathered in the room.

For Cheney, it was a huge moment. Although he was cautious in his public statements, as he had been in New Windsor, Cheney thought that capturing Saddam would mark the beginning of better times in Iraq.

As the Bush administration entered its fourth and perhaps final year, Cheney had many reasons to be pleased. The American military had fought and won two wars with a combination of superior technology, creative planning, and moral certitude. The United States had liberated some 50 million people from oppressive, despotic regimes, and the Bush administration was making good on its commitment to guide these delicate nation-states to some form of democracy.

In Afghanistan, al Qaeda's camps had been destroyed, the Taliban had been removed from power, and Hamid Karzai—a leader committed to Afghan self-government—had been confirmed as head of state by the country's *loya jirga* process. National elections were scheduled for sometime in 2004.

In Iraq, Saddam Hussein was in jail and his two sons were dead. Many leaders of the former regime had been captured or killed, too; and at least initially leaders in Iran and Syria were very nervous.

What's more, proliferators of weapons and state sponsors of terrorism throughout the world understood that the United States was deadly serious about fighting and winning the global war on terror. George W. Bush had said that the United States would treat sponsors of terror as terrorists themselves, and he certainly seemed to mean it. As if to confirm this sense, five days after Saddam Hussein had been captured in Iraq, Muammar Gadhafi of Libya publicly abandoned his WMD programs. (In private discussions about Gadhafi, Cheney took great delight in pointing out that the enigmatic Libyan leader did not choose to turn himself in to Kofi Annan and the UN.)

But there were problems, too. Afghanistan's drug trade was flourishing. International observers thought that Afghanistan was not yet prepared to hold its promised elections. And, of course, both Osama bin Laden and his chief deputy, Ayman al Zawahiri, remained at large.

In Iraq, the teams assigned to find and secure Iraq's stockpiles of WMD had little to show for their efforts. That failure was eroding domestic support for the war and beginning to chip away at the credibility of the U.S. government overseas. The insurgency, once understood as a small number of "dead-enders," was proving particularly stubborn in its efforts against U.S. troops and, increasingly, Iraqis themselves.

The vice president had grown concerned about the level of violence in Iraq over the eight months since the invasion. As usual, he said little in meetings of the National Security Council (NSC) and saved his advice for one-on-one meetings with the president. Occasionally, however, other participants at a meeting, or members of his staff, got a window into his thinking on an issue. Sometimes it would be the way he phrased a question or framed a point. At other times he would simply share his views in the course of a conversation. Those instances were rare, however, particularly in the regular video teleconferences the NSC and the principals committee conducted with Baghdad to assess progress and difficulties in postwar Iraq.

"Cheney's style, at least insofar as I saw it in those meetings, was to ask questions and not—he didn't really reveal his hand," says Bremer, the head of the CPA. "He did not very often say, 'This is what I think we should do.' It's not his style. His style is more Socratic."[2]

Cheney stepped out of character in a conversation with Bremer in early November, after a bloody end to October. On October 26, 2003, terrorists fired forty French and Russian antitank missiles at the al Rashid Hotel, temporary home to many senior officials of the CPA and, on that night, to Paul Wolfowitz, deputy secretary of defense. Wolfowitz narrowly escaped injury, but others in the building were not as lucky. There were several serious injuries, and Lieutenant Colonel Charles Buehring Jr., a senior communications adviser to Bremer, was killed.

The next day, insurgents conducted simultaneous bombings of four buildings in Baghdad: three police stations and the headquarters of the Red Cross. The attackers used decoy vehicles ahead of trucks carrying 1,000 pounds of explosives—all characteristics of attacks by al Qaeda. Less than a week later, insurgents brought down a Chinook helicopter near al-Falluja, killing seventeen U.S. soldiers and wounding several others. The number of daily attacks had tripled from twelve to thirty-six since the spring. There would be nearly as many casualties in October and November as there had been in the preceding four months combined.

On November 6, Bremer called Scooter Libby from Baghdad. Libby was not available, and Bremer was surprised when the vice president himself picked up the phone. Bremer shared his concerns about the deterioration of security.

"Mr. Vice President, in my view we do not have a military strategy for victory in Iraq," he said. "It seems to me that our policy is driven more by our troop rotation schedule than by a strategy to win." Bremer said he was particularly concerned about the talk of lowering troop levels and replacing American soldiers with poorly trained Iraqis. "The impression may well be growing among the insurgents that we won't stay the course," he warned.

According to Bremer's notes from the call, Cheney agreed. "I've been asking the same question," he said. "What's our strategy to win? My impression is that the Pentagon's mind-set is that the war's over and they're now in the 'mopping up' phase. They fail to see that we're in a major battle against terrorists in Iraq and elsewhere."[3]

Cheney doesn't remember the conversation with Bremer. But his concern is consistent with the recollections of his staff. The vice president regularly asked why the bureaucracies—the CIA and the Pentagon—were not devoting more resources to understanding the insurgency and coming up with new ways to defeat it. He was rarely happy with the answers.

The vexing problems of security contributed to the political difficulties of postwar Iraq. But there were other reasons for those political problems, Cheney thought, including some that he traced back to decisions he had supported more than a decade earlier.

"The Shia had been treated for centuries as second-class citizens, governed by the Sunnis, and in recent decades the Baathists, under Saddam Hussein. They had been encouraged, in '91, to rise up, and did, and were slaughtered for their troubles. Nobody came to their assistance," Cheney says.[4]

Many Iraqis assumed that the United States refused to remove Saddam Hussein after the first Gulf War because it wanted him to remain in power. Americans, in that view, were the willful enablers of the man who brutalized them for decades. "I think there are many Shia who still, to this day, aren't convinced we're going to stay the course, that we're going to get the job done."[5]

Earning their trust would be critical to the United States' success in Iraq. From the days immediately following the 9/11 attacks Cheney had spoken of the importance of "getting the locals into the fight." Sometimes he meant this literally—the Northern Alliance battling the Taliban and al Qaeda in Afghanistan. At other times it was figurative—the willingness of the Saudi regime to get serious about fighting Islamic radicals following a bombing in Riyadh in May 2003.

In Iraq, Cheney thought it important to establish Iraqi political legitimacy as soon as possible. Before the war, planners at the Pentagon had discussed bringing a government-in-waiting to Iraq to run the country after Saddam Hussein had been removed from power. The idea met stiff resistance at the State Department and the CIA, which worried that the Iraqi people would be skeptical about leaders handpicked by the United States and comprising mainly Iraqi exiles. Cheney understood their concerns, but the idea still had a certain crawl-before-you-walk appeal. A provisional Iraqi government, even an imperfect one, could help convince Iraqis that the U.S. government was serious when it promised to send a liberating force, not an occupying force.

If Cheney was skeptical about the CPA while it was operating, Bremer never saw that. "He was generally supportive of my view on how we had to go forward. I didn't get a sense that he was sort of working in a different direction. I don't know—there's all kinds of stories around all the time about how the people on his staff were working with the neocons at the Pentagon," he says. "I don't have any firsthand knowledge of that. It's possible. I never saw it from Cheney. He was always very supportive of what we were trying to do."[6]

Despite his cautionary public comments, Cheney shared the optimism of many who believed that capturing Saddam Hussein would prove demoralizing to the insurgents and might prove to be a military and political turning point in Iraq.

"My view, and thinking, is that there were milestones along the way that I would have expected would have moved things more in our direction, such as capturing Saddam Hussein," he says. It would be a temporary triumph.

Later that spring, Cheney's communications staff met at the vice president's residence to prepare for a busy month to come. Among the first items on the calendar was an appearance on April 5, the opening day of the baseball season, in Cincinnati. Cheney had been invited to the recently constructed Great American

Ballpark to throw out the first pitch in a game between the Cincinnati Reds and the Chicago Cubs. He had declined many similar offers, but the potential for good press at a harmless event in a swing state made this one too good to pass up.

As a bonus, one staff member jokingly explained to Cheney, Nick Lachey—a native of Cincinnati—would sing the national anthem before the game and would be accompanied by his girlfriend, the pop singer Jessica Simpson. Cheney thought Simpson's name sounded familiar. He asked his staff: "Is that the soldier who was captured in Iraq?"

As it turned out, neither Jessica Lynch nor Jessica Simpson would be at the ballpark that day, but Nick Lachey was among the 43,000 baseball fans who saw Cheney pitch to Reds catcher Jason LaRue. Before the game, the Reds' owner, Carl Lindner, a major donor to the Republicans, gave Cheney a tour of the facilities that included brief stops in the clubhouses of both his team and the Cubs.

The Reds' clubhouse was tense because of the VIP visitor. The ballplayers reacted awkwardly to Cheney's entourage—which included his Secret Service detail, his staff, and a horde of journalists. Several of the ballplayers spoke only broken English, and one staff member had the sense that few of them followed politics closely. Still, they seemed to be awestruck in the presence of one of the world's most powerful men.

The tension was exacerbated because several of the men following Cheney had only recently given up their own dreams of playing professional baseball. They took in the scene like eight-year-olds—mesmerized by the trappings of the clubhouse and captivated by the big-league ballplayers.

The Reds stood next to their lockers and waited their turn to shake Cheney's hand. Cheney was characteristically efficient. There was very little chitchat: a quick handshake, a nice-to-meet-you, and Cheney moved down the line. Then he came to Sean Casey, the Reds' hulking first baseman.

"Nice to meet you," Cheney said as he reached for Casey's hand. A mischievous grin crept across Casey's face as he began to pump Cheney's hand up and down.

"Big time."

Cheney gave Casey a friendly version of his crooked smile, and nodded his head in approval.

Four days later, on April 9, 2004, Cheney departed from Andrews Air Force base for Asia and a long-planned tour of the region. He would visit Japan, China, and South Korea. His priority was North Korea. As a potential candidate for president in 1994, Cheney had declared that North Korea posed the greatest threat to the United States. The same year, the Clinton administration negotiated an "Agreed Framework," in which the United States assisted North Korea with energy in exchange for a pledge from North Korea to stop work at two nuclear reactors suspected of being used to develop nuclear weapons. Cheney thought that this agreement rewarded North Korea for bad behavior and made the United States look weak.

The trip had been a long time coming. In January 2002, President Bush included the rogue North Korean regime in the "axis of evil," along with Iraq and Iran. Later that year, in October, the U.S. envoy James Kelly had traveled to North Korea to confront the regime of Kim Jong Il with fresh evidence that its nuclear programs were designed to produce weapons. Virtually everyone expected the North Koreans to dismiss the United States' findings and downplay their own nuclear ambitions. Instead, the North Koreans not only confirmed the existence of the programs, they boldly announced that they had long since abandoned the agreement of 1994 requiring them to halt their weapons program.

The United States declared the violations unacceptable, withdrew offers of humanitarian aid and energy assistance, and sought to recruit allies to pressure the North Koreans. But the North Koreans' admission—coming, as it did, just days before the U.S. Congress voted to authorize the war in Iraq and less than a month prior to the midterm elections of 2002—created a potential crisis for an administration focused on other issues. In short order, Cheney's staff stepped up its planning for a trip to Asia in early 2003.

But that trip was postponed by the Iraq War and the outbreak in Asia of the SARS virus, which Cheney's doctors be-

lieved would present risks to his health. By April 2004, when
Cheney finally set off for Asia, much had changed, and none of it
for the better. The failure to find weapons of mass destruction in
Iraq, the growing insurgency, the diminishing domestic support
for the war, a crumbling international coalition—all conspired to
weaken the hand of the United States as it sought regional support for an aggressive posture toward North Korea.

Cheney started the trip with three days in Japan, in a nod to
the "allies first" philosophy of the Bush administration. In To-
kyo, he thanked Prime Minister Junichiro Koizumi for Japan's
support in the war on terror. Then it was off to Beijing.

There, Cheney met with the four most powerful members
of the Chinese government: President Hu Jintao; Vice President
Zeng Qinghong; Prime Minister Wen Jiabao; and Jiang Zemin,
the former president, still powerful as head of the Central Military Commission. The meetings touched on economic development, trade, intellectual property, political reform, and regional
stability. But the two parties differed in their views of the most
important issue. For the Chinese, it was Taiwan; for Cheney, it
was North Korea.

On April 13, the *New York Times* had published a front-page
article about a Pakistani scientist named A. Q. Khan, considered
the father of Pakistan's nuclear program, who had confessed in
February to selling nuclear weapons expertise and equipment on
the black market. According to the story, Khan told Pakistani interrogators that he had visited North Korea and had been shown
three nuclear devices.[7]

Cheney used this article to press his Chinese counterparts on
the potential threat presented by North Korea. He was particularly aggressive in his talks with Jiang Zemin. Seated at a long
conference table—nearly a dozen Americans on one side, the
Chinese on the other—Cheney told Jiang that President Bush
remembered fondly their time together at Bush's ranch in Crawford. Cheney then moved quickly to North Korea. "The main
source of technology for uranium enrichment was Mr. A. Q.
Khan of Pakistan," Cheney explained, referring to the rogue
scientist who had created a worldwide network of clandestine
nuclear trading. "After he developed Pakistan's capability, Khan

began selling for his own purposes, particularly to Libya and North Korea. Our concern is that time is not necessarily on our side, that North Korea may continue to use this technology to further develop their capabilities. One of the greatest threats we all face is the proliferation of those technologies."

Jiang replied that he had followed the developments in North Korea closely and that China and the United States had common goals. Then he abruptly changed the subject. "Where is Mr. Stephen?" Jiang asked, his eyes scanning the American side of the table. Stephen Yates, Cheney's deputy national security adviser and a vocal critic of China in the 1990s, tentatively raised his hand, taken aback at the unexpected attention. "I know that Mr. Stephen speaks Chinese," Jiang declared. Yates acknowledged that he did, in fact, speak Chinese. Jiang then promptly returned to the discussion of North Korea.

"It is not an overstatement that sometimes I have found some shortcuts to access information," he said, noting that he had "made calls directly" to the North Korean leadership. Then, oddly, he downplayed the strength of those ties. "President Bush told me, 'Just tell them what to do.' It's not that simple," said Jiang, turning to Bush's representative in China, Ambassador Clark Randt. "I'm sure the ambassador has a better understanding of this issue."

Cheney's next meeting was with Hu Jintao, the Chinese president. Meetings between national leaders are often closed, to guard against leaks of sensitive information. In this case, Cheney's aides had warned him to assume that the room was equipped with listening devices enabling others to hear everything that was being discussed.

Cheney's meeting with Hu included just three participants on each side. The Americans were Cheney, Scooter Libby, and Victoria Nuland, a national security adviser to the vice president. The others, from both the Chinese and American sides, were led to a separate holding room for refreshments.

After waiting just ten minutes, the Chinese officials abruptly left the holding room without a word. The confused Americans didn't know whether to stay or go. After a tense interval, two of Cheney's staffers made their way down a long hallway in search of their suddenly elusive hosts. As they walked, they heard a muffled discussion in the distance. "It sounded like Charlie Brown talking," says one. "Wah, wah, wah." They followed the sound into a large room about halfway down the hall, where they found their missing Chinese counterparts huddled around a speaker listening to the supposedly sensitive meeting between Cheney and Hu. "Even the waitstaff was listening in," recalls one of the Americans.

Later, when Cheney's party returned to their hotel in Beijing, Cheney asked Yates about the odd exchange with Jiang Zemin.

"What the hell was that about?" asked Cheney.

"I think he likes me," Yates deadpanned.

"Um, I don't think he did that because he likes you." Cheney laughed.

When he departed for Asia, Cheney doubted that China was willing to pressure North Korea. The trip did nothing to persuade him that he had been mistaken.

Back in Washington, the special prosecutor's investigation continued. Even after the appointment of Fitzgerald, Libby continued his efforts to discredit Wilson. Libby regularly assigned sensitive research to junior staffers in the vice president's office. He also asked those staffers to reach out to journalists covering the case to challenge their reporting. Libby's efforts intensified with the publication of Wilson's book *The Politics of Truth* in early May 2004.

Part memoir and part polemic, *The Politics of Truth* received a flood of media coverage: NBC's *Dateline*, *Larry King Live*, Fox News's *On the Record with Greta van Susteren*, CNBC's *Capital Report*, *Charlie Rose*, several CNN shows, and even *The Daily Show with Jon Stewart*. On *The Today Show*, Wilson rated an interview with Katie Couric, who asked, "Why do you believe your wife was brought into this? Was it a simple payback, in your view, by the Bush administration for refuting claims that were made in the State of the Union address?"[8]

Wilson used the book to accuse Scooter Libby of involvement in the unmasking of his wife. He speculated that Libby was "quite possibly" the official who exposed his wife, a charge picked up in a piece in the *New York Times* about the book.

"Former Ambassador Joseph C. Wilson IV says in a new book

that he believes the White House official behind the disclosure
of his wife's identity as an undercover C.I.A. officer was 'quite
possibly' I. Lewis Libby, Vice President Dick Cheney's chief of
staff," wrote Richard Stevenson and David Johnston. "But Mr.
Wilson offers no firm evidence to support his assertion, and the
White House has denied it."[9]

Libby seized on the article and asked the communications
staff to distribute it to other reporters covering the Plame affair.
As Wilson made the rounds on television chat shows, Libby's
campaign became a top priority in meetings of the communications staff, so prevalent that one staff member even gave it a
name, the "No Evidence Project." The other staple of the project
was Tenet's statement of July 11, acknowledging that the CIA
sent Wilson on its "own initiative."

The only public hint of the ongoing campaign came on
May 4, when Keith Olbermann spoke with Wilson on Olbermann's show. "You do know that they are still going after you,
right?" he asked Wilson. "We promoted the fact that you would
be on this show tonight. Today we received three separate copies
of the same e-mail with talking points from the White House. . . .
Are you surprised by that?"

"No, I'm not surprised at all," Wilson responded. "I tell you,
this administration has tried to manage and direct the news from
the very beginning. As I point out in the book, they have made
the lives of journalists very unpleasant. One journalist said he
was afraid to go to print because he might end up in Guantá-
namo, which I take to be a metaphor for being cut out."[10]

Kellems didn't want to talk to reporters about the Wilson
case and didn't want Libby to ask others on the communications
staff to do it either. He was particularly concerned about Hannah Siemers, a communications assistant in her mid-twenties,
because she had been working in the vice president's office during the spring and summer of 2003, when the activity concerning Wilson started. Kellems became so anxious that he took the
problem to Addington.

"We're being encouraged to respond to reporters and to respond in a certain way,"[11] Kellems recalls telling Addington,
whom he described as the "rock" in Cheney's office in the Plame

matter. Kellems says that Addington told him to refuse Libby's orders. They would fall back on formal language denying comment: *The matter you're asking about is related to an ongoing investigation; if you're interested in any information on this, call the special prosecutor's office.*

Libby's campaign should have received a boost with the release of the report by the Senate Select Intelligence Committee (SSIC) on prewar intelligence on Iraq. Point by point, the SSIC rebutted Wilson's claims. Did the vice president send him? No. Had his wife recommended him? Yes. Did he challenge the authenticity of the forged documents? Not possible. Did he debunk the claim that Iraq had been seeking uranium from Niger? No.

On the last two points, the SSIC's report was particularly conclusive. Wilson could not have pointed out problems with the forged documents, the committee found, because the U.S. government did not yet have those documents in its possession at the time of his trip. And not only did Wilson's report fail to debunk the original reporting on Iraq and Niger; on balance the CIA believed that his information made the original intelligence seem more credible. Although State Department analysts thought Wilson's report supported their skepticism on Iraq and Niger, "for most analysts, the information in the report lent more credibility to the original Central Intelligence Agency reports on the uranium deal," read "Conclusion 13" of the SSIC's report.[12]

But for the most part, journalists ignored or buried the committee's critique; Wilson's further discrediting would be embarrassing to the news outlets that had, for more than a year, relied on his information for stories about the Plame case. And by the time the report was released, the story had moved beyond Wilson's claims to Fitzgerald's investigation and its likely outcome.

Reporters had also begun to devote considerable time and resources to the presidential campaign of 2004. There had long been whispers that Bush would look for someone to replace Cheney on the Republican ticket. By early summer, with the national convention just two months away, a potential early retirement for Cheney was the talk of the chattering classes. Paul Gigot, editor of the editorial page at the *Wall Street Journal*, acknowledged that speculation about dumping Cheney was prevalent. "It's out there," he said. "The buzz is out there."[13] The editor of the *Des Moines Register*, who described himself as an admirer of Cheney's, wrote an open letter to Cheney asking him to step down.[14] The conservative writer Debra Saunders recommended replacing Cheney with John McCain,[15] who did his part to squelch such speculation by praising Cheney for doing a "fine job" as vice president.[16]

One reason for the chatter was that Cheney had become a focal point of the Democrats' attacks on the Bush administration. In particular, Bush administration critics sought to link Cheney with no-bid contracts awarded to Halliburton, his former employer, since he had become vice president. Halliburton had a long history of no-bid contracts, including many for its overseas work during the Clinton administration. But that context was missing from much of the news coverage.

Also missing from most of the attacks was any evidence that Cheney had been involved in any discussions about Halliburton's contracts. That appeared to change, temporarily at least, with a story published in *Time* in early June. *Time* had obtained an e-mail written by an unnamed official in the Army Corps of Engineers about a contract for a project named "Restore Iraqi Oil." The e-mail noted that Undersecretary of Defense for Policy Douglas Feith had signed off on the contract "contingent on informing WH [White House] tomorrow. We anticipate no issues since action has been coordinated wVP's [Vice President's] office."[17]

In a conference call with reporters on behalf of John Kerry's campaign, Senator Patrick Leahy of Vermont went after Cheney. "It raises the real question, can the American people trust their government to do the right thing? We have very real rules here."[18]

As a substantive matter, the issue soon faded after six officials at the Pentagon who were involved in the contract told a congressional hearing that they had seen no evidence of involvement by Cheney's office before the contract was awarded, and his office had simply been notified after the fact.[19]

But as a political issue, Halliburton would live on. On June 21, the Democratic National Committee (DNC) kicked off what it called "Halliburton week." The chairman of the DNC, Terry McAuliffe, lashed out at the vice president. "I think the public is very concerned," he declared. "They don't want to see a sitting vice president receive personal financial gain by the awarding of these contracts."[20]

In reality, Cheney received only deferred payments that were unaffected by Halliburton's performance. On January 18, 2001, two days before he took the oath of office, he and his wife signed an irrevocable "gift of trust agreement," in which they divested "themselves of any and all economic benefit" from the stock options. As part of the agreement, Cheney had agreed to donate any profits from his stock options—then valued at about $8 million—to charity. A detailed study by the nonpartisan Annenberg Public Policy Center at the University of Pennsylvania later concluded: "Cheney's deferred payments from Halliburton wouldn't increase no matter how much money the company makes, or how many government contracts it receives," and, "It is clear that giving up rights to the future profits constitutes a significant financial sacrifice, and a sizeable donation to the chosen charities."[21]

Later on June 21, "Halliburton week" continued with a conference call with reporters featuring the Democratic senators Frank Lautenberg and, once again, Patrick Leahy. The two men discussed their concerns about no-bid contracts awarded to Halliburton by the Pentagon. They promised a congressional investigation.

The next day, Cheney traveled to Capitol Hill in his role as president of the Senate. Though Congress was not yet in session, the senators had gathered on the floor for an official photograph. Cheney had been told about Leahy's comments and initially shrugged them off. "They trashed me over Halliburton for the umpteenth time," he recalls. "I was used to that. Nothing new there."[22]

But what happened next clearly affected him more, touching off a fracas that would quickly become a minor legend inside the Beltway.

"I was in the well of the Senate on the Republican side and Leahy came over and put his arm around me. And he didn't kiss me but it was close to it. It was like we're sort of the best buddies in the world. And it pissed me off, frankly. So I flashed. And I told him—"[23]

Just as he begins to repeat his controversial comment to Leahy verbatim, Cheney thinks better of it, laughs loudly, and offers a redacted version:

"I dropped the F-bomb on him."[24]

"I think he was just having a bad day," Leahy said. "I was kind of shocked to hear that kind of language on the floor."[25]

News of the outburst surprised some of Cheney's friends. They had seen him keep cool under far worse attacks. Merritt Benson, Cheney's former state representative and fishing buddy, remembers one particularly harsh editorial board meeting in Wyoming in the 1980s. Cheney, he says, withstood one unfair accusation after another without so much as raising an eyebrow. Benson, however, was fuming. "We got out to the car and I asked him how he could sit there through that. He said, 'You never, ever let those people get to you. Or then they win.'"[26]

In his long public career, Cheney had only one notable outburst. In 1987, he called the House speaker Jim Wright "a heavy-handed son of a bitch," after Wright reneged on a deal he had struck with the Republican leadership.[27]

But Cheney's friends expected him to defend himself against attacks on his integrity, and he had always disliked the my-good-friend-from-Wyoming-phoniness of Washington. "I wasn't surprised that he told Leahy to shove off," says Ron Lewis, a friend from high school. "I was just surprised that he used that language."[28]

Looking back on the incident, Cheney has no regrets. "It was out of character from my standpoint, I suppose. But what can you say. . . . It was heartfelt."[29]

Six days later, on June 28, 2004, the CPA in Iraq handed power to an interim Iraqi government in a secret ceremony announced only after the transfer was completed. Although the handover marked the official end of American rule in Iraq, it would not mean an end to the violence that had beleaguered the country for more than a year.

For Cheney, the handover did not come soon enough. He

had been concerned that a long-term occupation of Iraq by the United States would breed resentment among the Iraqi populace, and by the summer of 2004 his concerns had been realized.

In retrospect, he says, the mechanism for U.S. governance in postwar Iraq was a failure. "I think we should have probably gone with the provisional government of Iraqis from the very outset, maybe even before we launched. I think the Coalition Provisional Authority was a mistake, wasted valuable time."[30]

Cheney's skepticism about the CPA was never apparent to the man who ran it. "I always felt that he was an ally," says Bremer. "I never felt that he was working against me either behind my back or in any other way. I must say my impression of him as a man is that he is very straight. He does not have a lot of curveballs.

"He's one of the people in the administration who probably best understands the relationship of what we're doing in Iraq to the war on terrorism. At least judging from his public statements he understands it very well. Says it, articulates it better than anybody else—than Powell or Rumsfeld or Rice. He really gets it."[31]

The summer of 2004 brought into sharp focus the debate over Iraq and its place in the war on terror. In July, came the report of the 9/11 Commission, a ten-person panel established to investigate the attacks of that day. And that same month, the Senate Select Intelligence Committee issued its report on the prewar use of intelligence on Iraq. The findings were bound to be discussed and debated at length, and the fact that they were released shortly before a presidential election resulted in a predictable political scrum about their meaning.

Democrats argued that the Iraq War had been a distraction from the war on terror; the White House and its allies maintained that the war had been central to that broader fight. Cheney made the administration's case most forcefully, a fact that brought eye-rolls from the more cautious White House communications team. That didn't stop him.

In the spring, a 9/11 Commission staff statement reported that its investigation had not turned up evidence of a "collaborative relationship" between Iraq and al Qaeda. The news media jumped on this as proof that the Bush administration had exaggerated claims of an Iraq–al Qaeda connection as it made its case for war. "Panel finds no Qaeda-Iraq Tie," blared a headline in the *New York Times*.

In its final report, released on July 22, the 9/11 Commission further narrowed its finding, concluding only that its investigation could not prove a "collaborative operational relationship" between Iraq and al Qaeda. Once again, news reports played the finding as an authoritative rebuttal to the Bush administration's prewar claims that Iraq and al Qaeda had a long relationship.

But the body of the report offered several troubling new pieces of evidence of what it called "friendly contacts" between Iraqi regime officials and senior al Qaeda leaders. And at a press conference held to release the report, 9/11 Commission co-chairman Thomas Kean confirmed the existence of Iraq–al Qaeda ties. "There was no question in our minds that there was a relationship between Iraq and al Qaeda," he declared.[32]

In its report on Iraq and terrorism, the SSIC included numerous excerpts of CIA reports that seemed to bolster the Bush administration's assertions regarding the relationship between Iraq and al Qaeda. A declassified CIA report from January 2003, called "Iraqi Support for Terrorism," concluded:

> *Iraq continues to be a safe haven, transit point, or operational node for groups and individuals who direct violence against the United States, Israel, and other allies. Iraq has a long history of supporting terrorism. During the last four decades, it has altered its targets to reflect changing priorities and goals. It continues to harbor and sustain a number of smaller anti-Israel terrorist groups and to actively encourage violence against Israel. Regarding the Iraq–al Qaeda relationship, reporting from sources of varying credibility points to a number of contacts, incidents of training, and discussions of Iraqi safe haven for Osama bin Laden and his organization dating from the early 1990s.*[33]

The SSIC report also noted that the Iraqi Intelligence Service (IIS) "focused its terrorist activities on western interests, particularly against the U.S. and Israel." According to the SSIC report:

The CIA summarized nearly 50 intelligence reports as examples, using language directly from the intelligence reports. Ten intelligence reports, [redacted] from multiple sources, indicated IIS "casing" operations against Radio Free Europe and Radio Liberty in Prague began in 1998 and continued into early 2003. The CIA assessed, based on the Prague casings and a variety of other reporting, that throughout 2002, the IIS was becoming increasingly aggressive in planning attacks against U.S. interests. The CIA provided eight reports to support this assessment.[34]

And yet the media coverage of the Senate report focused on its conclusion that there was little evidence of an "established, formal relationship" between Iraq and al Qaeda—something neither the Bush administration nor Dick Cheney had ever claimed. A *New York Times* story reported that CIA analysts had "largely discounted" the Bush administration's views on a long relationship, failing to mention the declassified letter that George Tenet had released before the war, noting "growing indications of a relationship" between Iraq and al Qaeda and senior-level contacts between the two "going back a decade." Many news articles misportrayed the reports as a corrective to claims Cheney had made about the relationship.

But the misleading reporting on Cheney's claims about Iraq and al Qaeda was nothing new. Back in March 2004, *Washington Post* intelligence beat reporter Dana Priest accused Cheney of making a "definite link between 9/11 and Iraq" in "all those appearances on national broadcasts." Priest was almost certainly referring to Cheney's discussions of intelligence that lead hijacker Mohammed Atta had met in Prague with an Iraqi intelligence officer several months before the 9/11 attacks.

In a December 9, 2001, appearance on *Meet the Press*, Cheney responded to a question from Tim Russert about media reports on the contact between Atta and the Iraqi.

"Let me turn to Iraq," said Russert. "When you were last on this program, September 16, five days after the attack on this country, I asked you whether there was any evidence that Iraq was involved in the attack and you said no. Since that time, a couple of articles have appeared which I want to get you to react to. The first: 'The Czech interior minister said today that an Iraqi intelligence officer met with Mohammed Atta, one of the ringleaders of the September 11 terrorist attacks on the United States, just five months before the synchronized hijackings and mass killings were carried out.'"

Cheney responded: "It's been pretty well confirmed that he did go to Prague and he did meet with a senior official of the Iraqi Intelligence Service in Czechoslovakia last April, several months before the attacks."

Russert followed up with several questions about Iraq's long history of harboring terrorists. "If they're harboring terrorists, why not go in and get them?"

Cheney was careful to distinguish between the indisputable proposition that Iraq provided safe haven to terrorists and the allegation that Iraq had a role in 9/11. "Well, the evidence is pretty conclusive that the Iraqis have, indeed, harbored terrorists. That wasn't the question last time we met. You asked about evidence of their involvement in September 11."

One line from Cheney's answer — "pretty well confirmed" — would be used against him for years to come as evidence Cheney's willingness to push the intelligence beyond what he was getting from the CIA. The intelligence community, critics would claim, had dismissed reports of a meeting.

But one week after Cheney told Russert that the contact between Atta and Ahmad al Ani had been "pretty well confirmed," an unnamed intelligence official confirmed the report, telling the *New York Times* "there was definitely one meeting."[35]

While some intelligence analysts came to view the alleged meeting with skepticism, George Tenet was not one of them. Tenet wasn't shy about sharing his views on the subject, even with reporters. Before the Iraq War, speaking as a "senior intelligence official" during a background interview with the *Washington Post* editorial board, Tenet explained that while there were some gaps in the intelligence, he believed Atta had met with an Iraqi intelligence officer. Although the newspaper had reported analysts' skepticism about the meeting on several previous occasions, Tenet's comments as a "senior intelligence official" went unreported. Even as CIA analysts raised doubts about the Prague encounter, Tenet continued to believe that the meeting had taken place. As the questions persisted, Cheney went to Tenet directly. "George, you're telling me this about Atta," Cheney said. "Is that still right?" The CIA director told Cheney that he continued to believe that the meeting had taken place. "That came directly out of the agency," he says. "George Tenet gave that to me."[36]

Cheney used it again on September 8, 2002, in another appearance on *Meet the Press*. Russert asked the vice president again about his assertion from a year earlier that the U.S. government did not have evidence of Iraqi involvement in 9/11.

I want to be very careful about how I say this. I'm not here today to make a specific allegation that Iraq was somehow responsible for 9/11. I can't say that. On the other hand, since we did that interview, new information has come to light. And we spent time looking at that relationship between Iraq, on the one hand, and the al Qaeda organization on the other. And there has been reporting that suggests that there have been a number of contacts over the years. We've seen in connection with the hijackers, of course, Mohamed Atta, who was the lead hijacker, did apparently travel to Prague on a number of occasions. And on at least one occasion, we have reporting that places him in Prague with a senior Iraqi intelligence official a few months before the attack on the World Trade Center. The debates about, you know, was he there or wasn't he there, again, it's the intelligence business.

Russert pressed him further. "What does the CIA say about that? Is it credible?"

"It's credible. But, you know, I think a way to put it would be it's unconfirmed at this point."

After the war, Tenet rejected claims that the meeting had been disproven. In an exchange at a hearing of the Senate Intelligence

Committee on February 24, 2004, nearly a year after the beginning of the Iraq War, Michigan Senator Carl Levin pressed Tenet to disavow the meeting. Tenet stubbornly refused.[37]

"Was the Intelligence Committee's assessment—what is the Intelligence Committee's assessment of whether or not 9/11 hijacker Mohamed Atta met with Ahmed al-Ani, an alleged Iraq intelligence officer in Iraq in April of 2001. What is your assessment?"

"Sir, I know you have a paper up here that outlines all that for you," Tenet responded. "It's a classified paper. My recollection is we can't prove that one way or another. Is that correct?"

Levin persisted. "The *Washington Post* says that the CIA has always doubted that it took place. Is that correct?"

"We have not gathered enough evidence to conclude that it happened, sir. That's just where we are analytically in the paper that—"

"It's not correct then that you doubt that it took place?"

"Sir, I don't know that it took place. I can't say that it did."[38]

Tenet's characterization did not much differ from Cheney's claim that the reporting had been "credible" but "unconfirmed." But this didn't matter. Reporters, including several on the intelligence beat, used Cheney's refusal to rule out the alleged meeting to accuse him of making false claims that Iraq was behind the 9/11 attacks.

In an interview on the subject during the contentious summer of 2004, Cheney chose his words carefully.

"I think it is important to the public that there be dialogue to make sure to make a distinction on the one hand, push on the question of whether or not there was Iraqi participation and support for what al Qaeda did in attacking the United States on 9/11, and what we've said is that we've never been able to prove that, we've been unable to confirm it," Cheney said. "The second proposition is between Iraq and al Qaeda and Iraqi intelligence services over a longer period of time and there we have said yes, there was, and we have been able to confirm that. . . . I think people sometimes get sloppy and sort of kluge together those two issues—well, there was no proof that Saddam Hussein was involved with 9/11, therefore there is no relationship. That's not accurate. There was a relationship. But we have not been able to confirm that there was any involvement in 9/11."[39]

Cheney, asked about critics who say he is the one who is sloppy about making the distinction: "I try not to be. It's possible that I am sloppy on occasion, but I try not to be."[40]

The claims that Cheney exaggerated the Iraq–Al Qaeda connection would dog Cheney for years to come, as critics, led by Senator Carl Levin of Michigan, continued to accuse him of manipulating intelligence. But Paul Bremer, the CPA administrator in Iraq, supports Cheney. On the broad question of the connection between Iraq and al Qaeda, Bremer says, "I don't think there's any question about that." Bremer had access to intelligence in Iraq immediately after it was collected, including information gleaned from captured Iraqi intelligence documents. "This idea that there was no relationship between al Qaeda and Saddam is just wrong," he says with an incredulous laugh. "Flat wrong on the record."

The presidential campaign intensified as the national conventions drew near. On July 16, 2004, Cheney and his staff flew to Minnesota for a rally with the talk radio host Sean Hannity at the Minneapolis Convention Center. En route aboard Air Force Two, the staff gathered in the aisles for a short birthday celebration for deputy press secretary Randy DeCleene. Cheney rarely comes to the back of the plane—that's where reporters sit—but he made an exception for the presentation of the cake. He looked relaxed, wearing a light blue button-down shirt and no tie. Cheney stood with his arms extended above his head and spread wide, the palms of his hands resting on the overhead luggage compartments on either side of the aisle. Speaking in front of the traveling journalists and the staff members, the vice president thanked DeCleene for his hard work and wished him a happy thirtieth birthday.

DeCleene appreciated the gesture. He had been on Cheney's staff for just four months, and having the vice president say nice things about him in front of the reporters he worked with on a daily basis would be helpful.

DeCleene's actual birthday was the following day and after

he had finished his work in Minneapolis, he caught a commuter flight to Chicago, his hometown, for a celebration with friends. It was a memorable night, or at least a festive one.

Four days later, on July 21, Cheney's staff gathered at the vice president's residence at the Naval Observatory for a cookout. The staff ate and drank alongside the pool on the west side of the large house. DeCleene was helping himself to a hamburger when someone behind him called his name. It was Cheney.

DeCleene usually knew in advance when he would have face time with the vice president. He would sometimes spend hours preparing thirty-second briefings for Cheney, rehearsing his lines again and again until he had memorized what he would say. Not this time.

"How was your birthday?"

DeCleene recalls being impressed that Cheney, in the midst of nonstop campaigning, remembered his birthday and even more impressed that he thought to ask about it. Perhaps it was this feeling of familiarity that made DeCleene comfortable with an answer so candid it surprised him even as it passed his lips.

"I think today is the first day that I'm not hung over," he said, exaggerating.

After a split second of awkward silence, Cheney offered a sympathetic grin and walked away.

"I was mortified," DeCleene says. "I spent the next two days wondering why the hell I told the vice president of the United States that I was hung over."[41]

Cheney continued in his role as the Bush campaign's attack dog, snarling after every Democratic misstep, particularly when John Kerry said anything that could be interpreted as soft on national security. He saved the sharpest attacks for the national stage provided by the 2004 Republican National Convention in New York City, which began in late July.

"Even in this post-9/11 period, Senator Kerry doesn't appear to understand how the world has changed. He talks about leading a 'more sensitive war on terror,' as though Al Qaeda will be impressed with our softer side," Cheney said with a dispassionate, almost clinical tone that took the sting out of his biting words. "He declared at the Democratic Convention that he will forcefully defend America after we have been attacked. My fellow Americans, we have already been attacked and, faced with an enemy who seeks the deadliest of weapons to use against us, we cannot wait for the next attack. We must do everything we can to prevent it and that includes the use of military force."

Cheney ripped Kerry for his votes on Iraq. "Although he voted to authorize force against Saddam Hussein, he then decided he was opposed to the war, and voted against funding for our men and women in the field. He voted against body armor, ammunition, fuel, spare parts, armored vehicles, extra pay for hardship duty, and support for military families. Senator Kerry is campaigning for the position of commander in chief. Yet he does not seem to understand the first obligation of a commander in chief and that is to support American troops in combat."

Shortly before sunrise on September 3, the morning after the convention ended, Cheney's motorcade departed from the St. Regis Hotel and cruised through the awakening city to La Guardia Airport. Many of the people they encountered along the way expressed their feelings about the Republican invasion of their town — some of them annoyed by the inconvenience of it all and others hostile to the politics. In any case, it was a far different reception from the one Cheney had gotten when he visited after the 9/11 attacks.

As the campaign wore on, the candidates were reminded frequently of the power of language. Carefully wrought talking points could slide away without impact, whereas an offhand remark could come to occupy the conversation for weeks at a time. At a rally in Des Moines on September 7, 2004, Cheney made a statement that, depending on who you asked, was either a straightforward assessment of an opponent's policy or an unforgivable bit of political fortune-telling: "It's absolutely essential that eight weeks from today, on November 2, we make the right choice. Because if we make the wrong choice, then the danger is that we'll get hit again and we'll be hit in a way that will be devastating from the point of view of the United States, and that we'll fall back into the pre-9/11 mind-set, if you will, that in fact these terrorist attacks are just criminal acts, and that we're not really at war."

Hearing the speech, Scooter Libby at once anticipated a problem. He wanted the official White House transcript of the event rushed to Cheney's plane as soon as it was finished. That long sentence, he knew, could be interpreted two different ways depending on the listener's understanding of "the danger." Was the danger that a candidate with a "pre-9/11 mind-set" would treat terrorist attacks as criminal acts? Or was it that the "wrong choice" in the election would actually increase the likelihood of devastating terrorist attacks? The first interpretation was well within the bounds of acceptable campaign rhetoric; the second, as they would soon discover, was not.

The journalists traveling with the vice president typically paid scant attention to the stump speech, which they had heard so many times they could deliver it from memory. But Cheney's warning about the consequences of the "wrong choice" was enough to arouse their interest. On Air Force Two, reporters were brimming with questions. Anne Womack, a Cheney press aide, walked to the back of the plane to try to clarify his comments. Womack insisted that Cheney's meaning was clear: a President Kerry would respond to future terrorist attacks by treating them as crimes, not acts of war, and such a response would be dangerous.

Reporters were skeptical. The consensus among the group was that Cheney had said that the "wrong choice" in the election would bring more attacks. They wanted to know whether Cheney was backing away from his remarks. Womack walked to the front of the plane and returned after a brief meeting with Cheney and his top staff. "The vice president stands by his statement," she said.[42]

More confusion. Did he stand by his statement as Womack interpreted it, or as the reporters interpreted it?

To complicate matters, the White House stenographer assigned to Cheney's speech interpreted the comments—and punctuated them—in a manner consistent with the way the reporters heard them. Libby ordered the transcript changed, but the stenographer resisted, strengthening the position of the traveling press corps.

Cheney and his team lost the argument. The lead in a story in the next day's *Los Angeles Times* was typical of the coverage.

"Vice President Dick Cheney suggested Tuesday that electing the Democratic presidential ticket would make the United States more vulnerable to attack."[43] Commentators roundly criticized Cheney for exceeding the bounds of proper campaign rhetoric, and although his staff spent hours arguing for his interpretation, the vice president's refusal to back away from the statement made the problem worse.

Two days after the original comments in Iowa, Laura Meckler, a reporter for the Associated Press traveling with Cheney, filed a dispatch from a stop in Cincinnati. At the end of the story she wrote: "The vice president argued that the United States cannot view terrorism as it views ordinary crime and said the nation's future depends on its response to the threat. But he did not repeat explosive comments from earlier in the week when he suggested that a vote for Democrat John Kerry would increase the likelihood of a terrorist attack."[44]

Cheney read Meckler's story, and he was annoyed. In his view, the story simply ignored the attempts of his staff to clarify his meaning. The vice president told Anne Womack that he wanted Meckler kicked off the plane.

It was a move fraught with peril. Removing a reporter from the plane was not likely to be a quick or clean fix. Other reporters would certainly take up their colleague's cause in private. Cheney's staff worried that the flap would be reported in the press coverage of the vice president's travels and urged him to reconsider.

"I don't have time for this shit," he said, in a tone that reflected weariness rather than anger. He agreed to let her remain on the plane for the duration of the trip, but said: "I don't want her on the plane again." As it turned out, the AP did not assign Meckler to cover Cheney for the rest of the campaign, much to the delight of Cheney's staff.

At least the Associated Press could send another reporter. The *New York Times* wasn't so lucky. In a story written for its front page two weeks after the incident with Meckler, the reporter Rick Lyman detailed the difficulties of trying to cover Cheney without a ride on Air Force Two.

"[T]here is not a seat for me," wrote Lyman. "Nor has there been a seat for the previous two *New York Times* reporters sent to cover the vice president. I am told not to take this personally. Nor, I am told, is this intended as a slight against the paper, which normally maintains a seat (paid for handsomely) on all campaign planes, presidential and vice-presidential." Some colleagues, Lyman continues, suspected that Cheney's "antipathy toward the newspaper" might explain the slight.[45]

One of those colleagues may have been Lyman's boss, the executive editor, Bill Keller. Earlier that summer, in an e-mail to Cheney's communications director Kevin Kellems, Keller said he had been thinking about Cheney's "antipathy for the *Times*," and "the attitude some officials in this administration have toward the *Times*." He wrote:

> *I know there are some people who actually believe that the* Times *has a partisan or ideological "agenda"—the favorite word of critics on the right AND the left. There are even a few people who think the news coverage and the editorial page operate in lockstep as part of a liberal cabal. The vice president is much too sophisticated and experienced, I suspect, to really believe that. I won't pretend that reporters' stories are never shaped by liberal bias (more accurately liberal assumptions about the world) but I think those instances are relatively rare, and I fight to filter them our* [sic] *and deplore them when they get into the paper. But that's not an "agenda." I'm pretty sure your boss understands that when reporters inject a "gotcha" attitude into their prose, their agenda usually has nothing to do with bias or conviction, it has to do with getting their stories on the front page and making their competitors envious. That, plus an eagerness to prove they have not been "spun": when it comes to dealing with politicians and public officials, there is often a reluctance to seem naïve, credulous, "in the pocket" of people they cover. All of this sometimes leads reporters to write tendentious language or to pump up little facts into doubtful stories, and it's part of our job to edit out those excesses. This was just as true in the Clinton Administration (which also had a lot of* Times*-haters at the top) as it*

is in this administration. So it doesn't entirely explain why
some officials at senior levels of this administration regard
the Times *with such scorn.*

Keller posited two other possibilities: "that our critics in the
administration attack us because it is red meat for the base"; and
"that the attacks are simply designed to knock us off balance and
perhaps move us to write more sympathetically. . . . I suspect there
are some people who believe if they scream 'Liberal agenda! Liberal agenda!' loud enough, we'll overcompensate by writing stories
they like. Or, if they make a show of not inviting us to briefings,
not including us on official trips, then we'll behave as they like."

The most distressing possible explanation for the tension,
wrote Keller, would be if "some officials have just given up hope
of being understood or accurately reflected in the pages of the
Times." Although reporters for the *Times* can still obtain information, Keller continued, "what's much harder to get, in an administration that chooses to shut us out, is understanding. . . . I
have always believed one of the most difficult tasks for a newspaper is to fight oversimplification and caricature, and to strive
for understanding—of nuance, of character, of motivation, of
ideas and ideals. Our job is not to 'support' our leaders, not to
buy in to any administration, Democrat or Republican, but our
job should be to figure out what they believe and why, and how
all of that shapes the policies they make."

Keller once again conceded that ideological biases of the establishment press cut against conservatives like Cheney.

I think the media generally—including the Times, *but also*
most other mainstream newspapers, news magazines and
television—has a more difficult time doing this with con-
servatives than with liberals. The reasons for that are prob-
ably both too obvious and too complicated to go into here,
but it's been a pet cause of mine at the Times *to make a*
special effort to understand and portray conservatives in a
way that is not necessarily flattering, but three-dimensional,
and that an honest conservative would regard as fair.

Keller finished his letter contrasting the coverage of Deputy Defense Secretary Paul Wolfowitz and Attorney General John Ashcroft. Wolfowitz spent time with Keller for a long piece he wrote for the *New York Times* magazine. The payoff for Wolfowitz, said Keller, was that "a considerable number of NYT readers found themselves reassessing the man, or at least understanding him at a different level." Ashcroft, by contrast, would never open himself up and "he still lives in the land of stereotype."

His message was not subtle: so does Dick Cheney.

On September 4, Cheney flew to Roswell, New Mexico, for a campaign rally. Roswell is known, to the extent it is known at all, as the world headquarters for research on unidentified flying objects and alien life. "The streetlights are little aliens, and they have the alien museum," says Senator John McCain, who joined Cheney for the event. "It's all about the alien invasion. It's incredible."[46]

Cheney didn't know it, but his visit to Roswell—the third since he reentered public life—was a big deal to the world's ufologists. They believe Cheney's experience at the Pentagon made him one of the few U.S. government officials to have been briefed on the UFO phenomenon. He was in a unique position to share the truth about UFOs and extraterrestrials, if he were so inclined.

Two events heightened this intrigue. The first took place at a rally for Bush and Cheney in Springdale, Arkansas, on July 28, 2000, just three days after Cheney was announced as the vice presidential candidate. A UFO enthusiast named Charles Huffer approached then-Governor Bush and sought a commitment. "Half the country believes UFOs are real. Would you finally tell us what the hell is going on?"

"Sure I will," said Bush.

Huffer motioned to Cheney. "This man knows. He was secretary of defense."

Minutes later, Bush was standing next to Cheney when Huffer made a second pass. Bush gestured to his running mate and told Huffer that Cheney had a new assignment. "It'll be the first thing he will do," Bush promised facetiously. "He'll get right on it."[47]

Many in the UFO community missed the candidate's sarcasm

and would come to speak of this exchange as "Bush's promise." Would it be just another broken promise from a politician seeking the votes of ufologists and their supporters? Word spread quickly throughout the UFO community. Cheney is on it. After he took office, they waited eagerly for news, but the vice president seemed preoccupied with the Energy Task Force and other matters. So on April 11, 2001, the UFO researcher Grant Cameron followed up in a phone call to Cheney, who was appearing on the *Diane Rehm Show* on National Public Radio.

"Since the statement made by George Bush last July, there is a vicious rumor circulating in the UFO community that you've been read into the UFO program. So my question to you is, in any of your government jobs, have you ever been briefed on the subject of UFOs, and if you have, when was it and what were you told?"

"Well," said Cheney, "if I had been briefed on it, I'm sure it was probably classified and I couldn't talk about it."

The host followed up. "Is there an investigation going on within this administration, Mr. Vice President, as to UFOs?"

"I have not come across the subject since I've been back in government, oh, like since January twentieth," Cheney said. "I've been in a lot of meetings, but I don't recall one on UFOs."[48]

The UFO community seized on Cheney's allusion to classified information, and its hope turned to bitterness. "He was like the secretive and calculating Cheney of old," Cameron complained.

Cheney's reticence may have dampened the enthusiasm of Roswell's ufologists. The Republicans attending his rally with John McCain in September 2004, however, were still excited. Polls showed that New Mexico was probably one of the states that would determine the winner if the election was close, so the Republican National Committee called in Senator John McCain. Cheney would rally the base; McCain would appeal to independents. Two for one.

These rallies often attract the most enthusiastic party activists and campaign volunteers. The result is a media-savvy campaign organizer's dream: a boisterous crowd, roaring its approval for everything that comes out of the candidate's mouth. Even by those standards, the gathering in Roswell was exceptional. "Everybody's enthusiastic," McCain says, lifting his hand in front of his mouth as he makes a crowd-goes-wild sound effect. "Dick starts reading, and he's reading pretty fast."

After nearly every sentence that Cheney spoke, even the throwaway lines, the audience would interrupt with some kind of reply.

At one point, Cheney said: "You may have noticed I now have an opponent in this campaign."

The crowd booed loudly.

At another: "We will never seek a permission slip to defend the United States of America."

They chanted: "U-S-A!"

Cheney said that Kerry voted against funding the war in Iraq as he set up what would become a signature attack on Kerry.

The audience broke into boos and cries of "Flip-flop!"

"Only twelve members of the United States Senate opposed the funding that would provide those resources for the troops. Only four senators voted for the use of force and against the resources our men and women needed. Only four. Senators Kerry and Edwards were two of those four."

More boos.

"At first Senator Kerry said he didn't really oppose the funding. He both supported and opposed it. And said—"

The crowd interrupted: "Flip-flop! Flip-flop!"

"And I quote, 'I actually voted for the $87 billion before I voted against it.'"

The audience clapped and laughed.

"Lately, he's been saying he's proud that he and John Edwards voted no, and he explained his decision was 'complicated.' But funding American troops in combat should never be a complicated question."

"Cheney! Cheney! Cheney!"

McCain sensed that Cheney was getting frustrated with the continual interruptions. The vice president tried a joke.

"Are you guys busy for the next fifty-nine days?"

The audience roared.

"We might take you with us."

"Four more years!" they chanted.

Then Cheney, feigning frustration, shouted over their deafening applause: "Do you want to hear this speech or not?"

Cheney delivered the line with a smile. But McCain found it telling. "I mean, we've got to be very honest," he says. "Dick doesn't like campaigning. Let's just tell the truth here. . . . He never shirked his duty, but I'm not sure that was his favorite pastime."[49] The crowd naturally responded to the vice president's needling: "Cheney! Cheney! Cheney!"

The vice president traveled extensively during October, attacking Democrats, drawing fire that might otherwise be trained on President Bush, and preparing for his upcoming debate against John Edwards on October 5.

The pressure on Cheney was immense. One week earlier, Kerry had dominated the first presidential debate with Bush. Kerry had displayed a solid grasp of the issues and handled himself well. Bush, by contrast, seemed aloof and annoyed. Although Bush's campaign staffers worked hard to spin the miserable performance as a win, they had trouble convincing themselves. Their own internal polling confirmed the flash surveys taken that night: Bush had lost and lost badly.

For months the polls had been tight, but the debate had given Kerry momentum that he did not get from his lackluster convention. And now he seemed to be pulling away. The Bush campaign's pollster, Matthew Dowd, told Cheney that he had no idea how far the numbers would tumble. It was Cheney's job, Dowd told him, to stop the slide.

The preparation for the debate was intense, as it had been in 2000. Again, the campaign asked Representative Rob Portman to work as Cheney's opponent. As he had with Joe Lieberman, Portman assigned a communications staffer to tape everything John Edwards said on the campaign trail and in media appearances. Portman listened to the tapes whenever he had a few spare minutes—as he traveled, as he commuted to and from work—until he had Edwards down.

Cheney studied his briefing books nightly in preparation for the first of several mock debates. In the first session, Portman was ready with arguments using Edwards's exact language. When he couldn't remember a fact or name, he simply made one up—something he had identified as a habit of Edwards's on the campaign. The debate ended with a dismaying result: Portman won.

Cheney's performances improved in the next sessions, but not enough. His aides speculated about his troubles. Their leading theory was that Cheney had been so immersed in the minutiae of national security that he was having trouble adjusting to the campaign. His struggles came mostly on domestic policy issues like Medicare and education.

On October 4, Cheney and Portman went fishing. Cheney gave direct orders to his staff: "Leave me alone." He needed to clear his head. The eighteen-hour days campaigning and endless study of his briefing books had cluttered his usually orderly mind. Time on the river could take care of that.

"It sort of pushes whatever other thoughts you've got out of your mind. If you're worried about the tax bill or the deficit or some other piece of governmental business, you're not paying attention to the fly. You're not doing all of those things you need to do in order to be able to fish well."

And for Cheney, there's no point in fishing if you're not going to fish well. "He's very competitive about it," says Portman. The vice president, Portman continues, "caught more fish than I did, but I caught the bigger fish. He's constantly slapping the fly down on the water. He's a good fisherman. Maybe not perfect from a technical perspective, but a very good fisherman. His fly was down on the water twice as much as mine. I gave him some grief about catching the bigger fish."[50]

There was no talk of the coming debate.

"We went fishing," says Cheney. "Rob loves to fish and I love to fish. I took him out on the South Fork. And we had a great day. But literally, it was going fishing."[51]

Back in Jackson, his staff received a copy of a story distributed by the Knight Ridder News Service. This chain owns some of the largest daily metropolitan newspapers in the country, and its Washington bureau had won plaudits from media critics for its skeptical coverage of the Bush administration's case for war. At the White House, many people viewed Knight Ridder's reporters as a useful megaphone for leaks from groups within the intelligence community that opposed Bush administration policies. This story fit that pattern.

"A new CIA assessment undercuts the White House's claim that Saddam Hussein maintained ties to al Qaeda, saying there's no conclusive evidence that the regime harbored Osama bin Laden associate Abu Musab al-Zarqawi. The CIA review, which U.S. officials said Monday was requested some months ago by Vice President Dick Cheney, is the latest assessment that calls into question one of President Bush's key justifications for last year's U.S.-led invasion of Iraq."[52]

The article framed the revelation in political terms. "Questions about whether the president and other officials overstated the intelligence about Iraq and omitted contradictory information and analysis are now at the center of the campaign debate over Iraq policy."[53]

Other news outlets quickly picked up the story. Cheney's debate team scrambled to come up with a response to the inevitable questions. They reviewed a copy of the full analysis by the CIA. It was bad. Some of the new information seemed to contradict prewar claims that Zarqawi was working with the Iraqi regime. But the analysis was not as devastating as the reporting implied. And one curious detail buried in the news stories deserved more attention: Saddam Hussein had personally ordered the release of two Zarqawi associates, detained by lower-level Iraqi intelligence officers.

The vice president returned from his day of relaxation to find his staff discussing possible responses. One option would be to declassify the entire analysis. President Bush had given Cheney declassification authority in March 2003. By making the assessment available to the public, the staff could provide a context for the findings hyped by the media. But as a matter of policy, Cheney was reluctant to declassify anything. And declassification had the potential to compound the problem if journalists focused not on the impropriety of the original leak but on Cheney's reaction to it.

The group looked for another line of defense, or perhaps even a way for Cheney to counterattack. They perused the study on prewar intelligence on Iraq issued in July by the Senate Select

Intelligence Committee (SSIC). Edwards, they noted, had been
on the committee and affixed his signature to its final report.
Some of its findings supported the administration's claims that
Iraq was a longtime supporter of terrorists, including al Qaeda.
Edwards, despite his endorsement of these findings and his prewar accusations that Iraq supported terrorism, had lately taken
to disclaiming any relationship between Saddam Hussein and Islamic radicals. The group decided that a quick "hypocrisy pop"
would be more effective than a lengthy discourse about the nature of terrorism against the West.

The prep session lasted late, until just after midnight, when
Cheney, still wearing jeans and hiking boots from his fishing excursion earlier in the day, abruptly ended it.

"Hey, guys," he said. "I have a debate tomorrow."

Scooter Libby told his boss that the responsibility of making
the Bush administration's case on Iraq—the most important issue in the campaign—would fall to him.

Lynne Cheney shot Scooter a look of disapproval. This was
exactly the kind of "bark off" feedback Cheney ordinarily liked.
But coming the night before the most important political performance of his life, it had the potential to add to the pressure.

The debate took place at Case Western Reserve University
in Cleveland, Ohio, with Gwen Ifill from PBS as host. In pre-
debate negotiations between the campaigns, Bush and Cheney's
side had insisted that the two candidates be seated next to each
other in a discussion format. Edwards, a former trial lawyer, had
wanted the option of walking as he spoke, as he had in the courtroom. In the end, Cheney got his way. Kerry's campaign, in a
move designed as much for the media coverage it would generate as its likely impact on Cheney, seated Senator Patrick Leahy
directly in the vice president's line of sight.

On ABC News, Peter Jennings opened his coverage by emphasizing the potential importance of the event. Vice presidential
debates are not typically appointment viewing, he said, but the
contest between Cheney and Edwards was a notable exception.
"Vice President Cheney is regarded today as the most powerful vice president in history," Jennings said. "And ever since the
presidential debate last Thursday, which was a surprise to many

people—Mr. Kerry was judged to have won that one—interest in tonight's main event has really gone up."⁵⁴ According to an ABC News poll, 76 percent of those surveyed said they were going to tune in that night.

The first question of the debate went to Cheney, and as his debate prep team had anticipated, it was about the Zarqawi analysis. It was the first of two questions about the leaked CIA analysis. In his first response, Cheney spoke broadly about the war on terror, emphasized Saddam Hussein's record of supporting terrorism, and reminded voters that George Tenet had testified in public to a long relationship between Iraq and Al Qaeda.

It was a reasoned answer, but it was a dodge. So moments later, Ifill tried again. Cheney replied:

> *Gwen, the story that appeared today about this report is the one I asked for. I ask a lot of questions. That's part of my job as vice president. The CIA spokesman was quoted in that story as saying they'd not yet reached the bottom line. And there's still debate over this question of the relationship between Zarqawi and Saddam Hussein. The report also points out that at one point, some of Zarqawi's people were arrested. Saddam personally intervened to have [them] released, supposedly at the request of Zarqawi. Let's look at what we know about Mr. Zarqawi. We know he was running a terrorist camp, training terrorists in Afghanistan, prior to 9/11. We know that when we went into Afghanistan, that he then migrated to Baghdad. He set up shop in Baghdad, where he oversaw the poisons facility up at Khurmal, where the terrorists were developing ricin and other deadly substances to use. We know he's still in Baghdad today. He is responsible for most of the major car bombings that have killed or maimed thousands of people. He's the one you will see on the evening news, beheading hostages. He is, without question, a bad guy. He is, without question, a terrorist. He was in fact in Baghdad before the war. And he's in Baghdad now, after the war. The fact of the matter is that this is exactly the kind of track record we've seen over the years.*

The debate featured several sharp exchanges. On the biggest issue, Iraq, Edwards criticized the launching of the war and its execution. He hinted several times that Cheney had deliberately misled the public about the case for war, and he portrayed the vice president as oblivious to the difficulties on the ground.

Cheney, meanwhile, noted that both senators had voted in favor of the war, but against funding it and suggested that Edwards and Kerry had become critics of the war because criticism was expedient.

"These are two individuals who have been for the war when the headlines were good and against it when the poll ratings were bad. We have not seen the kind of consistency that a commander in chief has to have in order to be a leader in wartime and in order to be able to see this strategy through to victory. If we want to win the war on terror, it seems to me, it's pretty clear the choice is George Bush, not John Kerry."

The moderator picked up on Cheney's last point to ask about his comments two months earlier about the consequences of voters making "the wrong choice" in November.

"Are you saying that it would be a dangerous thing to have John Kerry as president?"

"I'm saying specifically that I don't believe he has the qualities we need in a commander in chief," Cheney replied. "Because I don't think, based on his record, that he would pursue the kind of aggressive policies that need to be pursued if we're going to defeat these terrorists."

As Cheney anticipated, Edwards came after him on Halliburton. But rather than targeting the vice president on his deferred compensation, Edwards attacked Halliburton's business practices, twice mentioning Cheney's willingness to do business in hostile countries. "This vice president has been an advocate for over a decade for lifting sanctions against Iran, the largest state sponsor of terrorism on the planet," said Edwards. "It's a mistake."

On domestic policy, the candidates talked past each other. Each offered numbers about the economy, tax cuts, jobs, and education. Depending on who had the floor: the economy was either growing or in trouble, taxes would be cut or raised, jobs were being created or lost, and kids were getting smarter or dumber.

When Ifill asked Cheney about gay marriage and his experience with Mary, he acknowledged his differences with Bush. Still, Cheney said, the president "sets the policy for this administration, and I support the president."

Edwards prefaced his remarks with a personal word for his opponent. "Let me say first that I think the vice president and his wife love their daughter. I think they love her very much. And you can't have anything but respect for the fact that they're willing to talk about the fact that they have a gay daughter, the fact that they embrace her. It's a wonderful thing."

Edwards went on to turn his achingly sincere gesture into a long discourse on gay marriage. Ifill turned back to Cheney.

"Mr. Vice President, you have ninety seconds."

"Well, Gwen, let me simply thank the senator for the kind words he said about my family and our daughter. I appreciate that very much."

"That's it?"

"That's it."

"OK, then we'll move on to the next question."

President Bush later gave Cheney high marks for his restraint. "They brought his daughter up for political purposes," Bush says. "Dick Cheney, who has been known to express himself quite vividly at times, because he's a passionate man, was very measured."[55]

Cheney's gratitude was, of course, no less ironic than his opponent's words of admiration for his family. But his tone went largely unnoticed. It would not be the end of the issue.

The consensus after the debate was that both candidates accomplished their objectives. On CNN, Candy Crowley reported that Kerry's campaign was "definitely not giddy" about his performance, but the campaign staff thought it was "good enough."[56] Edwards at least dispelled any lingering notions that he was just a pretty face.

Cheney's performance was strong, but not perfect. In an assault on Edwards's poor congressional attendance record, Cheney claimed that as the Senate's president officer he had never before met Edwards. He was wrong. He had met Edwards before, and Kerry's campaign produced pictures to prove it almost immediately. Still, the campaign's top advisers believed Cheney had done well enough to stanch the bleeding from Bush's first debate.

In a conference call two days later, Bush's campaign pollster Matthew Dowd confirmed those results and congratulated Cheney. "As you know," he told the vice president and senior communications staff, "we dropped off after last Thursday from five points up to two points now. The average of all polls in the last forty-eight hours shows us up two points, and key states are starting to tick back up gradually. The vice president stopped the slide and we maintained our lead."

After Cheney noted that he was surprised at the large television audience for the debate, Dowd mentioned that 7.5 million people had watched it on the Fox News Channel and 8 million had watched it on CBS. "Good," said Cheney. "We ought to do all we can for Fox News."

Bush performed much better in the two subsequent presidential debates than he had in the first one. At the third debate, though, in Phoenix, Arizona, his performance was overshadowed by a second attempt by Kerry's campaign to make an issue of Mary Cheney's sexuality.

Gay marriage had been an issue in the presidential campaign. Courts in Massachusetts had ruled that gay marriage was legal. The decision took a dormant political issue and made it active. In San Francisco, the mayor began to marry gay couples on the steps of city hall. Other states moved to codify marriage as a heterosexual institution, and President Bush announced his support for a constitutional amendment declaring that marriage is between a man and a woman.

As he had done in the campaign of 2000, Cheney publicly took a position at odds with the president.

Bush is hesitant to discuss his conversations about Mary Cheney. "I don't know if I want to lay this one out there or not," he says, but presses on. "I learned a lot about Dick Cheney because of his love for his daughter, Mary, for example. And his basic advice was, let the states deal with this issue of marriage, gay marriage. In other words, he really felt uncomfortable with a, I think, federalization of the issue. And I said, 'I appreciate your advice.'"[57]

But the president was concerned about the personal implications of his policy on Cheney and his family.

"My only ask was that if his daughter doubted my tolerance to her orientation that I would hope that he would help make it clear to Mary that this is a—I was just worried about—the reason I'd federalized the issue is because I was worried about the courts' defining the issue and that we'd end up with de facto marriage that was not traditionally defined, I guess is the best way to put it."[58]

John Kerry struggled to explain his position, too. Kerry had been opposed to gay marriage and also opposed a constitutional amendment banning gay marriage. It's a coherent position, but one that gave Kerry difficulty during the course of the campaign.

At the third and final presidential debate, the moderator, Bob Schieffer, asked the candidates whether they believe homosexuality is a choice. Bush went first, saying he didn't know. He reiterated his opposition to discrimination and reminded voters that he proposed a constitutional amendment defining marriage as a union of a man and a woman.

Then Kerry answered. "We're all God's children, Bob. And I think if you were to talk to Dick Cheney's daughter, who is a lesbian, she would tell you that she's being who she was, she's being who she was born as."

There was an audible groan in the press viewing room when Kerry made his comments. It was one of the few times during the course of the campaign that Lynne Cheney agreed with reporters. Mrs. Cheney was watching the debate with her husband, Liz, and several senior staffers in a large suite at their hotel in Corapolis, Pennsylvania. Lynne Cheney and Liz were visibly angry. Lynne announced that she would make Kerry pay. If the vice president was angry, he did not show it.

Focus-group testing of the debate showed that undecided viewers found Kerry's remarks insincere and inappropriate. In post-debate television appearances, his surrogates defended Kerry's remarks as heartfelt and apolitical.

"She is someone who's a major figure in the campaign," said

Mary Beth Cahill, Kerry's campaign manager, a comment that surely surprised John Kerry's personal aide. "I think that it's fair game, and I think she's been treated very respectfully."

Almost everyone not affiliated with Kerry's campaign disagreed.

"The reason both Senator Edwards and Senator Kerry have raised it is to implicitly accuse Bush and Cheney of hypocrisy," said Jon Meacham, the editor of *Newsweek*. "I wouldn't have done it. I think it's out of bounds."[59]

On a tense elevator ride down from the suite to a previously scheduled rally, Liz cautioned her mother against saying anything she might regret. Lynne accepted the advice but remained determined to go after Kerry. In the course of introducing her husband, she made good on her threats, castigating the Democratic nominee for his "cheap and tawdry political trick." She continued: "Now, you know, I did have a chance to assess John Kerry once more. And now the only thing I could conclude: this is not a good man," she said. "This is not a good man."

And the controversy continued. The vice president reiterated his wife's comments. "You saw a man who will do and say anything to get elected, and I am not just speaking as a father here, although I am a pretty angry father," he said during a campaign event in Fort Myers, Florida.[60] All of the back-and-forth did not help Democrats. Polls showed that most Americans thought Mary Cheney should remain out of the campaign.

The campaign was draining. Cheney established a rule that he would attend no more than three events a day. He had seen Gerald Ford try to pack his daily schedule with several events, sometimes in several cities. Such overextension made Ford more susceptible to making mistakes born of sheer exhaustion.

But compared with the previous three years, nonstop campaigning was almost a break. Those who worked closest with Cheney on the campaigns noticed some changes in his demeanor between 2000 and 2004.

"The difference between the two was almost a sad thing for me," says one of Bush's senior advisers, who worked closely with Cheney on the campaigns. "Because of 9/11, he had grown much more"—the official stopped midsentence, searching for a word—"serious. I'm not sure that's the word, but I don't want to be negative. He was less animated, less—he was more somber, more reflective about international challenges. You have to remember, in the days and weeks after 9/11, he had taken a lead role on this mix of defense-intelligence-homeland security stuff. Between September 2001 and the summer of 2004, when the campaign really got serious, he was leading a pretty serious and somber life. I just got a different sense of responsibility, a different weight on his shoulders. It took a lot out of him."

If Cheney had changed—or if he was tired—Bush did not see it.

"He hasn't changed. Now, if somebody is bearing the burden of responsibility heavily, I would have seen a change," Bush says. "Maybe I'm just so used to seeing him and the change was so subtle that I don't see it. But I don't see that. Nor has he ever complained to me about the burden of responsibility—you know, 'Oh, how stressful this is. Oh, how terrible this is, oh, how this, that, or this is.' "[61]

Bush agreed that such a display would be out of character for Cheney. "He's not a real emotional person around other people."[62]

Cheney was smiling broadly on November 2. He and George W. Bush had been reelected with 51 percent of the popular vote. The difference in the electoral college tally—the one that matters—was one state, as it had been in 2000. This time it was Ohio.

Republicans also increased their margins in both the House and the Senate. It was far from a landslide, but after four tumultuous years that included two wars and a series of economic problems, it was a win.

CHAPTER SIXTEEN

Dick Cheney: New Democrat?

O ne month after Bush and Cheney were reelected, the vice president traveled to Afghanistan to represent the U.S. government at the inauguration of newly elected President Hamid Karzai. His visit was partly ceremonial, of course—a typically vice presidential function—but given America's stake in Afghanistan's fledgling democracy, there would be important business to attend to as well.

In three years, the United States had helped to remake Afghanistan. The Taliban government had been deposed. Al Qaeda's camps, which had trained some 20,000 terrorists in the preceding years, had been destroyed, and many of the men who ran those operations had been killed or imprisoned. Now, the U.S. intelligence community was bullish about the prospects for stability. In 2004, a classified assessment by Zalmay Khalilzad, the U.S. ambassador to Afghanistan, noted a decline in extremist violence and concluded, "Taliban and other terrorist elements may be on their last legs." That assessment would prove a bit too rosy, but it reflected the optimism of the Afghan people. Young women could once again attend school. Celebratory kites, forbidden under the Taliban, filled the skies. Democratically elected leaders would shape the future.

Between the fall of the Taliban and Karzai's inauguration, some 4 million Afghan refugees returned to their home country. Most regional warlords and tribal leaders, many of whom had a history of antagonizing whoever was in power, had decided join the political process rather than fight it. Ten million Afghans registered to vote, and 80 percent of them showed up at the polls in a presidential election that was praised by outside observers as extraordinarily well run. Elections for the legislature were planned for the spring.

The outlook wasn't entirely sunny, though. One result of the Afghans' newfound freedom had been an explosion of the poppy trade. According to estimates by the CIA, Afghanistan's poppy crop had increased threefold since 2001 and now accounted for a disquieting percentage of its GDP. And al Qaeda's fighters and their Taliban counterparts had escaped to establish pockets of resistance in the mountainous northeastern region of the country. Others hid in plain sight throughout Afghanistan's larger cities and threatened the fragile peace. The country's legal system was often dysfunctional or, in many places, simply nonexistent.

Afghanistan's neighbors presented concerns, too. The new leaders believed that President Pervez Musharraf of Pakistan was not doing enough to control extremist elements on the border between Pakistan and Afghanistan. And Iran was moving aggressively to assert its influence in an Afghanistan that had positioned itself as a proud U.S. ally.

So Cheney's agenda would include both the symbolic and the substantive. Among the materials packaged in preparation for the trip was a confidential memo to Cheney prepared by the State Department that laid out the diplomatic objectives of the delegation.

According to the memo, Cheney's visit "provides an opportunity to convey the strong and enduring U.S. commitment to a stable and secure Afghanistan, stress the U.S. commitment to aiding in the counternarcotics effort, make clear our interest in a cabinet of integrity and holding firm on parliamentary elections on schedule in the spring, and inform Karzai of our readiness to negotiate a long-term partnership agreement with Afghanistan."

The last of these points was particularly sensitive. Karzai was looking for assurances from the United States of long-term support, both military and financial, and he believed that announcing such a deal would enhance his political power at home and in the region. Cheney was concerned that public discussion of permanent U.S. military bases in Afghanistan could have an unwelcome ripple effect in Iraq, where the first national election was scheduled in a little more than a month, by making new voters there wary of endorsing candidates sympathetic to American causes.

A memo from Victoria Nuland, Cheney's deputy national security adviser, discussed the situation. "There are many potential benefits to a long-term U.S. military base in Afghanistan, but [the Department of Defense] is concerned about making a binding treaty commitment." It further notes: "Following your inquiry, the interagency shares your concerns about potential negative spill-over for the Iraq election if we begin a process which could look like base negotiations before January 30."

Cheney's objective, then, was to convince Karzai of the Bush administration's intentions, but to keep him from making them public.

The trip would be long: fifty-two hours from departure to return, with only seven hours on the ground in Afghanistan. Cheney's delegation would travel most of the way on the modified Boeing 757 designated as Air Force Two when the vice president is on board. With its familiar blue-and-white legend—"United States of America"—across both sides, the aircraft looks like a slightly smaller version of the Boeing 747 used by the president.

Cheney travels in a private cabin equipped with a work desk, a sofa that folds out into a double bed, a flat-screen television, a DVD player, and a secure telephone and fax. He uses the space to work, to meet with his staff, and, on rare occasions, to conduct background briefings or interviews with the media.

The vice president's cabin is the second of five compartments on the plane. The crew occupies the first, located directly behind the cockpit. The third section, with large leather chairs that swivel to face one another, is reserved for the senior staff and the vice president's family. Behind that section are two lavatories and several rows of seats reserved for the midlevel staff. The rear section contains smaller seats for the Secret Service and members of the media.

As the flight path took him high over Iraq and Iran, two-thirds of the "axis of evil," Cheney, in a tan suede jacket, blue jeans, and cowboy boots, occasionally ventured out of his cabin to chat with his senior staff. But he spent most of the long trip reading in his private quarters. His personal aide carries a green canvas bag, with a reinforced leather bottom, for Cheney's books; it often has between fifteen and twenty books inside.

In Oman, Cheney and his entourage switched over to a military C-17, the same kind of plane he had used in trouble spots during his trip to the Middle East in 2002. The temperature was below freezing when the plane landed in Afghanistan just after sunrise. Cheney would spend a busy day there and then leave.

Within an hour, Cheney arrived at the heavily fortified compound in Kabul that houses Afghanistan's national government and was ushered to the "presidential palace." There, he congratulated Karzai on the upcoming inauguration, and the two men settled down to issues: from the drug trade and threats by terrorists to economic development and Afghanistan's often strained relationship with Pakistan.

After their meeting, Karzai and Cheney emerged to address reporters in a courtyard outside the presidential offices. The walls and windows of the building that served as the backdrop for the brief press conference were scarred with bullet holes, marks of the fighting that in the past decades had become routine. Karzai, wearing his usual flowing green silk coat and lambskin hat, thanked the United States for its role in liberating Afghanistan:

Whatever we have achieved in Afghanistan—the peace, the election, the reconstruction, the life that the Afghans are living today in peace, the children going to school, the businesses, the fact that Afghanistan is again a respected member of the international community—is from the help that the United States of America gave us. Without that help Afghanistan would be in the hands of terrorists—destroyed, poverty-stricken, and without its children going to school or

getting an education. We are very, very grateful, to put it in
the simple words that we know, to the people of the United
States of America for bringing us this day.

In his own remarks, Cheney congratulated Karzai on the victory in the election and pledged America's support of the new Afghan democracy. There was a brief pause at the end of Cheney's statement, as both men seemed unclear about the procedures for the question-and-answer session to follow. When Karzai looked to Cheney for direction, the vice president leaned away from the microphone and in a voice audible only to those standing nearby, reminded Karzai of the obvious: "You're in charge now."

The final preparations for the inauguration continued as the press conference ended. Two lines of Afghan soldiers assembled in front of the once elegant building where the inauguration would take place. They stood at attention in their olive green dress uniforms as an officer paced purposefully in front of them, stopping every so often to bark instructions in the face of an unlucky soldier. Behind them, a long red banner hung from two large pillars. The words were in English, in large gold lettering: "December 7th celebrates the decision of the Afghan nation." Whoever made the sign had run out of room, and so the word "nation" was written in much smaller letters, tucked underneath "Afghan."

Cheney and his wife, arriving first, were greeted at the entrance and escorted to their seats in the front row. Donald Rumsfeld had made the trip separately, and was seated just behind them. Karzai came minutes later in a black Mercedes with tinted windows. He accompanied a very frail King Mohammed Zahir Shah, who had ruled Afghanistan from 1933 to 1973 and until recently had lived in exile in Italy.

Karzai's inaugural address was interrupted frequently by applause, and several members of the audience were moved to tears as he pledged to secure the country and prepare it for parliamentary elections. (Karzai's speech was interrupted several times by the chirping of cell phones, and at least one foreign dignitary snored loudly as the new Afghan president spoke.)

Karzai told a story of an elderly woman from the Farah province who came to a polling station with two voters' cards:

She went up to an election worker and declared that she wanted to vote twice, once for herself, and again for her daughter who, she said, was about to deliver her child and unable to come to the polling station to vote. "We are sorry, but no one can vote for another person, this is the rule," the elderly lady was told. So she voted—for herself—and left the station. Later in the day, the election worker was shocked to see the elderly woman back, this time accompanying her young daughter to the polling station. Her daughter carried her newborn baby, as well as her voting card, which she used to cast her vote.

Cheney was clearly affected by the ceremony. Afterward, in his stateroom aboard Air Force One, he sipped from a Starbucks cup and discussed the significance of the moment. "Think about what's happened in that country, what change has brought," he said. Cheney walked through the modern history of Afghanistan—from the Soviet-Afghan wars, the power struggles of the 1990s, the Taliban rule, al Qaeda's safe haven, the state from which 9/11 was planned. "Today, we swore in the first democratically elected president in 5,000 years. I think most of us think of it in terms of 9/11 and the subsequent three years, but there's a lot more history to it than that."[1]

In 1987, on a trip to Pakistan with the Intelligence Committee, Cheney had met with a group of Afghan mujahideen leaders who had crossed the Khyber Pass to share dinner with him and his delegation. Now, nearly twenty years later, one of the men Cheney had dined with that night, Sibghatullah Mujaddedi, sat two seats to his left during the inaugural ceremonies. "In those days he was one of the leaders of the muj," Cheney recalled.[2] After the fall of the Taliban, Mujaddedi headed the loya jirga, the traditional decision-making body that was convened after the war to select a new leader. At Karzai's inauguration, he delivered the closing prayer.

"I think it's one more example of the power of the idea of democracy, self-government, and the right of people to elect their own leaders," Cheney said. "It's not just an intellectual construct or something you read about in a political science text. It's a very powerful ideal."[3]

A little more than a month later, on January 30, the Iraqis would take their own first steps toward democracy in the first national elections since the U.S. invasion. Almost immediately after the fall of Saddam, Iraqis throughout the country had spontaneously begun organizing themselves into local governing councils. But little had been accomplished on a national level. With continuing violence in the most populous areas and extraordinary difficulties in simply trying to determine who was eligible to vote, the American and international media were filled with calls to postpone the elections. Elections in Afghanistan had been delayed several times before being held successfully, and many of the loudest voices in Washington—and even some in Iraq—were urging a similar postponement. Cheney rejected those arguments. In his view, the expansive American political presence had already stunted the growth of Iraqi democracy. Delaying the elections would make matters worse.

In briefings on Capitol Hill, Cheney spoke of his experiences in El Salvador as a member of Congress. "He told us, 'I saw people coming out of the jungles to vote,'" says one Republican who attended one of the sessions. "And he told us they'd do the same thing in Iraq. He was absolutely right about that."

In the weeks before the election, Abu Musab al Zarqawi, the leader of al Qaeda in Iraq, had warned his superiors about the potentially devastating effects to their terrorist campaign if the Iraqi people embraced democracy. In response, the insurgents had publicly promised violence against those who came to the polling stations. Nonetheless, on January 30, 2005, Iraqis turned out in large numbers to vote for candidates for the Iraqi National Assembly. Fifty-eight percent of the registered electorate voted (a turnout about twenty points higher than that for the congressional elections of 2002 in the United States).

Images of the long lines at polling stations bounced from satellites to television screens across the globe. Iraqis who had voted successfully had their fingers marked with distinctive purple ink to prevent their voting again. The pictures of their stained fingers carried with them the hope that this would be a turning point in the gantlet of daily violence. As the results came in, Iraqis and Americans could hope, for once, that Zarqawi might have been right.

In the first twenty-two years of his career, few would have described Dick Cheney as an outspoken advocate of democratic reform. He had been skeptical about the prospects for democracy in the former Soviet Union. He had opposed overthrowing Saddam Hussein in the early 1990s, and had been content to deal with dictators in the Middle East because he had seen virtually no chance for serious reform in the region. Stability was paramount.

"Certainly up through my time as secretary of defense, I would not have been as staunch an advocate as I am now of the importance of the reform process in that part of the world," he acknowledges. "You used to be able to look at that region and, you know, we were interested in it because we got a lot of oil out of it. And that was important from the standpoint of the economy. But it didn't constitute any kind of direct threat to the United States. Americans weren't going to die based on who was governing Iraq or most other places over there. That changed on 9/11, in my mind."[4]

The long-term national security of the United States, Cheney came to believe, depends in part on changing the nature of governance in the region. "It's not enough to go and kill terrorists. It's not enough to go in and take down or harass those who sponsor terror," he says.[5] Encouraging democratic reform must be part of the solution.

Although Cheney often speaks the same words about democracy as the president, the two men came to their views for very different reasons. For Bush, democracy is aspirational and moral. When he speaks of changes in the Middle East, he often invokes his faith and describes freedom as God's gift to mankind.

For Cheney, the changes are primarily utilitarian. "It's not a romantic or idealistic notion," he says. "In many respects, it's a very pragmatic proposition." When Cheney answers questions about the United States' efforts in postwar Afghanistan and Iraq, he almost always uses the word "obligation."

> *I am a big democracy advocate. And I say that for a couple of reasons. Because on the one hand I think we have an obligation, we Americans, if we go in and take down a*

government to do the best we can to stand up a new one in its place that meets the standards and principles that we believe in. It's not enough to go and take out one bad guy and put our dictator in his place. There might have been a time in the past when you could do that. But again, you come back to 9/11. The theory that we pursued is that you've got to drain the swamp. You've got to create a set of conditions that no longer lend themselves to fomenting this terror. . . . Political reform is part of that. What the president's recommending is supporting the proposition that we can have a bigger impact on that part of the globe by supporting freedom and democracy.[6]

"It's more than just elections," he adds. But long-term success in the war on terror is unlikely "unless you plant those basic, fundamental seeds that will ultimately blossom into peaceful governments, democratically elected."[7]

Nonetheless, several of his colleagues say that Cheney is a cautionary voice on the prospects for democratic reform. They point out that he has dealt with antidemocratic rulers like Hosni Mubarak and the Saudi royal family for decades and understands that fundamental change will not come easily. Condoleezza Rice says that he is "clear-eyed about how hard it is."[8]

On this issue perhaps more than others, it can be difficult to tell how much Cheney allows his policy to be set by the man he works for. In an interview, Bush acknowledged that he himself was "passionate" about democratic reform, but deflected a question about whether Cheney comes to the issue with greater skepticism.

"I would ask him that."

He continues.

I think this: I think that Dick Cheney understands that . . . a realistic way to deal long-term with a totalitarian ideology is to propose, defend, and encourage a competing ideology that is hopeful. There's no question that he brings a kind of—a caution about—of being unrealistic about how fast

*democracy can take hold in certain societies, which I ap-
preciate. I'm confident to tell you that he is a believer in the
capacity of liberty to transform countries and regions.*

*Look, in all this, there are degrees of optimism in all the
debates we have. And if you really think about it, we're
dealing with countries that are on their way, one way or
the other, to freer societies, with the notable exception of
Syria, Iran, and North Korea.*[9]

Bush begins to explain that Cheney understands that liberty
means more than just political freedom; it means opening markets, too. Then, he gives up.

"I'm not sure how to answer the question of how he stands
relative to me."[10]

When the White House counselor Dan Bartlett mentions that
Cheney is often labeled a neoconservative, Bush agrees with the
observation that the label is "silly shorthand." Then he tries once
again to explain what his vice president thinks about democracy.

*I think Dick Cheney believes in liberty, but is, having
dealt in this foreign policy arena and as a congressman, defense secretary, now as the vice president, sees the tension,
is aware of the tension that the freedom agenda can put on
strategic friends and allies.*

*The question, the fundamental question is, how hard
does the president push. And one of the roles of a Dick
Cheney, and anybody else, is to say, push harder, or push
less hard, as I begin to calibrate.*[11]

While those who work with Cheney every day suggest he
is hardheaded and analytical about the challenges of reform in
the region, critics portray him as unrealistically sanguine about
defeating the insurgency that threatens the political progress in
Iraq.

In an appearance on *Larry King Live* on May 31, 2005, taped
at the vice president's residence, Cheney was asked how long
Americans would be fighting in Iraq. "The level of activity that

we see today, from a military standpoint, I think will clearly decline," he said. "I think they're in the last throes, if you will, of the insurgency."[12]

There were those who agreed with Cheney—the foreign minister of Iraq, Hoshyar Zebari, called his assessment "realistic" because "our intelligence is improving, our military have taken the offensive now, taking the fight to the insurgents, to the terrorists"[13]—but they were very few and very far between.

John McCain says that the comments made Cheney look "out of touch."[14] Eighty Americans had been killed in May, more than those killed in April (fifty-two) or in March (thirty-six). These numbers fluctuate and thus are an imprecise measure of the strength of the insurgency. But other indicators suggested that the insurgency was not in its last throes. Violence against the Iraqi people had increased dramatically since the beginning of the war and showed little sign of abating.

In Washington, the White House was criticized for failing to address the violence and, worse, for failing to see it. "President Bush's portrayal of a wilting insurgency in Iraq at a time of escalating violence and insecurity throughout the country is reviving the debate over the administration's Iraq strategy and the accuracy of its upbeat claims," wrote Jim VandeHei and Peter Baker on the front page of the *Washington Post*. "While Bush and Vice President Cheney offer optimistic assessments of the situation, a fresh wave of car bombings and other attacks killed 80 U.S. soldiers and more than 700 Iraqis last month alone and prompted Iraqi leaders to appeal to the administration for greater help."[15]

McCain says: "I puzzled over that, too, because [Cheney]'s so smart. He is really one of the brilliant minds. The only thing I can think of as a rationale for that is that he feels strongly that you have to present a united front. And to show chinks in the armor, then the swords will come out."[16]

When Cheney was asked about his comments a month later, he defended them as literally true. "If you look at what the dictionary says about throes, it can still be a violent period, the throes of a revolution," he said on CNN's *Wolf Blitzer Reports*. "The point would be that the conflict will be intense, but it's intense because the terrorists understand that if we're successful at accomplishing our objective—standing up a democracy in Iraq—that that's a huge defeat for them."[17]

Nearly two years later, Cheney says the political and historical context is important. "I was thinking about the political process and the fact that we had intelligence at one point attributed to Zarqawi that once the Iraqis got a democracy established he was going to have to move on and find another base to operate from." But he doesn't defend the claim. "It was obviously wrong."[18]

Nevertheless, as the insurgency rolled on unabated, the political process in Iraq was also moving forward. On October 15, 2005, Iraqi voters once again went to the polls in a national referendum on a proposed constitution. It was a document that included many provisions that Americans would find objectionable, including an embrace of Islamic law. But the Iraqis ratified it, with a voter turnout of more than 60 percent, In an otherwise skeptical editorial, the *New York Times* writers conceded they were "impressed" that Iraqis had trekked to the polls "in defiance of terrorist threats, to decide their constitutional future. They have exercised a basic democratic right that would have been inconceivable just a few years ago."[19]

As the Iraqi people were enduring an often baffling mix of successes and disasters, Cheney was experiencing something similar in Washington. On October 28, 2005, just two weeks after the voting on the Iraqi constitution, Scooter Libby left the White House complex trailed by a phalanx of photographers. At a courthouse not far from 1600 Pennsylvania Avenue, the special prosecutor, Patrick Fitzgerald, had just announced the grand jury's indictment of Libby on charges of committing perjury, making false statements, and obstructing justice.

As Libby walked from his office, his assistant Jennifer Mayfield followed closely. She carried a duffel bag full of his clothes: Libby's emergency bag, just like the ones he had insisted that others on the staff keep at work in the days after the 9/11 attacks.

The scope of the indictment would be limited to crimes allegedly committed after the leak of Plame's name, and Fitzgerald had said pointedly that his prosecution should not be taken as an indictment of the Iraq War. But there were indications that he subscribed to the theory that the White House had authorized the leak to strike back at Wilson. Although Fitzgerald had implied in a press conference announcing the indictment that Valerie Wilson's covert status had been compromised, he never charged anyone with the offense of having compromised it. If Fitzgerald's posture at the time of the indictment struck many at the White House as odd, it would seem stranger still a year later, when the public would learn what Fitzgerald and the Justice Department had known from the earliest days of the inquiry: that the leak about Wilson's wife came not as political payback from the White House, but as gossip from a top State Department official who had shared Wilson's skepticism about the Iraq War.

The consequences for Cheney were great. One of his closest allies and advisers was suddenly gone. The vice president soon promoted David Addington to chief of staff and asked his deputy national security adviser, John Hannah, to assume Libby's duties as his chief national security adviser. Both Addington and Hannah figured in the investigation of the leak. People had speculated for months—inaccurately, it turned out—that Hannah would be indicted or that he had cut a deal with Fitzgerald and agreed to provide testimony damaging to Libby and to others in the White House. And the indictment of Libby indicated that Addington had given investigators statements contradicting Libby's claims about how and when he learned the identity of Valerie Plame.

Within the White House, Addington and Libby were very different operators. Libby was much more engaging with other senior officials in the Bush administration, and his relationship with Cheney was friendlier and less formal. He was a staffer first, but he was also a friend. Addington, by contrast, struck even those who respected him as aloof and, at times, needlessly acerbic. If a situation could be handled through either conciliation or confrontation, the new chief of staff always seemed to choose the latter. Addington's relationship with Cheney was different, too. Although he had known Cheney longer than Libby had—for more than twenty years when he was named chief of staff—Addington saw himself as strictly an employee and their relationship as professional, not personal.

On the level of policy, Libby had been an enthusiastic supporter of the Iraq War and, within the administration, an aggressive spokesman for Cheney on issues related to Iraq. Addington had been more skeptical. Before the war, he had worried that invading Iraq would be a distraction from the broader war on terror, and afterward he felt that Donald Rumsfeld was more concerned with transforming the military than with winning the occupation.

But Addington was no dove. In the months after 9/11, he played a major role in shaping the administration's legal strategy for the long war ahead. When Alberto Gonzales arrived as White House counsel, fresh from a judgeship in Texas, he found that Addington had memorized elements of national security law that he himself had never before encountered. And when lawyers around town called their friends at the White House with suggestions on legal strategy, they were invariably told that Addington had already considered them. Virtually every legal decision made by the administration after 9/11 that was related to the war on terror had at least crossed Addington's desk.

Addington and Cheney would spend much of that fall quietly fighting for two of the most controversial programs adopted after September 11. Since the earliest days of the war on terror there had been broad disagreements about how the United States should regard captured al Qaeda terrorists. Civil libertarians and many Democrats argued that Geneva Convention protections applied to al Qaeda; Cheney, Addington and many of the lawyers at the Justice Department's Office of Legal Counsel (OLC), the office that advises the president on such matters, disagreed.

The war on terror, they maintained, is a different kind of war. Not only do al Qaeda combatants fail to abide by the Geneva Convention themselves, their primary targets are innocent civilians. According them legal status as prisoners of war, Cheney believed, would provide them with legal protections to which they are not entitled and could restrict U.S. interrogators from using aggressive techniques to extract information. And those interrogations would be critical to preventing another attack.

"We've got to spend time in the shadows in the intelligence world," he said just days after the 9/11 attacks. "A lot of what needs to be done here will have to be done quietly, without any discussion, using sources and methods that are available to our intelligence agencies, if we're going to be successful. That's the world these folks operate in, and so it's going to be vital for us to use any means at our disposal, basically, to achieve our objective."[20]

Although issues regarding interrogation and allegations that Americans were torturing prisoners arose in public from time to time, the debate mostly took place out of public view, in the offices of a small subset of government lawyers with knowledge of the law of war. That changed for good in the spring of 2004, when *60 Minutes* broadcast photographs showing American soldiers abusing Iraqi prisoners in Saddam Hussein's infamous Abu Ghraib prison. The Bush administration, already on the defensive over Iraq's missing WMDs, tried to explain that such abuses were not routine and had not been approved by the White House. But few would listen, and the disclosure of an August 2002 memo written by Assistant Attorney General Jay Bybee added to the growing furor. Bybee reiterated the position of the OLC that al Qaeda terrorists were not protected by the Geneva Convention and argued for a restrictive definition of "torture" that would ensure that the techniques reportedly practiced by some interrogators would not be so classified.

On June 22, 2004, the Justice Department released reams of previously classified documents in an attempt to demonstrate that the administration had thoroughly debated the laws that govern interrogations. It didn't work.

Cheney explained his position in an interview the following day. "Let me just state a couple of general propositions. I've seen the big-picture documents they have released. I have not yet had the opportunity to read them all. We had, I think, several concerns in the aftermath of 9/11. We had al Qaeda terrorists kill 3,000 Americans that morning. We were concerned about the possibility of follow-on attacks. We still are. We also, as we went into Afghanistan, encountered significant numbers of al Qaeda and the Taliban in Afghanistan that were either captured or killed. And the al Qaeda in particular didn't operate by the rules of conventional warfare. We looked at all of that, and the decision was made that they were not lawful combatants. They didn't wear uniforms or have badges; they didn't carry arms out in the open. They did in fact set out to kill civilians; and, by definition, they did not qualify as lawful combatants. Also, as terrorists, they were not represented as any state."

Cheney explained that because al Qaeda was not a state—as Geneva Convention signatories are—claims that those terrorists were protected by the conventions were illogical. "We have the Abu Ghraib problem in Iraq, but Iraq is different than Afghanistan. Iraq is different than al Qaeda, and it is important to remember that, because people captured in Iraq are treated as under the Geneva Convention. Different set of facts. We also need to be able to interrogate terrorists who potentially had information on the operation plans that the organization might have. Future attacks that they had that they were working on designed to kill Americans, and so that it was necessary to have a policy in place that on one hand was consistent with our basic laws and statutes concerning such ordinances: torture, for example, which we don't engage in, but at the same time allow for interrogation, that would give us the information needed to save American lives."[21]

By the following summer, Congress had gotten involved. Senator John McCain, a POW in Vietnam, was pushing a measure that would place restrictions on the techniques available to U.S. interrogators. Cheney met with McCain several times in an attempt to discuss the issue.

"He feels strongly about the need to have these kinds of techniques allowed in interrogation," says McCain. "Period. Dick thinks that the most important tool we've got is intelligence in the war on terror. I get that. My view was that some techniques should never be used because they then damaged our image with people in the world. And in my view, it made us no different from them or hard to tell the difference between ourselves and them. And my belief is that if you inflict enough physical pain on somebody, then they're going to tell you whatever it is you want to hear."[22]

Despite Cheney's lobbying against the measure, the McCain amendment passed the Senate on October 5, 2005, on a vote of

90 to 9. McCain says that while their negotiations were intense, they were always professional. "We had a difference of opinion. I won the vote. It's over."[23]

Cheney has said very little in public on the subject of interrogations, so as with other aspects of his vice presidency, his precise views remain something of a mystery. And those familiar with Cheney's views on the subject comprise a much smaller group than those who claim to know what Cheney thinks. Much of the reporting on the subject seems to assume that because the vice president opposes granting Geneva Convention protections to Al Qaeda detainees, he supports the widespread use of torture to obtain information from them. It is a caricature of Cheney's views, say those who have spoken with him at length, including McCain.

"He's straightforward, says he disagrees with you, he tells you why he disagrees with you, and you tell him why you disagree with him."

On December 15, 2005, a second highly controversial program became the subject of intense public scrutiny when it was exposed by the *New York Times* in an explosive story. Since shortly after 9/11, the National Security Agency (NSA), the government entity responsible for listening in on enemy communications, had broadened its mandate to include communications in which one of the parties was in the United States. According to the *Times:*

> *Months after the Sept. 11 attacks, President Bush secretly authorized the National Security Agency to eavesdrop on Americans and others inside the United States to search for evidence of terrorist activity without the court-approved warrants ordinarily required for domestic spying, according to government officials.*

The officials told the Times that the NSA listened in on "hundreds, perhaps thousands" of people in the United States. This monitoring had taken place without warrants.

> *The previously undisclosed decision to permit some eavesdropping inside the country without court approval was a major shift in American intelligence-gathering practices,*

particularly for the National Security Agency, whose mission is to spy on communications abroad. As a result, some officials familiar with the continuing operation have questioned whether the surveillance has stretched, if not crossed, constitutional limits on legal searches.[24]

The reaction in Washington was quick and divisive. Democratic leaders said that the program was plainly unconstitutional; Senator Barbara Boxer hinted that Bush's authorization of the program might be an impeachable offense. Republicans in Congress, with few exceptions, supported the White House. Representative Peter Hoekstra, chairman of the House Intelligence Committee, called the program a "no-brainer."[25] And a Republican senator, Orrin Hatch of Utah, said, "The White House certainly has some case law on their side and the inherent powers of the president on their side, and those two, I think, would cause any reasonable person to side with the executive branch."[26]

The Justice Department released a letter to the leaders of the intelligence committees defending the program. "There is undeniably an important and legitimate privacy interest at stake with respect to the activities described by the president," wrote Assistant Attorney General William Moschella. "That must be balanced, however, against the government's compelling interest in the security of the nation."[27]

Cheney was angry about the leak. As the article in the *Times* acknowledged, the program had disrupted at least two plots—one in the United States and one in Britain—and Cheney believed that the disclosure of the program would compromise its effectiveness, as would the many follow-up stories that were sure to come.

The reporters at the *Times* had, in fact, learned about the program more than a year earlier, and the White House had implored the paper to kill the story, or at least hold it. Bush met with the publisher, Arthur Sulzberger, in the Oval Office to make his case. After listening to the president and top intelligence officials, Sulzberger agreed not to publish.

Cheney had recused himself from the meeting, believing that, given his tense relationship with the *Times*, Sulzberger might be more inclined to view the program skeptically if the vice president were its chief defender. Cheney also sat out when Bush invited the *Times*'s leadership back to the White House a year later in another attempt to keep the *Times* from exposing the program. This one was unsuccessful.

The dispute centers on whether the Bush administration should have used procedures established by the Foreign Intelligence Surveillance Act (FISA) of 1978. Critics of the program contend that FISA requires the U.S. government to obtain court authorization to conduct the kind of eavesdropping practiced by the NSA program.

Cheney's view is simple: acts of Congress that interfere with the president's ability to carry out his functions as commander in chief violate the Constitution. Those inherent powers, Cheney believes, coupled with the authorization of force passed by Congress shortly after 9/11, place the NSA program on solid constitutional ground. He says that the NSA's lawyers and its inspector general back him up.

The NSA's lawyers had approved similar programs in the past. In April 1992, in the last days of George H. W. Bush's administration, Admiral William Studeman left his position as director of the NSA to become deputy director of the CIA. Cheney, who as secretary of defense oversaw about 80 percent of the U.S. intelligence community, including the NSA, asked his colleagues about appointing Michael McConnell to the post. Impossible, he was told. McConnell had only one star and had had that one for only seven months. In order to serve as the director of NSA he needed three. Cheney asked if he could simply elevate McConnell to three-star status and was told he couldn't. Why not? he asked. He got a number of answers, all of them variants of the same argument: it's just never been done that way before. Cheney didn't care. If Bush lost to Clinton in the 1992 presidential election, he would be out of a job, and the appointment would be one of his final major personnel decisions. He wanted McConnell in the position.

"Do it," he said. "Make it happen."

"So," McConnell says, "I went from a one-star to a three-star."[28]

McConnell stayed on for most of Bill Clinton's first term.

When the Clinton administration sought to continue the "war on drugs," calling international drug traffickers a threat to U.S. national security, lawyers in the NSA's Office of General Counsel believed their agency might be in a position to help.

In many cases communications to and from drug smugglers did not fall neatly into the two classes of permissible targets for intercepts: foreign powers or agents of foreign powers. If a drug kingpin in Colombia calls an associate in Miami to plot an assassination of an official from the Drug Enforcement Agency, can the NSA listen in? NSA lawyers believed it could.

McConnell considered himself professional and aggressive, and as such understood the need to collect as much good information as possible. To do so, he was willing to interpret laws broadly, but not to break them. He wanted to cover himself with a legal opinion from the Justice Department, so he sought the counsel of Attorney General Janet Reno and her staff. They never provided it.

McConnell won't talk about the details of those discussions, but acknowledges that the issue had come up during the Clinton administration. "The debate recently about spying on Americans" was nothing new, he says. "We have that debate every so often."[29]

This time, though, America was at war. In the days after 9/11, President Bush instructed the heads of the agencies that make up the U.S. intelligence community to review their rules and procedures to determine what more they might do protect the country. He told them to be aggressive. Unconventional thinking was encouraged.

On September 13, 2001, General Michael Hayden, then director of the NSA, spoke to his staff in Fort Meade, Maryland, and around the world by video. It was a pep talk. He told the staffers that they would be on the front lines of the new war and that their work would save lives. Hayden also acknowledged the tension between freedom and security and promised: "We are going to keep America free by making Americans feel safe again."[30]

Hayden responded to Bush's admonition by broadening the scope of communications that the NSA considered potentially valuable—something he could do under authority that he already had. So the NSA began to look more carefully at communications that involved a "U.S. person"—not necessarily a U.S. citizen, but anyone in the United States.

This distinction would create a thicket of thorny legal issues. In an appearance on Capitol Hill more than a year before 9/11, Hayden told lawmakers: "If, as we are speaking here this afternoon, Osama bin Laden is walking across the Peace Bridge from Niagara Falls, Ontario, to Niagara Falls, New York, as he gets to the New York side, he is an American person. And my agency must respect his rights against unreasonable search and seizure, as provided by the Fourth Amendment to the Constitution."[31]

As buildings in New York and Washington smoldered, nothing seemed more preposterous than protecting the rights of Osama bin Laden. So some of the changes were obvious. But Hayden went farther. He was concerned that in some important respects, the laws of the United States had not caught up with global technology, and he believed that there were other measures, more aggressive still, that his agency could take. The NSA's lawyers signed off on a series of programs for more rigorous monitoring of al Qaeda.

Along with the director of the CIA, George Tenet, Hayden brought the proposals to Cheney. The vice president approved, and the three men went on to Bush for his OK. They got it.

The next question was obvious. How much do we share with Congress?

Cheney's answer did not require much thought: very little.

He came to this view through long experience. As White House chief of staff, he had seen many leaks from Capitol Hill. Even as a member of Congress, Cheney had argued forcefully not only that the executive branch had a right to restrict access to sensitive information, but that in some cases it had an obligation to do so.

On April 5, 1986, when Cheney was in his fourth term as a representative from Wyoming, a bomb tore through a discotheque in West Berlin, killing an American soldier. The Reagan administration quickly identified Muammar Gadhafi of Libya as a sponsor of the attack and warned that Libya was encouraging follow-up attacks on American targets.

Ten days later, more than 100 American bombs rained down on Libya. The Reagan Administration described them as the first shots in a broader "war against terrorism."[32] When the Democrats in Congress complained that the Reagan administration was doing too little to keep them informed, Cheney made it clear that he had little sympathy. "I am satisfied that I know all I need to know at this point, and I would disagree with what we often hear from the Hill, the cry for consultation in advance: 'Let us in the decision, we want to share responsibility,'" he said on the *MacNeil/Lehrer NewsHour*. "I don't think you can have 435 members of the House participate in that decision."[33]

Cheney not only rejected the calls for greater openness; he pushed the Reagan administration to be more secretive. He strongly cautioned against sharing more evidence of Libya's complicity in the attack in Berlin. The short-term gain in domestic and international support, he said, was not worth the long-term cost of potentially compromising an intelligence source. The "paramount interest," he said, is protecting sources and methods. And Congress would simply have to trust the executive branch.

"There's no reason to lay out the details of the information. If the president of the United States reviews it and feels it's adequate, if senior administration officials, civilian and military, review it and feel it's adequate, if senior members of Congress who have access to it see it and feel it's adequate, there's no reason in the world to lay that out and in effect make it impossible ever again to take advantage of our capabilities in a particular area. I think there's been far too much discussion at this point of the nature of the information we may have. We elect people to make these decisions for us, and we ought to trust them to do that."[34]

These arguments would echo loudly twenty years later.

"I served in the Congress for ten years," Cheney said in an interview shortly after the NSA's program was exposed. "I do believe that especially in the day and age we live in, the nature of the threats of we face . . . the president of the United States needs to have his constitutional powers unimpaired, if you will, in terms of the conduct of national security policy."[35]

For a select group of insiders in Washington, the *Times*'s

bombshell came as no surprise. When the NSA's Terrorist Surveillance Program began, the White House, with Cheney's strong encouragement, decided to strictly limit the members of Congress who would be briefed on its activities. Only the chairmen and ranking minority members of the House and Senate intelligence committees would be notified about the wiretapping—a total of four people.

Cheney presided over the briefings, usually held in his office in the West Wing, but said very little. Michael Hayden, director of the NSA, narrated the program's activities to Richard Shelby, Porter Goss, Bob Graham, and Nancy Pelosi. David Addington attended the sessions, and on most occasions so did George Tenet. There were few questions, and there was widespread agreement that the program was important and ought to continue. Cheney and Hayden conducted similar briefings for the leaders of the intelligence committee on a regular basis.

On July 17, 2003, Senator Jay Rockefeller wrote to Cheney, by hand, to express concern about the program without identifying it. "Clearly the activities we discussed raise profound oversight issues. As you know, I am neither a technician nor an attorney. Given the security restrictions associated with this information, and my inability to consult staff or counsel on my own, I feel unable to fully evaluate, much less endorse these activities."[36] Though Rockefeller introduced his letter by saying that he wanted to "reiterate" his concerns, those familiar with the briefings say it was the first time he had expressed any misgivings about the program.

After senior officials at the Justice Department raised questions about the legality of certain aspects of the program in 2004, the White House widened the circle of those briefed to include some leaders of the House and Senate. Some people in the intelligence community thought this was good politics. Adding congressional leaders to the conversation would make Congress feel as though it had more "ownership" of the program.

At a briefing in the White House situation room for the expanded group of nine legislators, Hayden described the program and shared some intelligence intended to demonstrate the value of the information the NSA was getting. Then Cheney spoke up.

Dennis Hastert, who was speaker of the House at the time, recalls Cheney's words. "I remember him specifically saying, 'OK,
we need your understanding to go forward. Does anybody have
any objection? Do we need to do anything legislatively?' It was
a question Cheney asked. And everybody agreed: no, we don't
need to do this in legislation. We need to let our intelligence go
forward and do what they're doing. So he laid it out very specifically to everybody. I remember everybody was present at that
time."[37]

Senator Pat Roberts, a Republican from Kansas who attended
the briefings as chairman of the Senate Intelligence Committee,
said he never heard Rockefeller object to the program. In a statement after the NSA's program was first publicly disclosed, Roberts said: "On many occasions Senator Rockefeller expressed to
the vice president his vocal support for the program; his most
recent expression of support was only two weeks ago."[38]

"It was their unanimous recommendation that we continue
with the program and that we not seek legislative authorization,"
says another participant in the meeting. "Jay Rockefeller was sitting at the table."

So was Nancy Pelosi, who had participated in the early briefings in her capacity as ranking Democrat on the House Intelligence Committee and continued as minority leader. When the
program was exposed, Pelosi led the call for a congressional investigation into its legality and told reporters that she had repeatedly expressed concerns in briefings. "She knew about it," says
the official. "And they all thought it was a pretty good program.
It was only after it became public—thanks to the *New York
Times*—then they had to cover their fannies politically."

Questions about the NSA program followed Cheney as
he traveled throughout the Middle East after the third and final successful election in Iraq on December 15, 2005. The trip
would take him first to Iraq, then on to Afghanistan, Pakistan,
and Oman. On one of the last legs of the trip, Cheney called the
seven reporters traveling with him back to his cabin for a chat.
The vice president appeared relaxed, wearing a lightweight black
U.S. Army jacket, a blue button-down shirt, gray flannels, and
brown hiking boots. He offered his guests a beer, but didn't have

one himself, saying he had lots of work to do on the flight back to Washington. The overstuffed three-ring binder on his desk—perhaps six inches thick—suggested he wasn't kidding.

Although much of the session was a playful back-and-forth between Cheney and the reporters, the vice president turned serious when he was asked about the *New York Times*'s story about the NSA program. Richard Stevenson of the *Times* asked Cheney for his thoughts about executive power.

Cheney answered as if he'd been waiting three decades for the question. Reporters were nonplussed as he guided them, at considerable length, through the recent history of executive power. The Vietnam War and Watergate, he said, led to a power grab by the legislative branch that resulted in things like the War Powers Act and the Congressional Budget and Impoundment Control Act of 1974. He spoke of his time in the Ford administration and Congress.

"If you want reference to an obscure text," he said, "go look at the minority views that were filed with the Iran-contra committee, the Iran-contra report in about 1987. Nobody's ever read them. Part of the argument in Iran-contra was whether or not the president had the authority to do what was done in the Reagan years. And those of us in the minority wrote minority views that we—actually authored by a guy working for me, one of my staff people, that I think are very good at laying out a robust view of the president's prerogatives with respect to the conduct of especially foreign policy and national security matters."

Then as now, he continued, those powers are crucial to protecting the American people. "In the day and age we live in, the nature of the threats we face—and this was true during the cold war as well as I think is true now—the president of the United States needs to have his constitutional powers unimpaired, if you will, in terms of the conduct of national security policy."

Nedra Pickler, a reporter with the Associated Press, asked Cheney: "Do you not understand, though, that some Americans are concerned to hear that their government is eavesdropping on these private conversations?"

"What private conversations?" Cheney shot back.

"The private conversations between Americans and people overseas."

"Which people overseas?"

"You tell me."

Cheney was exasperated. "It's important that you be clear that we're talking about individuals who are al Qaeda or have an association with al Qaeda, who we have reason to believe are part of that terrorist network. There are two requirements, and that's one of them. It's not just random conversations. If you're calling Aunt Sadie in Paris, we're probably not really interested."

The vice president said that Bush administration policies—not luck or fate—are responsible for keeping the United States safe since 9/11. "You know," he said, "it's not an accident that we haven't been hit in four years."

It was an argument Cheney had wanted to use during the 2004 campaign. But the White House communications shop and Bush campaign advisers had argued that making such a claim was too risky. It was too stark. If the White House had made those claims and the United States were hit again, they cautioned, the chances for reelection would be blown.

Cheney disagreed. By failing to take credit for the success of their policies, the campaign was refusing to capitalize on what was their greatest success. Bush sided with his political strategists and Cheney held his tongue until the return trip from the Middle East. Although most of the interview was "on background," Cheney specifically directed that his discussion of the NSA program and his defense of executive power be placed on the record.

He ended with what may be the most forceful on-the-record defense of Bush administration national security policy yet.

There's a temptation for people to sit around and say, well, gee, [9/11] was just a one-off affair, they didn't really mean it. Bottom line is, we've been very active and very aggressive defending the nation and using the tools at our disposal to do that. That ranges from everything to going into Afghanistan and closing down the terrorist camps, rounding up al Qaeda wherever we can find them in the world, to an active, robust intelligence program, putting out rewards, the capture of bad guys, and the Patriot Act. . . . Either we're serious about fighting the war on terror or we're not. Either we believe that there are individuals out there doing everything they can to try to launch more attacks, to try to get ever deadlier weapons to use against [us], or we don't. The president and I believe very deeply that there's a hell of a threat, that it's there for anybody who wants to look at it. And that our obligation and responsibility, given our job, is to do everything in our power to defeat the terrorists. And that's exactly what we're doing.[39]

Cheney continued to defend the NSA's program back home. In speeches and television appearances, he challenged reports in the media that labeled the wiretapping "domestic surveillance" and suggested that the mere existence of the secret program was a scandal.

"It's not domestic surveillance," he said on *The NewsHour with Jim Lehrer.* "The requirements for this authorization to be utilized are that one end of the communication has to be outside the United States, and one end of the communication has to involve reason to believe that it's al Qaeda–related or affiliated or part of the al Qaeda network. Now, those are two very important and very clear-cut criteria, and for this presidential authorization to be used in this way, those two conditions have to be met."[40]

Michael McConnell, the man Cheney had chosen to run the NSA more than a decade earlier, happened to be watching the PBS newscast with his wife. Few people knew the NSA's operations and capabilities better than McConnell. And yet when he had listened to NSA's director Michael Hayden and Attorney General Alberto Gonzales describe the Terrorist Surveillance Program, he was more confused afterward than he had been before they started. Cheney was different. "He laid it out chapter and verse, as plain as could be," McConnell recalls. "He has such a way of making it simple and compelling."[41]

Cheney's views should not have been surprising to anyone who had been following his career even casually. And in the aftermath of 9/11, as he was working behind the scenes on programs like the Terrorist Surveillance Program, Cheney said publicly that the war would be long and the government would be aggressive.

"This is going to last for a long time. We are vulnerable as a society to these people who wish us ill and are willing to die in the effort, and so we're all going to have to make some changes and possibly accept some limitations we'd rather not accept, but it's necessary unfortunately in the time we live in."[42s]

officials had concluded that some of the emergency measures taken immediately after 9/11 had outlived their usefulness.

Cheney was not among them. He believed that the United States had to fight the war with the same urgency that had guided the decisions of September 12, 2001, and that the conflict in Iraq was central to that broader fight.

"It may be that part of the difficulty of having people accept what's going on in Iraq is that they don't perceive that it is an effort to do anything positive or affirmatively in terms of dealing with the war on terror that they don't believe exists," he says.[2]

"I mean, for me there's no question. Part of it, of course, is we deal in that all the time. We see on a regular basis the threat. We have a pretty good fix on what the enemy is attempting. We know they're seriously interested in trying to acquire a WMD or a nuke or a biological agent of some kind. . . . Some people think if we just walk away from Iraq everything will be fine, that it's the optional war, that you don't have to be there, that it's possible to retreat behind our oceans and be safe and secure; withdrawal from Iraq doesn't damage our interest in this wider conflict. And that may be in part because they don't believe there's a wider conflict. I know different. . . . It's so clear to me, I have trouble understanding why it is unclear to everybody else."[3]

As the pressures of the second term added up, Cheney often sought to escape Washington to clear his head. He spent much of each August at his home in Jackson, Wyoming, and frequently stole away for long weekends or even day trips to hunt or fish. As he had been since his childhood, Cheney remained a passionate outdoorsman. His first love was fly-fishing.

"One of the keys to understanding Dick Cheney," said Lynne Cheney, introducing her husband at the Republican convention in 2000, is fly-fishing. And when Nick Lemann of *The New Yorker* profiled Cheney in the early days of the Bush administration, he opened his piece on an imaginary river. "If I were entering a contest to win a dream date with Dick Cheney, here is what I would say: We would definitely go fishing. Not bait fishing, which is for amateurs, a category that does not include Cheney, but fly-fishing."[4]

Fly-fishing isn't the most efficient way to catch fish, but it's the purest. In some cases its practitioners catch more fish than bait fishermen. Often, they don't. But the rituals are their own reward. Its purveyors almost pride themselves on the inefficiency. The fact that it confounds others brings fly fishermen together in a special fraternity of fishing classicists. Cheney is a member.

When Cheney goes fishing with Bush in the president's stocked pond on his ranch in Crawford, Cheney stands out. "We float out there and he's firing a fly at these largemouth bass," says Bush. "We're chucking Bubba bait, and he's fly-fishing."[5]

On the bookshelves of the library at the vice president's residence in Washington, D.C., across from the shelf featuring works about and by vice presidents, are dozens of books about fish and fishing. The books range from the thick two-volume *Trout* by Ernest Schwiebert to the somewhat narrower *Minor Tactics of the Chalk Stream* by the legendary G. F. M. Skues. There are accessible, lifestyle works—*Fishless Days, Angling Nights*; *The Well-Tempered Angler*; *Trout Bum*—and there are books for the technician: *Caddisflies*; *A Modern Dry-Fly Code*; *Practical Dry-Fly Fishing*; *The Fly-Fisher's Entomology*; *Nymph Fishing for Larger Trout*. When Cheney's secretary, Debbie Heiden, notices that he seems tired or particularly overworked, she slips a fly-fishing magazine into his travel bag. This is something she has done since they first worked together at Halliburton.

That Cheney has fished on four continents is only one indication of what might rightly be called an obsession with fly-fishing. Another would be a trip he took in June 1998 to the Ponoi River in Russia, with his daughter Mary and his old friend John Robson. It is a journey no casual fisherman would make.

The trip took them from New York to Helsinki on commercial airliners. In Helsinki, they boarded a weekly flight to Murmansk, the largest town north of the Arctic circle, where they caught an old Soviet Union helicopter to complete the journey. On the helicopter, they rode with fifteen other anglers on bench seats for nearly three hours, their gear strapped in at their feet, stopping halfway to refuel from large drums of helicopter fuel that had been deposited earlier on a barren patch of tundra. It was a trip into what one outfitting company calls "one of the most logistically complex parts of the world."[6]

The camp was comfortable, but not luxurious, and they slept in sleeping bags on cots in canvas tents with wood reinforcements. The fishing, for Atlantic salmon, was extraordinary. Over the course of the week, each angler caught some eighty fish. Twenty-fish days were not uncommon. And because the sun never set, they could fish late into the night.

Cheney fishes as often as he can. When he is home in Jackson for a weekend, he will sneak away on the Snake River, some-times just for an hour or two, and even when the weather isn't ideal. He's been seen fishing with ragg wool gloves—fingertips cut out—with snow on the banks of the river.

Each summer, Cheney takes a trip with a small group of friends—usually eight—for a week that is a return to pre–vice presidential normality. The men take turns cooking and doing dishes. They fish all day and spend the night around a campfire telling stories, often the same stories they told the year before and the year before that. Cheney sips Johnny Walker Red and snacks from a jar of Planters dry-roasted peanuts. ("Yeah, and he doesn't share," laughs one friend.)

For years, these trips took the men to remote rivers in west-ern Canada, usually the Dean or the Babine, for some of the best fly-fishing in North America. The men fish for steelhead trout in icy waters surrounded by pristine wilderness. Words stream out of Cheney's mouth as he describes his favorite fish. "A steel-head is a magnificent fish. It's a sea-run rainbow that spawns in fresh water. It hatches out, spends maybe a couple of years in fresh water. And then goes to sea, just like Atlantic salmon. A couple years in the ocean, cruising the Pacific, grows to consider-able size and then comes back to fresh water. Probably the big-gest steelhead I've caught—a few in the twenty-pound class," he says, then clarifies, "twenty-pounds-plus. That's a big fish on a fly rod. They catch a few up there every year where we go, over thirty pounds. I've never caught a thirty-pounder. And it's very tough technical fishing. You might fish all day long and not have a strike, but, boy, once you've got one on it's just—it's an amaz-ing experience when you've got a twelve-, fifteen-pound steel-head on the end of your line, tail-walking down the river, putting up a hell of a fight. And you do it in some of the most beautiful country. If I had one fishing trip left in me I want to go spend a week on the Babine."[7]

If others want to talk politics or current events, Cheney doesn't mind. But he rarely brings up current events on his own and most of his contributions to the conversation come in the form of one-liners. The Secret Service agents are there, as they always are, to protect the man they know back in Washington by the call sign "Angler." In one campfire discussion, Cheney was asked how many Secret Service agents serve in his detail.

"You don't want to know," he responded.

The fishing is highly competitive, with a daily "biggest fish" contest. The winning days are quite memorable, judging from Cheney's vivid recollection of beating his friend Dick Scarlett for the "money fish" one day last summer. "There was a small stream coming in from the right, and a big, long, very still pool. Dick went up and fished a bit and didn't have any success and moved on up above. I can remember coming in below him and I'd seen a big brown cruising, heading up into the water, and I made a perfect cast with a little fly tied on, a nymph, and I nailed it and fought him for a while and landed him. It was a twenty-three-inch brown trout, which is a big brown. The technical aspect of it was fun and important. It was a beautiful fish. The best part of it was, Scarlett had just been right on that spot and he hadn't got anything. That was the money fish of the day."

During the reelection effort in 2004, the campaign's media strategists pushed Cheney to spend time with local editorial boards or do interviews with regional television networks to communicate with voters, avoiding the cynical Washington press corps. When Cheney's advisers recommended that he spend an afternoon fishing with a reporter from a cable network that covered the outdoors, Cheney scoffed. You're just trying to soften my image, he joked, and recommended that they go hunting instead. He had been correct, of course. His advisers were trying to "humanize" Cheney, and hunting wouldn't necessarily accomplish their objectives.

And after a fateful trip to south Texas in late winter 2006, he might have had a difficult time getting reporters to join him.

As the sun was setting late in the afternoon on Saturday,

February 11, Cheney and a small group of hunters at the Armstrong Ranch in Texas decided they would try to bag an additional quail or two before they retired for the day. They had spent part of the afternoon touring the ranch and had resumed hunting about two hours earlier. Oscar Medellin, an outrider, had found a covey of quail and radioed to the group. Just before they reached the first covey, they came upon a second and decided to hunt there first.

Of the five people traveling in ranch jeeps, only three got out to hunt: Cheney; Pamela Willeford, a friend of Bush's who was serving as U.S. ambassador to Switzerland; and Harry Whittington, a prominent Texas attorney and Republican fund-raiser. When the first covey was flushed, Whittington shot two quail off to his right. Cheney also got a bird. Whittington walked with a guide and a Labrador into the tall grass to retrieve them.

Cheney and Willeford made their way to the second covey, which was to their left. The two stood side by side—Willeford on the left, Cheney on the right—as they waited for the birds to take flight. About 100 yards directly behind them, one of the birds Whittington had shot was proving difficult to locate. After about five minutes of searching, Whittington left his guide and returned to the vehicles, where the outrider congratulated him on his "double." He then began to walk toward Cheney and Willeford.

"We got over there, a bird flushed, a single bird," Cheney says. "I was on the right, out to the west. The sun is just starting down on the horizon, and there's nobody out there. I'm the last one on the line—well, there are just two of us, basically. A bird flushed, flew out to the right, and I turned and fired. Unbeknownst to me, Harry had come up and was standing where nobody had been before. He was down, there was a bit of a fold in the land, so he was down not completely behind it but partly, up to about his waist, down in this fold in the land. The sun was behind him. So I didn't see him in time to not fire."[8]

Cheney watched his friend crumple to the ground. "The image of him falling is something I'll never be able to get out of my mind. I fired, and there's Harry falling. And it was, I'd have to say, one of the worst days of my life, at that moment," he recalled in an interview four days later.[9]

Cheney and the guide who had been searching for the missing bird rushed to Whittington, followed shortly one of the physician's assistants who travel with the vice president at all times. Whittington was bloodied and stunned. "It knocked him silly," said Katharine Armstrong, an owner of the ranch.[10] "Harry, I had no idea you were there," Cheney said.[11] Cheney's medical team treated Whittington on the ranch. He was then taken by ambulance to a local hospital and then airlifted by helicopter to the Christus Spohn Medical System hospital in Corpus Christi.

Several eyewitnesses noted that Whittington had not called out his new position, as hunting etiquette requires. But Cheney did not deflect responsibility for the accident. "Ultimately, I'm the guy who pulled the trigger that fired the round that hit Harry. And you can talk about all of the other conditions that existed at the time, but that's the bottom line. And there's no—it was not Harry's fault. You can't blame anybody else. I'm the guy who pulled the trigger and shot my friend."[12]

Neither Cheney nor his staff put out word of the shooting until the next morning, when Katharine Armstrong telephoned a reporter for the *Corpus Christi Caller-Times*, which broke the story early Sunday afternoon, about twenty hours after the incident occurred. Journalists in Washington were incensed. How, they wondered, is it possible that the vice president shoots someone and neither the White House nor the vice president's office bothers to release that information? Even reporters sympathetic to Cheney's penchant for secrecy on national security matters— and there are only a few—thought that Cheney erred by delaying the news.

Cheney defended the decision in an interview with Brit Hume of Fox News Channel, saying his first concern was for Whittington's health. When Katharine Armstrong suggested to him that they give the story to a local reporter who might know more about hunting, Cheney agreed.

"I had a bit of the feeling that the press corps was upset because, to some extent, it was about them," he said. "They didn't like the idea that we called the *Corpus Christi Caller-Times* instead of the *New York Times*. But it strikes me that the *Corpus Christi Caller-Times* is just as valid a news outlet as the *New*

York Times is, especially for covering a major story in south Texas."[13]

The story almost immediately became fodder for late-night comedians, who would tell hundreds of jokes about the incident. "Dick Cheney accidentally shot a fellow hunter, a seventy-eight-year-old lawyer," said Jay Leno. "In fact, when people found out he shot a lawyer, his popularity is now at 92 percent."

Jon Stewart called it "Dick Cheney Shot-a-Guy-in-the-Face-Gate." He said: "Vice President Dick Cheney accidentally shot a man during a quail hunt . . . making seventy-eight-year-old Harry Whittington the first person shot by a sitting veep since Alexander Hamilton. Hamilton, of course, [was] shot in a duel with Aaron Burr over issues of honor, integrity, and political maneuvering. Whittington? Mistaken for a bird."

On February 16, 2006, the day after his interview with Hume, Cheney attended a small meeting with President Bush and congressional Republicans to discuss Iraq. Before the meeting began, Mike Pence, from Indiana, a leader of conservatives in the House, brought up the issue everyone was afraid to mention. Pence, seated directly to the right of President Bush, addressed his words to Cheney, who was sitting quietly in a corner of the Roosevelt Room.

"Mr. Vice President, I just want to tell you that we in Indiana know what an accident is. And we understand that accidents happen."

Cheney was typically understated in his response. "Thanks, Mike."

The official meeting began, and Zalmay Khalilzad, the U.S. ambassador to Iraq, appeared on a large television screen to brief the members of Congress. As Khalilzad began to describe security in Iraq, Pence was startled by a hard whack on his arm.

"That was great of you to say that, Mike," said Bush, leaning over Pence's shoulder. "It was good for him to hear it. That was really great."

But as Whittington's condition improved and time provided emotional distance from the incident, Bush joined the rest of the world in teasing Cheney about his marksmanship. One month after the shooting, Cheney's blunder was the theme of the evening

at the annual Gridiron Dinner in Washington, D.C., an exclusive
gala that brings reporters and politicians together for an evening
of drinking and joking. "Dick, I've got an approval rating of 38
percent, and you shoot the only trial lawyer in the country who
likes me," said Bush. He continued, "When Dick first heard my
approval rating was 38 percent, he said: 'What's your secret?'"[14]

The president also mocked the extensive media coverage of the
incident. "Good Lord," he said, "you'd [have] thought he shot
somebody or something."

Bush made the same cracks in private. When a visitor to
the Oval Office casually remarked that he had spent time with
Cheney immediately after the vice president returned from hunting, Bush said, "Better be careful." And on it went.

Over the course of the next year, jokes about Cheney and
hunting were everywhere.

An episode of Fox's popular animated show *Family Guy* de-
picted Cheney shooting the main character, Peter Griffin, ten
times at close range on a hunting trip. ("Sorry," Cheney says.
"I thought you were a deer.") The shooting was also featured
on *Family Feud*. "Name something you might need if you went
hunting with Dick Cheney," said the host, John O'Hurley. (Number one answer: bulletproof vest.) And the Las Vegas Wranglers,
a minor-league hockey franchise, hosted a "Dick Cheney Hunting Vest Night," distributing blaze-orange vests that read "Don't
Shoot, I'm Human."[15]

Later that spring, Cheney gave the commencement speech
at Louisiana State University. He had been invited to speak by
his former colleague at the Pentagon and the White House, Sean
O'Keefe, who was serving as chancellor of the university. As they
drove to the campus, supporters and detractors greeted Cheney
in ways that registered their views. "Lots of folks waved," says
O'Keefe, "and most of them used all of their fingers." The speech
was largely well received, ending in two standing ovations. But
not everyone appreciated Cheney's presence. Christine Koch-
Harris, who received her doctoral degree in French studies,
mocked his hunting accident by wearing a hunting vest over her
gown and a bull's-eye affixed to the top of her cap. As Koch-
Harris walked directly in front of Cheney to cross the stage to

receive her degree, the vice president turned and whispered his reaction to O'Keefe. "Find out who she is," he said, feigning indignation. "Tell her I thought that was well done."[16]

Nearly a year later, Cheney flipped on *The Tonight Show* with Jay Leno, as he occasionally did to unwind before going to sleep. "Even in Washington, everybody's into Valentine's Day," Leno said. "In fact, today Vice President Dick Cheney shot Cupid in the face."

Cheney laughed about Leno's joke the next day. "It never goes away," he says, shaking his head ruefully.[17]

He was right. A week later the *Sydney Morning Herald* welcomed Cheney to Australia with a front-page cartoon showing a bloodthirsty Cheney carrying a shotgun.[18]

Three weeks after the shooting, the *Washington Post* highlighted the results of a CBS News poll that helped explain why Cheney had been jealous of Bush's 38 percent approval rating. The survey found that only 18 percent of the American public had a favorable view of the vice president.

"These must be sobering days for Vice President Cheney as he reflects on recent events from his secret Fortress of Solitude," wrote the *Post*'s columnist Richard Morin. "Iraq teeters on the brink of civil war. The Bush agenda is in tatters. And one of his friends is recovering from an accidental gunshot wound inflicted by Cheney on a hunting trip. A particularly unfortunate mishap, as we learned last week, because Cheney wounded one of the rarest birds in America: someone who actually likes the vice president."[19]

Morin compared Cheney's approval ratings with those of historically unpopular public figures. For instance, Cheney was less popular than Michael Jackson after accusations of pedophilia and O. J. Simpson after he was tried for the murder of his wife and Ron Goldman. Cheney, wrote Morin, was "less popular with Americans than Joseph Stalin is with Russians. In 2003, fully 20 percent said Stalin, blamed for millions of deaths in the former Soviet Union during the 1930s and 1940s, was a 'wise and humane' leader."[20]

The results of other polls conducted that spring were not quite as unfriendly to Cheney. A poll by NBC News/*Wall Street Jour-*

nal found that 30 percent of registered voters had positive feelings about Cheney and 17 percent said they were neutral. And a poll by CNN/USA Today/Gallup reported that 40 percent of adults approved of Cheney's job as vice president.

Still, the low numbers were hardly surprising. It had been a difficult stretch for the Bush administration. The White House was still smarting from perceptions that it had responded inadequately to hurricane Katrina. The botched nomination of Harriet Miers to the Supreme Court had bolstered charges of cronyism from critics on both the left and the right.

For many politicians, such abysmal numbers would trigger changes. The unprincipled would simply abandon public discussion of the positions that had become a drag on their popularity. At the very least, they might be expected to launch a public relations offensive designed to boost those numbers.

Not Cheney. Though he still kept a relatively low profile, when he did appear in public he remained a forceful defender of the Bush administration's most controversial policies. He also remained its most aggressive rhetorical combatant. And he never wavered in his conviction that he had been elected to advise the president, not talk to journalists.

"We're pretty much a seen-and-not-heard operation," says one of Cheney's senior staffers. "We're not interested in building Dick Cheney's image. Neither is Dick Cheney."

"It's not my job to go spin the press," Cheney says. "It's my job to give [the president] the best policy advice. And to a large extent I'm very cautious about how much I see the press. The extent to which I'm available, I'll do some of it when I can be helpful. When it makes sense for me to go out and do a Sunday show on subject X because it's part of a program and we've got a campaign under way for the next three weeks to mark the second anniversary of the war or whatever matter is, I do my part. But I don't do much at all these days by way of press guidance."[21]

Bush says it's a quality he appreciates. "One thing about Dick Cheney is, you don't have to worry about him taking a political opinion poll and then trying to fashion his policy to make him popular," says Bush. "We all want to be popular. I mean, deep down, anybody who is running for office wants to be liked, popular. But popular—making decisions to enhance your own popularity does the country an injustice."[22]

As the public grew more frustrated with the Bush administration, Cheney took the largest dose of the venom. The novelist Tony Hendra, a contributor to the liberal weblog the *Huffington Post*, once offered a Thanksgiving Day prayer for Cheney's death. "I give thanks O Lord for Dick Cheney's Heart, that brave organ which has done its darn-tootin' best on four separate occasions to do what we can only dream about. O Lord, give Dick Cheney's Heart, Our Sacred Secret Weapon, the strength to try one more time! For love hath no heart than that it lay down its life to rid the planet of its Number One Human Tumor."[23]

For three decades, Cheney didn't have to worry about enhancing his popularity. He had been well liked and highly regarded by his colleagues and constituents. His political opponents often spoke of his trustworthiness, and even journalists praised him as competent.

Had Dick Cheney changed? Had the times changed? Both?

Brent Scowcroft, Cheney's old friend and longtime colleague who had become a bitter critic of Bush administration policies, told *The New Yorker* that Cheney was a different person. "The real anomaly in the administration is Cheney. I consider Cheney a good friend—I've known him for thirty years. But Dick Cheney I don't know anymore."[24]

Bob Michel, Cheney's congressional mentor and longtime friend, says he is not sure whether Vice President Dick Cheney is the same man he had handpicked to succeed him as Republican leader more than a decade earlier. "Well, I don't know. I always thought of Dick Cheney as a conservative, like I am. I never thought of him as being a hothead with respect to going to war, so anxious to go to war to achieve an end. He was a mediator, I thought, much more so."[25]

And yet Michel raises his voice in anger when he talks about media coverage of Cheney and his tenure as vice president.

"There are millions of people out there who only are persuaded by what they read in the *Washington Post* or the *New York Times* and their characterizations of Dick. That just galls my hide, that some of these big news outlets that so many people—

far too many people rely on—picture my friend that way. It gets my goat."[26]

Lee Hamilton, Cheney's Democratic colleague, believes that the poisonous political environment is partly responsible for Cheney's poll numbers. "I think when you become vice president or president and you're put in that spotlight, you naturally become a polarizing figure. The other problem is that there are few Democrats who have personal contact with him today. If more people had personal relations with him, that would make a difference."[27]

Perhaps.

On September 11, 2006, the country would set aside its enmities and political strife to commemorate the fifth anniversary of the attacks that had set the course of a new and difficult era for the country and for this administration. President Bush was in New York for much of the day, so the vice president represented the administration in ceremonies at the White House and the Pentagon.

The day started with a service at St. John's Episcopal Church, one block from the White House, where Cheney had attended services on his first day in office nearly six years earlier. On this day, like that one, a cool rain fell. Streets were closed for blocks. Bomb-sniffing German shepherds inspected the perimeter of the church, a small yellow building. One by one, cabinet secretaries filed by the small pool of reporters and photographers, most wearing somber expressions: Secretary of State Condoleezza Rice; Attorney General Alberto Gonzales; Secretary of Housing and Urban Development Alphonso Jackson; Director of the Office of Management and Budget Rob Portman. Alone among them, Secretary of Education Margaret Spellings smiled, incongruously, for the cameras and waved as if she were making an entrance at the Academy Awards.

The vice president and Mrs. Cheney arrived shortly before the service began. The vice presidential limousine was led by a policeman in full rain gear riding a Harley Davidson and followed by three oversize black SUVs.

The service opened with readings from Gonzales; Reverend Kathleene Card, the wife of the White House chief of staff, Andy Card; and Commander David Tarantino of the navy. Tarantino, who had been in his office at the Pentagon when the plane crashed into it, had rescued a civilian trapped under debris in the Navy Command Center.

After the small congregation sang "Be Thou My Vision," Reverend Luis Leon offered his personal recollections of the day, five years earlier. "All of the female staffers coming down Sixteenth Street were running in high heels," he remembered. "And guys who haven't run since the high school track team were sprinting down the street." When he finished, the congregation joined him in singing "America."

From there, Cheney made his way back to the White House to participate in a moment of silence at 8:46 AM eastern standard time on the South Lawn, five years to the minute after American Airlines Flight 11 struck the North Tower of the World Trade Center. As the Marine Corps Band played "Amazing Grace," the White House staff formed two lines arcing out from the doors to the South Lawn. In one line stood well-groomed cabinet secretaries and their spouses, most of them wearing handsome black suits and coats. In the other, facing this collection of the nation's most powerful, were National Park Service groundskeepers who had worked until just moments earlier to prepare the grounds for the short ceremony. They were a disheveled bunch, with dirt visible on their olive green uniforms even from a distance. One man nervously combed his tousled hair with his hands after he removed his cap out of respect for the flag. They stood tall and seemed proud of what they had contributed. Journalists looking for a poignant moment on this solemn day had found one.

Cheney escorted his wife and the former British prime minister Margaret Thatcher down the aisle created by the two lines. Thatcher, in town for another event, had asked the White House to include her in the commemoration ceremonies. Cheney regarded the formidable "iron lady" as a heroine and was eager to accommodate her request. As the United States contemplated military action after Iraq's invasion of Kuwait, Thatcher had famously told George H. W. Bush not to go "wobbly" in his response. Now, at eighty, she seemed unsure of her footing as she walked slowly down the grass pathway, clinging tightly to Cheney's arm. Finally, everyone was in place.

Then, silence.

After one minute, a lone bugler played taps. The ceremony was over.

Cheney was rushed off to his waiting limousine, and his entourage—the Secret Service, staff, photographers, and reporters—scrambled to keep up. A long line of black vehicles hustled from the White House along the mall, past the Washington Monument and over the Potomac River, zipping past traffic left over from the morning rush hour that had been blocked to allow the vice president to pass.

At precisely 9:37 AM, a second solemn remembrance began, this one on the River Parade Grounds at the Pentagon. Many in the audience looked skyward as an airplane roared overhead on its departure from National Airport, just two miles away, a powerful, unintentional reminder of the attacks. A massive American flag was unfurled from the roof of the Pentagon, released by the men who had famously done the same thing five years earlier.

Peter Pace, chairman of the Joint Chiefs of Staff, spoke first. Rumsfeld followed, and Cheney came last. A light rain began to fall as the vice president opened his remarks. "The ones who were lost," he said, had begun their day "busy with life."

> *They had people who cared about them, people who depended on them, people who loved the sight of their face and the sound of their voice. They were unsuspecting of danger and undeserving of their fate. Each one of them had hopes and plans for the future. . . . From two miles away, an Army chief warrant officer, whose wife worked in this building, saw the fire and ran to the scene. He joined in the rescue effort, and stayed in the work even after learning his wife could not possibly have survived.*

Cheney stopped briefly. He was visibly moved.

> *We know of these and so many similar acts of courage and kindness on that terrible morning. Other stories, we will never know. Surely men and women here, and aboard Flight 77, were, in their last moments, holding and com-*

forting one another. And when we think of them, it will always be with a special feeling of empathy and sorrow.

As Cheney spoke these words, he looked out at the families of the ones who were lost. Many of them, holding and comforting one another, looked back at him, their faces streaked with tears. A burly Army Ranger stood alone next to the holding area for the press, weeping silently as he listened.

We will always understand the pain of their families and our nation will forever look with reverence upon their place—this place where their lives ended.

And then Cheney paused, his words tangled in his throat. He started to speak and then, choking back tears, stopped again. Reporters exchanged quick glances as if to confirm what they were seeing.

It was not a good cry, exactly. But from the man who had long evaded questions about how those attacks affected him personally, it was, at last, an answer.

Throughout its momentous first six years, the Bush administration had had an enthusiastic witness in Bob Woodward, the legendary reporter for the *Washington Post* who had installed himself in the journalistic firmament during the Nixon years by exposing the Watergate scandal. Since then, Woodward had written a stream of books, the last two of which—*Bush at War* and *Plan of Attack*—had given the public an inside glimpse of the normally impenetrable Bush White House. Now he was planning a third.

Cheney had cooperated with Woodward on his last two books, but only because Bush had ordered him to do so. The result in both cases was vintage Woodward, filled with the kind of "in-the-room" reporting that propels each of his books to the top of the best-seller lists. The first, *Bush at War*, covered 9/11 and its aftermath, and contained an overwhelmingly positive portrayal of the president. *Plan of Attack,* which examined the buildup to and execution of the Iraq War, was more critical. But it put the decisions that led to the Iraq War in their proper context, as much of the daily reporting on the war had failed to do.

In that sense, the first two books tracked closely with public opinion. Bush and Cheney were riding high in the polls after removing the Taliban in Afghanistan; by 2004, they remained just popular enough to win reelection.

Woodward's third book would be different. It had to be. Since *Plan of Attack* was published in October 2004, there had been history-making events in both Afghanistan and Iraq. But despite the political successes in Baghdad, the insurgency continued and sectarian clashes threatened to escalate into a full-scale civil war. In *State of Denial*, Woodward would explore what went wrong and who was accountable.

President Bush had decided not to spend time with Woodward for the book, and he had not asked Cheney to cooperate, either. When Woodward asked Cheney for interview time, Cheney declined. The vice president had always liked Woodward. "I'd worked with Bob before and enjoyed working with him," he says. "He's good at his craft." But Cheney's natural inclination was to turn down the request for an interview. Woodward persisted. "He came at me two or three times," Cheney recalls. "And I said: 'No, I'm not going to do it.' "[28]

For Cheney, the issue was settled. A short time later, though, former president Ford called Cheney with a request. Bob Woodward is doing a book on the Ford administration, he explained. I'm cooperating with him and I hope you will, too.

Cheney agreed. "I was happy to do that," he says. "Ancient history. I would do just about anything Ford asked me to do if there was any way I could. And, you know, it sounded like fun. I was happy to do it. So I gave Woodward a couple of hours."[29]

But, Cheney says, his cooperation was conditional. The content of the interview was to be embargoed until Woodward's book on Ford was published. "The ground rules were very clear: that he wouldn't use any of that until it was part of the book," Cheney recalls. "He agreed to all that, so I did two interviews with him."[30]

Cheney didn't think much more about the book until Thursday, October 2, one month before the midterms, when a staffer played him a short video CBS was running on the Internet to advertise Sunday's edition of *60 Minutes*.

Cheney could not believe what he heard. The interviews he

had given Woodward were being used in the promotion. They would later be used on the show.

Woodward and Mike Wallace discussed information from the interviews Cheney had given Woodward for the book about Ford.

"Cheney stunned Woodward by revealing that a frequent adviser to the Bush White House is former secretary of state Henry Kissinger, who served Presidents Nixon and Ford during the Vietnam War," said Wallace.[31]

Woodward: "He's back. In fact, Henry Kissinger is almost like a member of the family. If he's in town, he can call up, and if the president's free, he'll see him."

Cheney thought Woodward's discussion of their interviews constituted an obvious violation of their agreement. It would get worse.

Wallace: "Woodward recorded his on-the-record interview with Cheney, and here's what the vice president said about Henry Kissinger's clout."

The next voice to speak was Cheney's, from the tape of the interview: "Of the outside people that I talk to in this job, I probably talk to Henry Kissinger more than just about anybody else. He comes by, I guess, at least once a month and I sit down with him."

Woodward: "And the same with the president?"

Cheney: "Yes. Absolutely."

Woodward: "President Bush is, I understand, is a real—"

Cheney: "A big fan of his."[32]

They had played the tapes of his interviews with Woodward as if they'd been done for *State of Denial.* Cheney, furious, did something he almost never does. "I called him up." The vice president launched a low-volume tirade.

Woodward denied that he had violated their agreement. "He said, 'It's on the record. It's on the record. It's on the record. Everything is on the record.'"

Cheney was shocked. He thought Woodward was not being honest. "It's not true," Cheney says. "There were conditions under which we did those interviews." It was soon clear that Woodward was not going to apologize. There would be no middle ground, Cheney felt.

"I was so angry—finally ended up hanging up on him."[33]

The fact that Cheney chose to talk about something related to the current Bush administration, Woodward argues, made it something he could use.

"The discussions about Ford and the Ford administration, indeed, you know, not to be used. That was the agreement. But these are on-the-record interviews," he says.

"That had nothing to do with Ford, and we're having an on-the-record interview. It's like if he said, 'Oh, by the way, the following happened yesterday' or whatever. We're having an on-the-record interview. It's not, it's—nothing about Ford could be used until I do my Ford book. But as I said to him, when he called, I said: 'On the record is on the record.'"

"You know, it's clear it's for the Ford book but he brought up something about now. And he didn't say, 'Oh, yeah, off-the-record' or 'This is on background.'"

Woodward says he can understand why Cheney would be upset, but says he never thought about leaving it out.

"There was no question of my using it."[34]

Woodward explained the incident six days later on *Meet the Press*, and when he was done, the host, Tim Russert, summarized. "He thought he was talking to you for one project and you used it in another project."

"Well, exactly," Woodward responded. "But it had nothing to do with it, and it's clearly spelled out that it's an on-the-record interview. And so—now, what does he do instead of saying, 'Well, OK, I look at it this way, you look at it that way.' It's a metaphor for what's going on. Hang up when somebody has a different point of view or information you don't want to deal with."[35]

Cheney may have thought it was a metaphor. It was the meaning of the metaphor that was different. Over the course of his career, and in particular his vice presidency, Cheney had grown increasingly disillusioned with the news media. And now, he thought, Bob Woodward, the most highly regarded reporter of his generation, had violated the ground rules of their interview and had been dishonest with him about it afterward.

It was an augur of things to come. Woodward's book would be one of dozens criticizing the Bush administration released in the four months before the congressional elections of 2006.

There were *Fiasco: The American Military Adventure in Iraq*; *Imperial Life in the Emerald City: Inside Iraq's Green Zone*; *Hubris: The Inside Story of Spin, Scandal, and the Selling of the Iraq War*; *The Greatest Story Ever Sold: The Decline and Fall of Truth from 9/11 to Katrina*; *Blind into Baghdad: America's War in Iraq*; *The End of Iraq: How American Incompetence Created a War without End*; *Conservatives without Conscience.* As the election approached, authors promoted their books in a steady stream of media appearances, in effect reinforcing the main argument of Democratic candidates around the country.

President Bush gave few indications that he was ready to make significant adjustments to his strategy in Iraq or among those charged with executing it. In an interview one week before the election, Bush told reporters that he supported Donald Rumsfeld "strongly" and thought his Secretary of Defense was doing a "fantastic job."

Cheney campaigned extensively for Republican candidates across the country, appearing at 160 events over the election cycle of 2006. He stumped mostly for candidates in conservative districts where turnout would determine the outcome. Despite his efforts, the Republicans lost six seats in the Senate and twenty-seven in the House. The Democrats had retaken control of both chambers.

One day after the election came an announcement that stunned Washington: Donald Rumsfeld was resigning. The news broke just before lunch on November 8. Rumsfeld had spoken with Bush and spent the better part of his morning informing a few of his most loyal staff members at the Pentagon.

It was an abrupt reversal for Bush. Shortly before the election, the president said that he expected Rumsfeld to remain in his job until the end of the Bush administration. "After a series of thoughtful conversations," Bush said, in announcing the change, "Secretary Rumsfeld and I agreed that the timing is right for new leadership at the Pentagon."

The Republicans on Capitol Hill were furious. Many of them believed that Rumsfeld had become a totem for the increasingly difficult war in Iraq. And Bush's unwillingness to fire him, once a sign of loyalty, had become further evidence of the president's stubbornness. Clay Shaw, one of the Republicans defeated in the Democratic sweep, told reporters that he would have won if Rumsfeld had been removed before the election. "My first impression was that the actual votes I would have needed would have been there," Shaw told the *Miami Herald.* "I think the Republicans would have been a little more energized."[36]

If Rumsfeld had been a convenient scapegoat for sour politicians, it would soon be clear that Bush saw him as an obstacle to a change of course in Iraq. The president had begun to consider expanding the number of troops deployed, and Rumsfeld, who had begun the war as an exercise in using a smaller, faster army, had long opposed such a plan. On the day Rumsfeld offered his resignation, Bush was asked about retooling his strategy. His response was terse. "Well, there's certainly going to be new leadership at the Pentagon."

Within a month of Rumsfeld's departure, Bush and that new leader, the former CIA director Robert Gates, had decided on significant changes to the U.S. strategy in Iraq. The new plan included adding some 21,000 additional combat troops in Baghdad and Anbar province and more aggressive rules of engagement for soldiers on the ground.

Cheney thought such an adjustment was overdue: as Bush says, "He was a more-troops man."[37] But the vice president was not happy that his longtime friend and mentor had been fired.

Cheney was asked about Rumsfeld's departure three weeks after it was announced. It was the last question at the end of a three-hour interview. "Anything you could say on the resignation of your good friend?" Rumsfeld was still at the Pentagon and Cheney at first refused to address the sensitive subject.

"Probably not." After a moment, he revised his answer.

"Someday, but—you can ask that next time. Let me think about it."

Cheney thought about it for one second.

"He is my good friend, and I think one hell of a defense secretary. Maybe the best we ever had. And that opinion is unchanged by any events that have happened in recent times."[38]

Three weeks later, on a podium outside the Pentagon at a full honor review ceremony to mark Rumsfeld's departure, Cheney dropped the qualifier. "Don Rumsfeld is the finest secretary of defense this nation has ever had," he said.

Cheney spoke before the president. His speech lasted ten minutes. For someone so intimately familiar with presidential protocol, it was an obvious violation. When a vice president and a president share a stage, the vice president, if he speaks at all, keeps his remarks brief.

"In a lifetime, one meets only a few people of such caliber and character, and so my first association with Don Rumsfeld was one of life's great turning points, both professionally and personally," he said. "On the professional side, I would not be where I am but for the confidence that Don first placed in me those many years ago. And on the personal side, it's enough to say that I have no better friend and ask for none."

And with Bush sitting just feet away, Cheney said: "I've never worked harder for a boss, and I've never learned more from one, either."[39]

In an interview in the Oval Office four days earlier, Bush acknowledged that his vice president was less enthusiastic than others in the administration about Rumsfeld's departure.

"I agree with that," says Bush, "because there was a personal relationship."[40]

For more than two years, Bush had sided with the vice president—and against several other senior advisers, including Andy Card and Condoleezza Rice—in deciding to allow Rumsfeld to remain in his job. The election-time discussion between the two men was not the first occasion when Cheney had tried to keep his boss from firing his friend.

But Bush had decided it was time to make a change. And unlike several previous discussions he had with Cheney about the secretary of defense, this one had ended with Bush firmly committed to replacing Rumsfeld. The conversation between Bush and Cheney had been intense. "He listened very carefully, and— he listened very carefully," says Bush. "He thinks Don Rumsfeld was a great secretary of defense, as do I. And I think in this case, he really felt like his friend ought to stay the years if that's what Donald Rumsfeld felt like doing himself."[41]

Bush contends, improbably, that Cheney simply misunderstood the situation and that Rumsfeld had gone freely. "It was a mutually arrived-at decision," he insists, "genuinely so. And I don't think Dick necessarily viewed it that way."[42]

Cheney's personal loyalty to Donald Rumsfeld reached back nearly thirty years. Throughout the Bush administration Cheney and Rumsfeld spoke almost daily and socialized at least once a week. They had plenty of differences over policy—"They don't always see eye to eye on things, not by any means," says Condoleezza Rice—but they remained close.[43]

In order to defend his old friend, Cheney would have to violate one of his cardinal rules. He would have to admit publicly that he had differences with his president. Cheney had done this once in six years, and then only because it was an issue that directly affected his family. He was prepared to do it a second time.

On January 10, 2007, President Bush announced that he would send additional combat troops to Iraq. The White House asked Cheney to promote the changes later in the week, on *Fox News Sunday*. The show's host, Chris Wallace, would conduct the interview. To prepare, Cheney got together with his staff and his elder daughter, Liz, for a "murder board" session. An aide fired one tough question after another at the vice president.

Then: Did you agree with President Bush's decision to replace Donald Rumsfeld as secretary of defense?

"Absolutely not," Cheney replied without elaborating. His answer surprised the small group with him, but it was the answer he was determined to give if Wallace asked, even at the risk of angering his boss. But the story was a month old, and Wallace never asked this question.

As Cheney's longtime mentor was ushered out of the Bush administration, one of his protégés was asked to join. In November, Michael McConnell, who had been working on intelligence issues in the private sector since resigning from the NSA in 1996, was asked to consider joining the Bush administration as the nation's top intelligence official, the director of national intelligence (DNI).

When McConnell was first approached, there was not yet a vacancy. John Negroponte, former U.S. ambassador to Iraq and the first person to hold the position, was still serving when McConnell received a call from the White House to gauge his interest. McConnell was told that President Bush wanted to move Negroponte to the State Department and was eager to hire someone with a background in intelligence for the job.

McConnell was honored to be asked, but had serious reservations. He had been unimpressed with many aspects of the Bush administration and its conduct of the war on terror, particularly what he felt was a politicized use of intelligence in the lead-up to the Iraq War.

"All of these current players, Secretary Rumsfeld, Vice President Cheney, and the President," McConnell said in an interview in late November 2006, "what's come through for me as a citizen— [I'm] no longer [on] active duty so I can say these things—they had first and foremost very strong political convictions. My sense of it is their political faith and convictions influenced how they took information and interpreted [it], how they picked up and interpreted outside events.

"As a former intel pro—I use that term meaning careerist, not that I was any more professional than anyone else—when you don't like the answer and you set up your own thing, you tend to get the answer you want. You hire people who think like you do or want to satisfy the boss. . . . I've read much more about the current set of players and they did set up a whole new interpretation because they didn't like the answers. They've gotten results that in my view now have been disastrous."[44]

So he called Colin Powell, his former boss on the Joint Chiefs of Staff. Powell's views on dealing with Cheney and Rumsfeld were no secret. He had cooperated with Bob Woodward on each of Woodward's increasingly critical books; and his former chief of staff, Larry Wilkerson, had been a go-to quote for reporters seeking a critical voice within the Bush administration.

"I called General Powell and we had a long discussion about it," says McConnell. "And I'm not really close to him but it was a big decision, and I thought I'd ask him for his advice. So he sort of played back for me what it was like. When I thought about it as an intel guy who's got to speak truth to power . . . but if power won't listen . . ." McConnell's voice trails off. "I don't know if the vice president is being enormously loyal to the president—trying to carry the president's water. Or, I don't know. I just don't know."[45]

He adds, "You can grade me as a Cheney fan in several dimensions. Watching him in action as secretary of defense—briefing him, learning from him, trying to get him information—man, he was impressive. I was a huge fan. . . . Now, the subsequent decisions about how we've gone about Iraq and the invasion and sort of the underhanded control that Secretary Rumsfeld asserted and so on. On a personal political level, as a citizen looking at this, that didn't go down too well with me. Remember, I did the game plan that said if we went without a game plan we'd end up with a mess, that's years and years and years ago, and then here we are and it's played out that way. Weapons of mass destruction, which the independent group set up by Rumsfeld and the Pentagon say . . . were there, they were wrong."[46]

It was not, of course, an independent group set up by the Pentagon that concluded Iraq had weapons of mass destruction; it was the consensus of the U.S. intelligence community and the world. But McConnell's comments, made shortly before he accepted Bush's offer to run that community, reveal just how decisively the White House has lost the public relations battle over the intelligence failure on Iraq's WMD.

There was more fallout to come.

Shortly before noon on March 6, 2007, the vice president took calls from his wife and elder daughter in his office in the West Wing. The all-news cable channels were reporting that a verdict in the trial of Cheney's former chief of staff, Scooter Libby, was imminent. Alone in his office, Cheney watched as the news anchors shared in real time the information as it was passed from their reporters in the courtroom. Libby was found guilty on four out of five counts of lying and obstructing justice.

It was devastating personal news. Cheney was close to Libby. He knew Libby's wife and his two young children and was well aware of how they had suffered, first through the investigation, then through the indictment, and finally through the six-week trial.

Cheney left the White House to attend the Senate Republican policy lunch on Capitol Hill. Several senators approached Cheney to tell him that they were thinking about Libby or to express their anger about the verdict. The somber lunch opened with a prayer that included Scooter Libby and his family. After it was through, Cheney returned to the White House and got back to work.

If Cheney had long been angry about what he saw as political investigation, facts that had come to light the previous summer upset him even more. In August 2006, the *Nation*'s David Corn—the journalist who first raised the possibility that Valerie Plame was covert and suggested that the Bush administration had deliberately exposed her—reported a fact that complicated the anti-Bush administration narrative. In *Hubris: The Inside Story of Spin, Scandal, and the Selling of the Iraq War*, Corn and *Newsweek* reporter Michael Isikoff produced a significant scoop: they revealed that the original leaker was not Scooter Libby or someone in the Bush White House or even a strong supporter of the Iraq War. It was Deputy Secretary of State Richard Armitage.

The original leak, Corn and Isikoff conceded, complicated the prevailing narrative.

Armitage had shared the information with Bob Woodward, who did not use it, and Bob Novak, who did. In an interview on June 13, 2003, made public as part of the Libby trial, Woodward asked about Joe Wilson in an interview with Armitage.

"But it was Joe Wilson who was sent by the agency," Woodward said. "I mean that's just—"

"His wife works in the agency," Armitage replied.

"Why doesn't that come out? Why does that have to be a big secret?"

"Everyone knows it."

"Everyone knows."

"Yeah," Armitage confirmed. "And I know [expletive] Joe Wilson's been calling everybody. He's pissed off because he was designated as a low-level guy, went out to look at it. So he's all pissed off."

"But why would they send him?"

"Because his wife's a [expletive] analyst at the agency," Armitage repeated.

"It's still weird."

"It's perfect. This is what she does. She is a WMD analyst out there."

Only after Novak wrote in October 2003 that he had gotten

the news from someone not regarded as "a partisan gunslinger," Armitage claims, did he realize that he was Novak's source.

"Well, I was reading the newspaper column again of Mr. Novak, and he said he was told by a non-partisan gunslinger," Armitage explained on CBS News. "I almost immediately called Secretary Powell and said, 'I'm sure that was me.'"

Armitage then confessed his role to FBI investigators. "I told them that I felt that I was the inadvertent leak." Neither Armitage nor Powell told the White House.

But the FBI investigation—launched under political pressure just days earlier after the leak revealing that the CIA had referred the case to the Justice Department—continued apace. When Patrick Fitzgerald started his own investigation three months later, he had already known that Armitage leaked Valerie Plame's identity to Novak. He asked Armitage not to discuss his role. "The special counsel, once he was appointed, asked me not to discuss this, and I honored his request."[47]

Armitage was never prosecuted for the leak, a fact Libby defenders take to mean that Fitzgerald was never able to prove that Plame was covert. Fitzgerald also chose not to file charges against Armitage for his failure to tell investigators that he also leaked Plame's identity to Woodward.

The outcome further damaged the already deteriorating relationship between Cheney and Powell.

Whether Powell himself talked with reporters about his relationship with Cheney or his subordinates spoke on his behalf, in books and in news accounts the public learned that Powell no longer cared much for the vice president. Missing from most of those accounts was any characterization of Cheney's thoughts on Powell.

Cheney will acknowledge that there was some tension with his former colleague, but otherwise refuses to talk about him. "The relationship clearly soured in the current administration," he says.[48]

Most of the trial had focused on a dispute about the exact moment at which Libby had learned Valerie Plame's identity, and from whom he had learned it. Libby had told investigators that he believed he had first heard about Plame from reporters and specifically mentioned a conversation with Tim Russert, host of NBC's *Meet the Press.* But Russert contradicted Libby, saying he did not know Plame's identity when he spoke with Libby on July 10, 2003.

A host of Bush administration officials, intelligence officers, and reporters provided testimony that often cast doubt on the claim of the defense that Libby had simply misremembered the details of the Plame controversy in the whirlwind of activity that his high-level jobs required. The defense had shown that many of those who claimed to remember conversations with Libby about Wilson and Plame had memory issues of their own. But Judge Reggie B. Walton refused to allow the defense to call memory experts to support their claims about Libby's recollections. Testimony from important witnesses for the prosecution—including David Addington, Cheney's former communications adviser Cathie Martin, and the former White House spokesman Ari Fleischer—helped convince the jury that Libby had indeed known Plame's identity—and that she was employed by the CIA—before his conversation with Russert. And some jurors would later say that they found Russert's testimony, in which he denied Libby's version of events, particularly persuasive.

Although Fitzgerald had said, in announcing his indictment of Libby in October 2005, that the trial would be about the narrow issue of whether one man lied or obstructed justice during the course of a federal investigation, his closing statement made it clear that he did not believe his own words.

"What is this case about?" Fitzgerald asked the jury. He wondered if it was about "something bigger" than the charges he brought against Scooter Libby.

"There is a cloud over the vice president," Fitzgerald proclaimed. "And that cloud remains because this defendant obstructed justice."[49]

It was a gift to the Washington press corps.

"From the start, the case was only marginally about Libby," wrote Michael Duffy of *Time* in a cover story, "Cheney's Fall from Grace." Duffy continued, "What was really on trial was the whole culture of an Administration that treated the truth as a relative virtue, as something it could take or leave as it needed. Everyone knows now that Bush and Cheney took the country into a deadly, costly and open-ended war on flimsy evidence of weapons of mass destruction. Yes, Congress went along. And yes, the public on balance supported it. But no one was more responsible than the Vice President for pushing the limits of the prewar intelligence that did all the convincing. And when former ambassador Joseph Wilson questioned the credibility of that intelligence—and the motives that helped polish it—it was Cheney who led the fight to bring him down."[50]

In story after story, reporters simply elided the facts that complicated this preferred story line: that the claims about Iraq's WMD came not from Cheney's fevered imagination, but from the "key judgments" of a document reflecting the consensus of the U.S. intelligence community; that the original leak of Valerie Plame's identity came not from the office of the vice president but from the State Department; that Joseph Wilson built his personal indictment of Cheney on a claim that not only wasn't true but could not have been true; and that Cheney and Libby had a compelling interest in ensuring that those fabrications were corrected.

As much as Cheney has tried to ignore the harshest of the criticism, he concedes that the sustained attacks about the administration's use of intelligence have not only registered but have changed the way he performs his job.

"Frankly, I'm gun-shy now," he says. "Because of the fact we've gotten hammered so often and so many times. And you get to the point where you're cautious about who you talk to and what you ask for."[51]

Isn't that a detriment to national security?

"Yes, it is. It should never be politicized to the extent that it has been. I think it's outrageous for the Senate Democrats, and people like Carl Levin, for example—and this is on the record—to continually hammer away at this notion and allege that [we were] trying to warp [intelligence]. This is all after the fact. He signed the intelligence committee bipartisan report that said there was nobody trying to coerce or cow the intelligence community into producing a particular product."[52]

How does this change the way you do your job?

"You're just more cautious. There have been, as well—the president has talked about it—leaks out of the community that appear to be timed to influence elections." Cheney makes it clear that he believes the problems are not with the entire intelligence community. "You have to be very careful here not to cast aspersions at the community. The vast majority of folks are thoroughgoing professionals, care very deeply, believe completely in what they're doing, and just want to get it right. But it doesn't take many to try to—it starts to break down the element of trust that needs to exist between the elected leadership and the intelligence professionals who are going to provide what the leaders need in order to be able to make intelligence judgments and decisions."[53]

Those intelligence judgments and those policy decisions, Cheney believes, are as important today as they were immediately following the attacks on September 11, 2001.

> *The notion that somehow we've got to get across to people is they just cannot think of this as a conventional war. This is not Desert Storm. It's not Korea. It's not World War II. This is a struggle that's going to go on in that part of the world for decades. I don't know that you're going to be involved in Iraq for decades; I don't want to say that. But just think about it.*
>
> *We just have to have people understand that and understand that the alternative is not peace. The alternative is not we go back to the way the world was before 9/11. You can't turn back the clock.... There's always a possibility that maybe the next president you elect decides they don't want to continue the policy and so they adopt the other approach, the one that failed before 9/11. And I think to some extent the terrorists are betting that they can run out the clock on the Bush-Cheney administration and that it will be easier for them in the future because they won't face the kind of determined action that this administration has taken to take them on—to take the fight to them, to put in place first-rate defenses here at home, to do all those things we've done that have kept us safe and secure for the last five years.*[54]

And the war goes on.

NOTES

INTRODUCTION

1. Mark O. Hatfield, *Vice Presidents of the United States, 1798–1993*. Washington, D.C.: U.S. Government Printing Office, 1997.
2. Steve Tally, *Bland Ambition: From Adams to Quayle — The Cranks, Criminals, Tax Cheats, and Golfers Who Made It to Vice President*. New York: Harcourt, Brace, Jovanovich, 1992.
3. Kenneth T. Walsh, Kevin Whitelaw, Angie C. Marek, and Jim Stanford, "The Man behind the Curtain," *U.S. News and World Report*, October 13, 2003, p. 26.
4. Author interview with Condoleezza Rice, August 15, 2006.
5. Judy Keen, "Cheney Says It's Too Soon to Tell on Iraqi Arms," *USA Today*, January 19, 2004.
6. Author interview with Vice President Dick Cheney, August 31, 2005.
7. Michael Medved, *The Shadow Presidents: The Secret History of the Chief Executives and Their Top Aides*. New York: Times Books, 1979, p. 346.
8. Author interview with Norma Fletcher, August 11, 2005.
9. Author interview with Mary Matalin, n.d.
10. Dick Cheney, commencement address, Natrona County High School, Casper, Wyo., May 27, 2006.

CHAPTER ONE: THE WEST

1. Author interview with Vice President Dick Cheney, August 31, 2005.
2. "The War of the Rebellion: A Compilation of the Official Records of the Union and Confederate Armies," Series 1, Vol. 38 — In Five Parts, "Operations in Northern Georgia, etc. May 1–September 8, 1864, Relating

Especially to the Atlanta Campaign," Serial No. 72. Washington, D.C.: Government Printing Office, 1891.
3. Cheney interview, August 31, 2005.
4. Ibid.
5. Author interview with Vic Larson, June 26, 2006.
6. Cheney interview, August 31, 2005.
7. Ibid.
8. U.S. Census, *Wyoming Population of Counties by Decennial Census: 1900 to 1990.*
9. Edna Kukura and Susan True, *Casper: A Pictorial History.* Norfolk/ Virginia Beach, Va.: Donning, 1986, p. 137.
10. Ibid., p. 148.
11. Jean Mead, *Casper Country: Wyoming's Heartland.* Boulder, Colo.: Pruett, 1987, pp. 134–138.
12. Author interview with Norma Fletcher, August 11, 2005.
13. Ibid.
14. Cheney interview, August 31, 2005.
15. Author interview with Mark Vincent, August 11, 2005.
16. Cheney interview, August 31, 2005.
17. Author interview with Darla Howard Burris, August 12, 2005.
18. Author interview with Tom Fake, April 28, 2006.
19. Cheney interview, August 31, 2005.
20. Author interview with John Castle, June 26, 2005.
21. Cheney interview, August 31, 2005.
22. Castle interview, June 26, 2005.
23. Author interview with Ron Lewis, August 11, 2005.
24. Cheney interview, August 31, 2005.
25. Interview with Tom Fake, April 28, 2006.
26. Interview with Cheney, August 31, 2005.
27. Nicholas Lemann, "The Quiet Man: Dick Cheney's Discreet Rise to Unprecedented Power," *New Yorker,* May 7, 2001.
28. Ibid.

Chapter Two: To Yale and Back

1. Author interview with Tom Fake, April 28, 2006.
2. Author interview with Dennis Landa, April 19, 2006.
3. Nicholas Richer, "Professor Recalls Living with Cheney," *State News,* February 11, 2006.
4. Author interview with Jim Little, April 25, 2006.
5. Author interview with Landa.
6. Cheney interview, August 31, 2005.
7. Author interview with Peter Cressy, April 19, 2006.
8. Author interview with Rees Jones, May 12, 2006.
9. Little interview, April 25, 2006.

10. Ibid.

11. Author interview with Ned Mason, June 25, 2006.

12. Ibid.

13. Little interview, April 25, 2006.

14. Cheney interview, August 31, 2005.

15. Ibid.

16. George W. Bush, commencement address, Yale University, New Haven, Conn., May 21, 2001.

17. Cheney interview, August 31, 2005.

18. See http://www.thesmokinggun.com/archive/cheneydwi1.html.

19. See http://www.thesmokinggun.com/archive/cheneydwi2.html.

20. Author interview with Joe Meyer, August 12, 2005.

21. Nicholas Lemann, "The Quiet Man," *New Yorker*, May 7, 2001.

22. Meyer interview.

23. Cheney interview, August 31, 2005.

24. Ibid.

25. Author interview with Dick Cheney, April 27, 2006.

26. Meyer interview.

27. Cheney interview, August 31, 2005.

28. Ibid.

29. Ibid.

30. Melissa Healey, "Cheney Courts Support as Nomination Hearings Begin," *Los Angeles Times*, March 15, 1989.

31. George C. Wilson, "Cheney Believes Gorbachev Sincere," *Washington Post*, April 5, 1989, p. A12.

32. Cheney interview, April 27, 2006.

33. Ibid.

34. David Maraniss, *They Marched into Sunlight: War and Peace, Vietnam and America October 1967*. New York: Simon and Schuster: 2003, p. 113.

35. Richard B. Cheney and Aage R. Clausen, "A Comparative Analysis of Senate and House Voting on Economic and Welfare Policy: 1953–1964," *American Political Science Review* 13 (Spring 1970).

36. Cheney interview, August 31, 2005.

37. Author interview with Joseph Tydings, June 26, 2006.

38. Ibid.

39. Ibid.

Chapter Three: Choosing Government

1. Cheney interview, August 31, 2005.

2. Letter to the author from Jeffrey Biggs, Director, American Political Science Association, Congressional Fellowship Program, June 1, 2006.

3. Cheney interview, August 31, 2005.

4. Ibid.

5. Donald Rumsfeld, hearing before the Committee on Labor and Public Welfare, U.S. Senate, 91st Congress, 1st session, May 13, 1969.

6. Gerald S. Strober and Deborah H. Strober, *The Nixon Presidency: An Oral History of the Era.* Dulles, Va.: Brassey's, 2003, p. 117.

7. Interview with Christine Todd Whitman, July 5, 2006.

8. Robert F. Kennedy Performance Project; transcript of speech at Floyd County Courthouse, 1968.

9. Interview with Donald Rumsfeld, June 1, 2006.

10. Eve Edstrom, "OEO Overides [sic] Nunn's Veto," *Washington Post,* October 8, 1969, p. A8.

11. Tom Loftus, "Treva Turner Howell, 1923–2004; Breathitt Political Heir Dies Following Surgery," *Louisville Courier-Journal,* September 4, 2004, p. 1B.

12. Edstrom, "OEO Overides."

13. Rumsfeld interview, June 1, 2006.

14. Leroy F. Aarons, "Client Sues OEO to Regain Funds," *Washington Post,* January 18, 1970, p. 1.

15. Ibid.

16. Allen J. Matusow, *Nixon's Economy: Booms, Busts, Dollars, and Votes.* Lawrence: University of Kansas Press, 1998, p. 150.

17. Murray N. Rothbard, "The President's Economic Betrayal," *New York Times,* September 4, 1971, p. 21.

18. Rumsfeld interview, June 1, 2006.

19. James L. Rowe Jr, "Rumsfeld Leads New Boards," *Washington Post,* October 24, 1971, p. L1.

20. Cheney interview, April 27, 2006.

21. Memo from Cheney to Nixon, November 15, 1971.

22. Richard M. Nixon, November 15, 1971, in *Public Papers of the Presidents,* p. 1110.

23. James L. Rowe Jr., "Pay Board Criticized for Lack of Control," *Washington Post,* December 7, 1971, p. D6.

24. Memo from Cheney to Ken Cole, December 27, 1971.

25. Author interview with Vice President Dick Cheney, April 27, 2006.

26. Ibid.

27. Ibid.

28. Cheney interview, August 31, 2005.

29. Cheney interview, April 27, 2006.

CHAPTER FOUR: THE FORD YEARS

1. Philip Shabecoff, "McDonald's Told to Reduce Prices," *New York Times,* June 3, 1972.

2. Ford's debate-preparation videotape, September 19, 1976, Ford Library.

3. Dick Cheney's remarks, March 21, 1986, University of Virginia, Miller Center: "The Ford Presidency: Twenty-Two Intimate Perspectives of Gerald R. Ford."

4. Ibid.

5. Cheney interview, August 31, 2005.

6. Author interview with Donald Rumsfeld, June 1, 2006.

7. Cheney interview, August 31, 2005.

8. Michael Medved, *The Shadow Presidents: The Secret History of the Chief Executives and Their Top Aides*. New York: Times Books, 1979, p. 337.

9. Cheney interview, April 27, 2006.

10. Cheney's remarks, March 21, 1986.

11. Gerald R. Ford, *A Time to Heal*. New York: Harper and Row, 1979, p. 228.

12. Cheney interview, April 27, 2006.

13. Ford's "talking points" for November 15, 1974, senior staff meeting.

14. Ron Nessen, *It Sure Looks Different from the Inside*. Chicago: Playboy, 1978, p. 32.

15. John Hersey, *The President*. New York: Knopf, 1975, p. 7.

16. Cheney, memorandum for the president, undated.

17. Memo from Cheney to Jim Conner, January 16, 1975.

18. Seymour Hersh, "Huge CIA Operation Reported in U.S. Against Antiwar Forces, Other Dissidents in Nixon Years," *New York Times,* December 22, 1974, p. 1.

19. Nessen, *It Sure Looks Different*, p. 55.

20. Cheney's memo, December 27, 1974.

21. Frank J. Smist Jr., "Seeking a Piece of the Action: Congress and Its Intelligence Investigation of 1975–1976," in *Gerald R. Ford and the Politics of Post-Watergate America,* Vol. 2, ed. Bernard J. Firestone and Alexej Ugrinsky. Westport, Conn.: Greenwood, 1993, p. 463.

22. Cited in Kathryn Olmsted, "Reclaiming Executive Power: The Ford Administration's Response to the Intelligence Investigations," *Presidential Studies Quarterly* 26 (3), Summer 1996, pp. 725–737.

23. Seymour M. Hersh, "Submarines of U.S. Stage Spy Missions Inside Soviet Waters," *New York Times,* May 25, 1975, p. 1.

24. Cheney's memo, May 29, 1975.

25. "U.S. Is Said to Spy on Soviet Subs," *Washington Post,* May 25, 1975, p. A2.

26. Memo from Cheney to Rumsfeld, May 31, 1975.

27. Cheney interview, April 27, 2006.

28. Author interview with Vice President Dick Cheney, August 22, 2006.

29. Author interview with Ron Nessen, August 15, 2005.

30. Richard L. Madden, "Ford Says Indochina War Is Finished for America," *New York Times,* April 24, 1975, pp. 1, 19.

31. Nessen, *It Sure Looks Different*, p. 110.

32. Memo from Cheney to James Conner, January 22, 1975, Staff Secretary Files, White House memos, Cheney 1, Box 7.

33. Memo from Cheney to Jerry Jones, January 27, 1975, Cheney 6, Box 10.

34. Memo from Cheney to Rod Hills, July 14, 1975, GRF Library.

35. Ron Franck, "Cheney: Political Violence Is a Burden Mankind Must Carry," *Laramie Boomerang,* April 1, 1981.
36. Phyllis C. Richman and Lynne Cheney, "The People's Choice," *Washington Post*, March 9, 1975, p. 212.
37. Aaron Latham, "The Sunday Morning Massacre: A Murder-Suicide?" *New York,* December 22, 1976, p. 45.
38. John W. Finney, "Schlesinger Finds Arms Cuts 'Savage,'" *New York Times,* October 21, 1975, p. 13.
39. Cheney interview, April 27, 2006.
40. Ibid.
41. Nessen, press conference transcript, November 4, 1975.
42. Nessen files, GRF Library.
43. Charles Mohr, "New Chief Assistant," *New York Times*, November 5, 1975, p. 20.
44. Lou Cannon, "Stepping Out of Rumsfeld's Shadow," *Washington Post*, November 6, 1975, p. A3.
45. Author interview with Joe Meyer, August 12, 2005.
46. Jim Little, letter to Cheney, courtesy of Jim Little.
47. Ibid.
48. Cheney interview, August 31, 2005.
49. Nessen interview, August 15, 2005.
50. Cannon, "Stepping Out."
51. Medved, *Shadow Presidents*.
52. Notes from Cheney files, GRF Library.
53. Dick Cheney's remarks, "The Ford White House: A Miller Center Conference," University of Virginia, April 23, 1977, p. 76.
54. Nessen interview, August 15, 2005.
55. Interview with Dick Cheney, *Face the Nation,* CBS News, January 4, 1976.
56. Ibid.
57. Dick Cheney's remarks, "The Ford White House," p. 80.
58. Memo from Cheney to Robert Hartmann, Bob Hartmann Papers. Box: White House Speech Files. Folder: "1/19/76 First Draft Dick Cheney's Comments," GRF Library.
59. Cheney interview, April 27, 2006.
60. Medved, *Shadow Presidents*, p. 342.
61. Cheney interview, April 27, 2006.
62. Medved, *Shadow Presidents*, p. 341.
63. Cheney interview, April 27, 2006.
64. Ibid.
65. Ibid.
66. Nessen, p. 230; see also, Medved, p. 343; James Main, *Rise of the Vulcans: The History of Bush's War Cabinet.* New York: Penguin Books, 2004, p. 73.
67. Cheney interview, April 27, 2006.
68. Ibid.

69. Ibid.
70. Ibid.
71. Rowland Evans and Robert Novak, "Evans-Novak Political Report," No. 266, August 20, 1976, p. 1.
72. Videotape of Ford's debate preparation, Session 1, GRF Library.
73. Cheney interview, April 27, 2006.
74. Ibid.
75. Ibid.
76. Ibid.
77. James M. Cannon, *Time and Chance: Gerald Ford's Appointment with History.* New York: HarperCollins, 1994, p. 408.
78. Ibid., p. 409.
79. Cheney interview, April 27, 2006.
80. Interview with Jim Naughton, n.d.
81. Cheney interview, April 27, 2006.
82. Naughton interview.
83. Samuel Kernell and Samuel L. Popkin, eds., *Chief of Staff: Twenty-Five Years of Managing the Presidency.* Berkeley: University of California Press, 1986, p. 188.
84. Robert T. Hartmann, *Palace Politics: An Inside Account of the Ford Years.* New York: McGraw Hill, p. 283.
85. Jerry Jones, interview with Stephen Wayne, December 17, 1976.
86. Bob Teeter, interview with David Horrocks, May 5, 1997.
87. Gerald R. Ford, *A Time to Heal: The Autobiography of Gerald R. Ford.* New York: Harper and Row, p. 442.
88. James E. Connor, memo, November 2, 1976.

CHAPTER FIVE: ON THE BALLOT
1. Author interview with Vice President Dick Cheney, August 9, 2006.
2. Author interview with Vice President Dick Cheney, August 22, 2006.
3. Ibid.
4. Cheney interview, August 9, 2006.
5. Ibid.
6. AP, "Roncalio Retiring after This Term," *Washington Post*, September 19, 1977, p. A8.
7. Cheney interview, August 9, 2006.
8. Ibid.
9. Author interview with Kathy Morton, n.d.
10. Cheney interview, August 9, 2006.
11. Ibid.
12. Bruce McCormack, "Cheney Will Run," *Casper Star-Tribune*, December 15, 1977.
13. Dick Cheney for Congress Committee, American Heritage Center, University of Wyoming.

14. Cheney interview, August 9, 2006.
15. Interview with John Vandel, August 9, 2005.
16. Author interview with Bob Gardner, August 16, 2006.
17. Cheney interview, August 9, 2006.
18. Author interview with Mark Vincent, July 15, 2006.
19. Cheney interview, August 9, 2006.
20. Author interview with Vice President Dick Cheney, April 27, 2006.
21. Cheney interview, August 9, 2006.
22. "Cheney Back Home in Casper," *Casper Star-Tribune*, No. 181, June 30, 1978, p. 1.
23. Cheney interview, August 9, 2006.
24. Ibid.
25. Author interview with Bill Bagley, October 4, 2006.
26. Wyoma Haskins, "Congress/Cheney vs. Bagley," *Branding Iron*, Special Issue: Voter's Guide, November 2, 1978.
27. Bagley interview, October 4, 2006.
28. Ibid.
29. Cheney interview, August 9, 2006.
30. Ibid.
31. Author interview with David Gribbin, July 6, 2006.
32. Ibid.
33. George F. Will, "Politics—As Steiger Practiced It," *Washington Post*, December 7, 1978, p. A23.
34. Ibid., p. 107. James Reston, "Steiger of Wisconsin," *New York Times*, December 8, 1978, p. A29.
35. Ibid., p. 2. David S. Broder, "'Stupendous Steiger,'" *Washington Post*, December 10, 1978, p. C7.
36. Cheney interview, August 9, 2006.
37. Ibid.
38. Author interview with Bill Thomas, October 27, 2006.
39. Ibid.
40. Mary Russell, "New House Republicans Try to Break Out of Stereotype," *Washington Post*, February 18, 1979, p. M1.
41. Gribben interview, July 6, 2006.
42. Ibid.
43. Ibid.
44. Bill Sniffin, "Skiers, Cutters and the Homeliest Stewardesses," *Lander Journal*, February 21, 1980.
45. Cheney interview, August 9, 2006.
46. Bernard Horton, "Cheney," *Cheyenne Eagle*, May 18, 1980.
47. Cheney interview, August 9, 2006.
48. William Branigan, "Iranians Seize U.S. Mission, Ask Shah's Return for Trial," *Washington Post*, November 5, 1979, p. A1.
49. President Jimmy Carter, Remarks to the nation, January 4, 1980.

50. Douglas MacEachin, "Predicting the Soviet Invasion of Afghanistan: The Intelligence Community's Record," Center for the Study of Intelligence, CIA.

51. James M. Flinchum, "Backing Reagan, Cheney Sees No Way Carter Can Win," *Cheyenne Tribune*, April 7, 1980.

52. Interview with Representative Richard B. Cheney, *Pinedale Roundup*, April 17, 1980.

53. Ibid.

54. Cheney interview, August 9, 2006.

55. UPI, "Judge: Contract Was 'Illegal, Improper and Non-Binding,'" March 6, 1984.

56. "Write-In Candidate against Cheney Eyed," *Cheyenne Tribune/UPI*, July 18, 1980.

57. Ibid.

58. Ibid.

59. Dennis Curran, "Rogers Wins Wyoming Democratic Primary," Associated Press, September 10, 1980.

60. "Write-In Candidate."

61. Bob Messenger, *Casper Star-Tribune*, No. 255, September 11, 1980.

62. Ibid.

63. "Iowa Is Heard From," Editorial, *Washington Post*, January 23, 1980, p. A22.

64. Adam Clymer, "Ford Declares Reagan Can't Win," *New York Times*, March 2, 1980, pp. 1, 17; Wayne King, "Reagan Challenges Ford to Join Him in Primary Trail," *New York Times*, March 3, 1980, p. A1.

65. Harry F. Rosenthal, "Ford—By April 1 You'll Know," Associated Press, March 4, 1980.

66. Lou Cannon and Edward Walsh, "While Ford Maneuvers, Reagan Fumes," *Washington Post*, March 11, 1980, p. A3.

67. Cheney interview, August 9, 2006.

68. Ibid.

69. Ibid.

70. Ibid.

71. Ibid.

72. Haynes Johnson, David S. Broder, Lou Cannon, Bill Peterson, Martin Schram, and Felicity Barringer, "The Republicans in Detroit: The Cement Just Wouldn't Set on GOP's Alliance; Reagan-Ford: The Dissolving of a Tenuous Alliance," *Washington Post*, July 18, 1980, p. A1.

73. Ibid.

74. Cheney interview, August 9, 2006.

75. Ibid.

76. "The Republicans in Detroit."

77. Cheney interview, August 9, 2006.

78. Cheney interview, April 27, 2006.

79. Cheney interview, August 9, 2006.
80. Cheney interview, April 27, 2006.
81. *Almanac of American Politics.* Boston: Gambit, 1982, p. 1209.
82. Cheney interview, August 9, 2006.
83. Author interview with Bob Michel, October 19, 2006.
84. Ibid.
85. Ibid.
86. "The New House Leaders: Bipartisan Compromisers," *National Journal,* December 13, 1980, p. 2136.
87. Howard Fineman, "Congress Gropes to a Close," *Newsweek*, December 22, 1980, p. 26.
88. "Inside Report," *Christian Science Monitor,* December 12, 1980, p. 2.
89. Ronald Koven, "U.S. Envoy Says Hostage Accord Sets 'Very Dangerous Precedent,'" *Washington Post,* January 22, 1981, p. A21.
90. "Iranian Situation—Cheney Says—'Let Dust Settle,'" Associated Press/ *Sheridan Press*, January 27, 1981. Associated Press, "Cheney: Be Wary of Revenge," *Casper Star-Tribune,* No. 25, January 25, 1981, pp. A1, A3.
91. Hedrick Smith, "Reagan Putting His Stamp on U.S. Policies: At Home and Abroad, a Change of Course," *New York Times*, January 30, 1981, p. A11.

Chapter Six: Leadership

1. Author interview with Vice President Dick Cheney, August 9, 2006.
2. Author interview with Leon Panetta, October 23, 2006.
3. Transcript, "Revitalizing America: What Are the Possibilities?," American Enterprise Institute (Forum 49), December 9, 1980.
4. James A. Baker III, *The Politics of Diplomacy: Revolution, War, and Peace, 1989–1992.* New York: Putnam, 1995, p. 22.
5. Ibid., pp. 22–23.
6. James A. Baker III, *Work Hard, Study . . . and Keep Out of Politics! Adventures and Lessons from an Unexpected Public Life.* New York: Putnam, 2006, p. 137.
7. Ibid.
8. Cheney interview, August 9, 2006.
9. Author interview with Jim Steen, January 20, 2007.
10. Edward Walsh, "$500,000 Ad Push Set for Next Week on Reagan Tax Bill," *Washington Post,* July 24, 1981, p. A2.
11. Tom Raum, "Republicans Start Media Blitz," Associated Press, July 23, 1981.
12. Ronald Reagan's remarks at meeting of House Republican Conference, July 24, 1981, *Public Papers of the Presidents: 1981,* p. 658.
13. Dick Kirschten, "The Domestic President," *National Journal* 13, no. 44, October 31, 1981, p. 1953.
14. Cheney notes, courtesy Jim Steen.

15. Newt Gingrich, testimony before the House Permanent Select Committee on Intelligence, August 11, 2004.
16. Robert Coram, *Boyd: The Fighter Pilot Who Changed the Art of War.* New York: Back Bay/Little, Brown, 2004, p. 355.
17. Angus Deming, "A Vote against Israel," *Newsweek*, June 29, 1981, p. 40.
18. Author interview with Vice President Dick Cheney, August 9, 2006.
19. Ibid.
20. Cheney interview, August 9, 2006.
21. Author interview with Merritt Benson, August 10, 2005.
22. Ibid.
23. Ibid.
24. Ibid.
25. Cheney interview, August 9, 2006.
26. Ibid.
27. Adam Clymer, "How to Keep Winning the West Is a Challenge," *New York Times*, March 7, 1982, p. 2.
28. Cheney interview, August 9, 2006.
29. Ibid.
30. William Chapman, "West's Docile Conservatives Rebel against Big Oil in Wilderness," *Washington Post*, February 7, 1982, p. A21.
31. Benson interview, August 10, 2005.
32. Ibid.
33. Ibid.
34. Ronald Reagan, Address to the nation, October 25, 1983.
35. Steven V. Roberts, "O'Neill Criticizes President; War Powers Act Is Invoked," *New York Times*, October 29, 1983.
36. Dan Balz and Thomas B. Edsall, "The Invasion of Grenada; GOP Rallies around Reagan; Democrats Divided on Grenada," *Washington Post*, October 26, 1983.
37. Quoted in Jacob K. Javits, "War Powers Reconsidered," *Foreign Affairs* (Fall 1985), p. 137.
38. Balz and Edsall, "The Invasion of Grenada."
39. Ward Sinclair, "Student Evacuees Return, Praise U.S. Military," *Washington Post*, October 27, 1983, p. A1.
40. Ibid.
41. Philip J. Hilts, "565,000 Jam ABC's Phone Lines," *Washington Post*, October 30, 1983, p. A18. See also Barry Sussman, "Reagan's Talk Gains Support for Policies," *Washington Post*, October 30, 1983, p. A1.
42. Ronald Reagan, Address to the nation, October 27, 1983.
43. Bernard Weinraub, "100 Bodies Are Reported Found at a Training Camp in Grenada," *New York Times*, November 7, 1983, p. A1.
44. Hedrick Smith, "O'Neill Now Calls Grenada Invasion 'Justified' Action," *New York Times*, November 9, 1983, p. A1.
45. Weinraub, "100 Bodies Are Reported Found."

46. Ibid.
47. Transcript, "War Powers and the Constitution," American Enterprise Institute, Forum 61, December 6, 1983.
48. Ibid.
49. A subsequent edition was published in 1996 and includes a chapter on Newt Gingrich and the Republican takeover of Congress in 1994.
50. Richard B. Cheney and Lynne V. Cheney, *Kings of the Hill: How Nine Powerful Men Changed the Course of American History*. New York: Touchstone, 1983, p. xii.
51. Eric Redman, "Master of the House," *Washington Post*, July 17, 1983, p. 11.
52. Ken Huff, "In Politics and Now in Print, Wyoming's Dick and Lynne Cheney Go a Country Mile for Each Other," *People,* June 27, 1983, p. 82.
53. Author interview with Dennis Hastert, January 27, 2007.
54. Cheney interview, August 9, 2006.
55. Author interview with Mark Vincent, July 15, 2006.
56. Cheney interview, August 9, 2006.
57. Author interview with David Gribbin, July 6, 2006.
58. Author interview with David Addington, August 10, 2006.
59. Interview with Richard Kerr, "The Darkside," *Frontline,* PBS, January 25, 2006.
60. Author interview with Lee Hamilton, September 2, 2006.
61. Ibid.
62. Ibid.
63. Author interview with Vice President Dick Cheney, November 21, 2006.
64. Baker, *The Politics of Diplomacy*, p. 24.
65. Gloria Borger, "Revolt of the 'Robots,'" *Newsweek,* December 23, 1985, p. 26.
66. Author interview with Jack Kemp, November 29, 2006.
67. Author interview with Bob Michel, October 19, 2006.
68. Dick Cheney, speaking on the Tax Reform Act of 1986, 99th Congress, 2nd session, *Congressional Record* 132 (128), September 25, 1986.
69. Michel interview, October 19, 2006.
70. Donald Rothberg, "Tax Law's Many Heroes—and No Political Losers?" Associated Press, August 20, 1986. David T. Cook, "White House Makes Final Tax-Reform Push," *Christian Science Monitor,* December 9, 1985, p. 1.
71. Jack Kemp, "Supply-Sider for Treasury Secretary," *Seattle Post-Intelligencer*, August 2, 2000.
72. *Green River Star,* September 1, 1987.
73. Author interview with Al Simpson, August 18, 2006.
74. Cheney interview, August 22, 2006.
75. Simpson interview, August 18, 2006.
76. Ibid.
77. Cheney's confirmation hearing, March 14, 1989.
78. Cheney interview, November 21, 2006.

79. David S. Broder, "Good News Cheney," *Washington Post*, March 15, 1989, p. A23.
80. Dick Cheney, "Making Appropriations for Aid to Nicaragua," 99th Congress, 1st session, *Congressional Record* 131 (49), April 24, 1985.
81. Walter Pincus, "Shultz Protested Iran Deal; U.S. Reassured Iraq of Neutrality in Persian Gulf War," *Washington Post*, November 7, 1986, p. A1.
82. George P. Shultz, *Turmoil and Triumph: My Years as Secretary of State*. New York: Scribner, 1993, p. 816.
83. Cheney interview, August 9, 2006.
84. Cheney interview, November 21, 2006.
85. Helen Dewar and Edward Walsh, "Hill Sees Policy in 'Disarray'; Members of Both Parties Express Shock, Plan Probes," *Washington Post*, November 26, 1986, p. A1.
86. David S. Broder, "Secrecy Trips a President," *Washington Post*, November 26, 1986, p. A1.
87. Cheney interview, November 21, 2006.
88. Ibid.
89. Cheney interview, November 21, 2006.
90. Mary McGrory, "Hill to Reagan: All Is Forgiven," *Washington Post*, August 4, 1987, p. A2.
91. Sam Nunn's letter to Cheney, August 10, 1987.
92. Report of the Congressional Committees Investigating the Iran-Contra Affair, Minority Report, p. 438.
93. Ibid., p. 437.
94. Hamilton interview, September 2, 2006.
95. "The Best of Capitol Hill," *U.S. News and World Report*, December 21, 1987, p. 47.
96. Cheney interview, November 21, 2006.
97. Ibid.
98. Michel interview, October 19, 2006.
99. Ibid.
100. Cheney interview, August 9, 2006.

CHAPTER SEVEN: AT WAR

1. Author interview with Vice President Dick Cheney, August 9, 2006.
2. Ibid.
3. Ibid.
4. Ibid.
5. Ibid.
6. Ibid.
7. Audio recording provided by Jon Maltese, March 10, 1989.
8. Ibid.
9. Cheney interview, August 9, 2006.

10. David S. Broder, "Good News Cheney," *Washington Post*, March 15, 1989, p. A23.
11. Ibid.
12. U.S. Senate, hearing of the Senate Armed Services Committee, March 14, 1989.
13. Ibid.
14. Ibid.
15. Ibid.
16. Ibid.
17. Donna Cassata, "Senate Panel Opens Cheney Hearings, Swift Action Promised," Associated Press, March 14, 1989.
18. Cheney interview, August 9, 2006.
19. Ibid.
20. Ibid.
21. Ibid.
22. Ibid.
23. Ibid.
24. Ibid.
25. Richard Halloran, "Scramble On to Succeed Chairman of Joint Chiefs," *New York Times*, August 7, 1989, p. A12.
26. Colin Powell, *My American Journey.* New York: Ballantine Books, 1995, pp. 394–395.
27. Ibid.
28. Cheney's speech to World Affairs Council, Pittsburgh, Pa., October 30, 1990.
29. Andrew Rosenthal, "Latest Pentagon Report Lifts Shadow of Soviet Menace, a Bit," *New York Times*, September 28, 1989, p. A10.
30. David C. Morrisson, "Pentagon's Boss Still Untested?" *National Journal* 21, no. 37, September 16, 1989, p. 2267.
31. Cheney interview, August 9, 2006.
32. William Branigan, "As Observer, Carter Proves Acute," *Washington Post*, May 10, 1989, p. A23.
33. Author interview with John Murtha, December 4, 2006.
34. Thomas Donnelly, Margaret Roth, and Caleb Baker, *Operation Just Cause: The Storming of Panama.* New York: Lexington Books, 1991, p. 97.
35. Cheney interview, August 9, 2006.
36. Ibid.
37. Ibid.
38. Ibid.
39. See Bob Woodward, *The Commanders* (New York: Touchstone, 1991), and Michael R. Gordon and Bernard E. Trainor, *The Generals' War: The Inside Story of the Conflict in the Gulf* (New York: Back Bay Books, 1995), pp. 4–30.

40. Stewart Powell, "Cheney Warns Iraq against Any Threats to Persian Gulf," *Seattle Post-Intelligencer,* July 20, 1990, p. E4.
41. Author interview with Michael McConnell, November 22, 2006.
42. Ibid.
43. Interview with Dick Cheney, *The MacNeil/Lehrer NewsHour,* PBS, August 1, 1990.
44. Gordon and Trainor, *The Generals' War,* p. 29.
45. Cheney interview, August 9, 2006.
46. Ibid.
47. Ibid.
48. Ibid.
49. Woodward, *The Commanders,* p. 271; see also, Peter W. Wilson and Douglas F. Graham, *Saudi Arabia: The Coming Storm* (Armonk, N.Y.: M. E. Sharpe, 1994), pp. 112–113.
50. Cheney interview, August 9, 2006.
51. Ibid.
52. Ibid.
53. Ibid.
54. Ibid.
55. Ibid.
56. Henry Rowen, "Inchon in the Desert: My Rejected Plan," *National Interest,* Summer 1995.
57. Cheney interview, August 9, 2006.
58. Author interview with Vice President Dick Cheney, November 29, 2006.
59. Rick Atkinson, "U.S. to Rely on Air Strikes If War Erupts," *Washington Post,* September 16, 1990.
60. Ibid.
61. Ibid.
62. Author interview with Michael J. Duggan, November 12, 2006.
63. Cheney's press briefing, Department of Defense, September 17, 1990.
64. McConnell interview, November 22, 2006.
65. Author interview with David Gribben, July 6, 2006.
66. Donna Cassata, "Committee Raps Delay in Duggan Retirement," Associated Press, October 29, 1990.
67. Gribben interview, July 6, 2006.
68. *Meet the Press,* NBC, November 18, 1990.
69. Author interview with Porter Goss, October 26, 2006.
70. Interview with F. Michael Maloof, *Frontline,* PBS, January 10, 2006.
71. *Face the Nation,* CBS, November 25, 1990.
72. Ibid.
73. Dick Cheney, Defense Department briefing, January 16, 1991; "Cheney's Remarks on Attack on Iraq," *New York Times,* January 17, 1991, p. A17.

74. Anthony O. Miller, "Iraq Makes Unconditional Withdrawal Offer," UPI, February 15, 1991.
75. Gordon and Trainor, *The Generals' War*, p. 446.
76. Ibid, p. 449.
77. Benson interview, August 10, 2005.
78. Dick Cheney, address to the Discovery Institute, Seattle, Wa., August 14, 1992.
79. Al Gore, remarks, Washington, D.C., September 29, 1992.
80. Ibid.
81. Ibid.
82. Ibid.

CHAPTER EIGHT: PRESIDENT CHENEY?
 1. *Larry King Live*, CNN, January 27, 1993.
 2. John J. Glisch, "For Gays, a Fight to Serve Their Country," *Orlando Sentinel*, October 25, 1992.
 3. Memorandum for the secretary of defense, "Ending Discrimination on the Basis of Sexual Orientation in the Armed Forces," January 30, 1993.
 4. *Larry King Live*, CNN, January 27, 1993.
 5. Ibid.
 6. Mary Cheney, *Now It's My Turn: A Daughter's Chronicle of Political Life*. New York: Threshold Editions, 2006.
 7. ABC News, *This Week with David Brinkley*, August 4, 1991.
 8. *This Morning*, CBS, August 6, 1991.
 9. Michael Duffy, "That Sinking Feeling," *Time*, June 7, 1993.
10. Author interview with Vice President Dick Cheney, November 21, 2006.
11. Ibid.
12. Dick Cheney, 1993 Francis Boyer Lecture, American Enterprise Annual Dinner, Washington, D.C., December 8, 1993.
13. Gloria Borger, "Dick Cheney at the Starting Gate," *U.S. News and World Report*, October 25, 1993.
14. Morton M. Kondracke, "On Cheney's Trail: Will He Run in '96? Could He Win?" *Roll Call*, July 19, 1993.
15. Marianne Means, "GOP Eager Beavers Champing at Bit," *Seattle Post-Intelligencer*, July 4, 1993, p. E2.
16. David Broder, "Cheney Bides His Time," *Washington Post*, June 20, 1993, p. C7.
17. Dick Cheney, "Inadequate Strategy, Inadequate Resources: Bill Clinton's National Security Deficit," *CommonSense*, Fall 1994.
18. Dick Cheney address, "American Leadership in the New Security Environment," July 27, 1994.
19. Ibid.
20. George Rodrigues, "Can Former Defense Chief Fire Up GOP Supporters?," *Dallas Morning News*, December 18, 1994, p. 1J.

21. Author interview with Vice President Dick Cheney, n.d.
22. "Cheney Won't Run in '96," UPI, January 3, 1995.
23. Author interview with Joseph Duggan, n.d.
24. "Poll: Dole Still '96 GOP Frontrunner," UPI, January 2, 1995.
25. Author interview with Vice President Dick Cheney, November 29, 2006.
26. Kondracke, "On Cheney's Trail."
27. Author interview with Tom Cruikshank, December 4, 2006.
28. Ibid.
29. Author interview with Vice President Dick Cheney, August 22, 2006.
30. Jason Marsden, "Cheney Out of Politics—Mostly," *Casper Star-Tribune*, May 26, 1996.
31. Borger, "Cheney at the Starting Gate."
32. *Meet the Press*, NBC, September 16, 2001.
33. Patrick Crow, "U.S. Petroleum Firms Hit Hard by Washington's Unilateral Sanctions," *Oil and Gas Journal*, May 5, 1997, p. 37.
34. Dick Cheney, address to Collateral Damage Conference, Cato Institute, "Defending Liberty in a Global Economy," June 23, 1998.
35. Cheney interview, November 21, 2006.
36. Ibid.
37. Toby T. Gati, assistant secretary of state for intelligence and research, Testimony before the Senate Select Committee on Intelligence, February 5, 1997.
38. "Cheney Backs Iran Investment," *Hart's Middle East Oil and Gas* 2, no. 13, June 27, 2000.
39. Cheney interview, November 21, 2006.

Chapter Nine: Another George Bush

1. Author interview with Vice President Dick Cheney, August 22, 2006.
2. Wayne Slater, "Ex-President's Allies Generous to Governor," *Dallas Morning News*, February 21, 1998.
3. Ken Herman, "Bush to Take Step Toward 2000 Campaign," *Austin American-Statesman*, February 25, 1999, p. A1.
4. Bob Novak, *Inside Politics*, CNN, February 24, 1999.
5. Author interview with President George W. Bush, December 11, 2006.
6. Frank Bruni, "Bush Vows Money and Support for Military," *New York Times,* September 24, 1999, p. A22.
7. Dick Cheney, interview with Brit Hume, *Special Report with Brit Hume,* Fox News Channel, September 23, 1999.
8. Paul Bedard, "Veep Watch," *U.S. News and World Report,* October 4, 1999, p. 9.
9. Cheney interview, August 22, 2006.
10. Bush interview, December 11, 2006.
11. Cheney interview, August 22, 2006.
12. Ibid.

13. "GOP Veepstakes: Early List Includes Most of RGA," *National Journal's Hotline,* March 13, 2000; Paul West, "As Gore, Bush Sit Pretty, Aspirants Vie to Be No. 2," *Baltimore Sun,* March 11, 2000, p. 1A.

14. Cheney interview, August 22, 2006.

15. Ibid.

16. Glen Johnson, "Bush Seeks Help in VP Search," Associated Press, April 25, 2000.

17. Trent Lott, press conference, Washington, D.C., July 25, 2000.

18. Cheney interview, August 22, 2006.

19. Bush interview, December 11, 2006.

20. Cheney interview, August 22, 2006.

21. Dick Cheney's address to the World Petroleum Congress, Calgary, Canada, June 14, 2000.

22. Jeffrey Jones, "Cheney Urges End to Iran Sanctions," Reuters, June 15, 2000. See also Brent Jang, "U.S. Urged to Let Oil Firms Back into Iran," *Globe and Mail,* June 15, 2000, p. B11.

23. "Cheney Backs Iran Investment," *Hart's Middle East Oil and Gas* 2, no. 13, June 27, 2000.

24. Cheney interview, August 22, 2006.

25. *Inside Politics,* CNN, July 3, 2000.

26. Cheney interview, August 22, 2006.

27. Ibid.

28. Dick Cheney, remarks on the future of Azerbaijan, American Enterprise Institute, February 18, 1997.

29. For more on Lisa Myers's account, see *Meet the Press,* NBC, July 23, 2000.

30. Paul Begala, *Equal Time,* MSNBC, July 21, 2000.

31. Author interview with John McConnell, July 31, 2006.

32. Anne E. Kornblut, "Weeks of Wooing by Bush Persuaded Cheney," *Boston Globe,* July 26, 2000, p. A1.

33. Cheney interview, August 22, 2006.

34. "The Choice for a Running Mate; Excerpts from Statements by Bush and His Running Mate," *New York Times,* July 26, 2000, p. A18.

35. Cheney interview, August 22, 2006.

36. Dick Cheney, interview with Larry King, *Larry King Live,* CNN, July 25, 2000.

37. Richard L. Berke, "A Safe Pick Is Revealing," *New York Times,* July 26, 2000, p. A1.

38. Tom Daschle, remarks at media stakeout following Senate policy luncheons, U.S. Capitol, Washington, D.C., July 25, 2000.

39. Ron Hutcheson and Steve Thomma, *Fort Worth Star-Telegram,* July 26, 2000.

40. Chris Matthews, *Hardball,* MSNBC, July 25, 2000.

41. Joe Battenfeld and Andrew Miga, "Dems Take Aim at Cheney Record," *Boston Herald,* July 31, 2000.

42. For more, see Stephen F. Hayes, "Clinton Campaigned for Mandela's Jailer," *National Review Online,* July 31, 2000.
43. Cheney interview, August 22, 2006.
44. Ibid.
45. Bush interview, December 11, 2006.
46. Cheney interview, August 22, 2006.
47. Ibid.
48. Ibid.
49. Karen Gullo, Associated Press, November 7, 2000.
50. Author interview with Michael McConnell, November 22, 2006.
51. Stephen Ohlemacher, "Cheney Blasts Clinton Administration, Says U.S. Military Readiness Has Slipped," *Cleveland Plain Dealer*, October 13, 2000, p. 2A.
52. Ceci Connolly and Mike Allen, "Crises Take Precedence with the Candidates," *Washington Post*, October 13, 2000, p. A30.
53. Cheney interview, August 22, 2006.
54. Ibid.
55. Ibid.
56. Ibid.
57. Ibid.
58. Ibid.
59. *Larry King Live*, CNN, November 22, 2000.
60. James A. Baker III, *Work Hard, Study . . . and Keep Out of Politics! Adventures and Lessons from an Unexpected Public Life*. New York: Putnam, 2006, p. 361.

CHAPTER TEN: THE BUSH ADMINISTRATION

1. Author interview with Vice President Dick Cheney, August 22, 2006.
2. Ibid.
3. Ibid.
4. Ibid.
5. Peter Jennings, *Special Report: The 2001 Inauguration*, ABC News, January 20, 2001.
6. Tom Brokaw, *Special Report: Inauguration of George W. Bush*, NBC News, January 20, 2001.
7. Author interview with Vice President Dick Cheney, April 27, 2006.
8. Episode 494, season 26, *Saturday Night Live*, NBC, December 16, 2000.
9. Andrea Mitchell, *NBC Nightly News*, NBC, January 23, 2001.
10. In the same poll, conducted February 7–8, 2001, Bush's favorability rating was 60 percent; 28 percent of the respondents expressed an unfavorable view. A poll by Fox News/Opinion Dynamics, taken the same week, rated Cheney even higher: 61 percent favorable and 18 percent unfavorable.
11. Author interview with President George W. Bush, December 11, 2006.

12. Cheney interview, April 27, 2006.

13. Bush interview, December 11, 2006.

14. Cheney interview, April 27, 2006.

15. Cheney interview, August 22, 2006.

16. Ibid.

17. Dick Cheney, memo to Donald Rumsfeld, "Decontrol of Oil Prices," May 27, 1975.

18. Cheney memo to Donald Rumsfeld, "Decontrol of Oil Prices," May 27, 1975.

19. Author interview with George W. Bush, December 11, 2006.

20. Author interview with Andrew Lundquist, September 20, 2006.

21. Ibid.

22. David Sanger, "President Offers Plan to Promote Oil Exploration," *New York Times*, January 30, 2001.

23. Joel Connelly, "Cheney Blasts Energy Policy; In Spokane, He Joins Wave of Campaigners Stumping in Northwest," *Seattle Post-Intelligencer*, October 25, 2000.

24. Ron Hutcheson, "Bush Says Election Is Choice between Energy Costs, Energy Security," Knight Ridder Washington Bureau, October 14, 2000.

25. Lundquist interview, September 20, 2006.

26. Letter from Dick Cheney to the president, May 16, 2001.

27. Tom Curry, "Cheney's Win: More Than Legal," MSNBC, June 24, 2004.

28. Katharine Q. Seelye, "Nuclear Power Gains in Status after Lobbying," *New York Times*, May 23, 2001.

29. Spencer Abraham, remarks to the U.S. Chamber of Commerce, National Energy Summit, Washington, D.C., March 19, 2001.

30. Ibid.

31. John D. Dingell and Henry A. Waxman, letter to Honorable David Walker, April 19, 2001.

32. Frank Bruni, "Bush Has Tough Words and Rough Enunciation for Iraq's Chief," *New York Times*, December 4, 1999, p. A12.

33. Bill Clinton, remarks at the Pentagon, Arlington, Va., February 17, 1998.

34. White House press release, "Domestic Preparedness Against Weapons of Mass Destruction," May 8, 2001.

35. Nicholas Lemann, "The Quiet Man," *New Yorker*, May 7, 2001.

36. Author interview with Mark Vincent, July 15, 2006.

37. See Staff Statement no. 8, "National Policy Coordination," Eighth Public Hearing of 9/11 Commission, Washington, D.C., March 22–23, 2004; Final Report of the National Commission on Terrorist Attacks upon the United States, pp. 190–197.

38. "Accounting for the Cole Attack," Editorial, *Washington Post*, January 10, 2001, p. A18.

39. Richard A. Clarke, "Presidential Policy Initiative/Review—The al-Qaida

Network," Memorandum to Condoleezza Rice, January 25, 2001, National Security Archives, George Washington University.

40. Cheney interview, August 9, 2006.
41. Cheney interview, August 22, 2006.
42. Ibid.
43. Cheney interview, August 9, 2006.
44. Statement by the GAO comptroller David M. Walker, September 7, 2001.
45. Lundquist interview, September 20, 2006.
46. Ellen Nakashima and Dan Eggen, "White House Seeks to Restore Its Privileges," *Washington Post*, September 10, 2001, p. A02.
47. Ibid.

CHAPTER ELEVEN: SEPTEMBER 11, 2001

1. "Lamb on Barbecue Menu as U.S. Welcomes Howard," *Gold Coast Bulletin,* September 11, 2001.
2. Ed Anderson, "Tax Relief Still a Priority, Cheney Says"; *New Orleans Times-Picayune*, September 11, 2001, p. 3.
3. Jack Brammer, "Vice President Cheney Speaks to Southern Governors," *Lexington Herald-Leader*, September 11, 2001.
4. Anderson, "Tax Relief Still a Priority, Cheney Says."
5. Mark R. Chellgren, "Governors Will Be Receptive Audience for Cheney Energy Speech," Associated Press, September 10, 2001.
6. *NBC Nightly News,* NBC, September 10, 2001.
7. Author interview with John McConnell, July 31, 2006.
8. Ibid.
9. Author interview with Jennifer Mayfield, July 18, 2006.
10. McConnell interview, July 31, 2006.
11. Ibid.
12. Author interview with Vice President Dick Cheney, August 9, 2006.
13. *Newsweek*, "The Day That Changed America," December 2001.
14. Ari Fleischer, *Taking Heat: The President, the Press, and My Years in the White House*. New York: William Morrow, 2005, p. 42.
15. Cheney interview, August 9, 2006.
16. Author interview with Dennis Hastert, n.d.
17. Cheney interview, August 9, 2006.
18. Laura Blumenfeld, "Norman Mineta: Shipping Out," *Washington Post*, June 29, 2006, p. A25.
19. Cheney interview, August 9, 2006.
20. Who first gave the "shoot-down" order has been one of the enduring controversies of 9/11. According to the report of the 9/11 Commission, there is no record of this phone call: "Among the sources that reflect other important events of that morning, there is no documentary evidence of this call, but the relevant sources are incomplete." Cheney

and Rice say they remember the conversation with Bush, and Bush described it in his interview with the commissioners. Cheney's military aide remembers the vice president speaking with the president just after entering the shelter, but did not hear the conversation. The 9/11 Commission reports that Josh Bolten, who was then deputy chief of staff, had not heard the call and later suggested to Cheney that he ought to call Bush for approval of the shoot-down order. Bolten disputes this. He says that he suggested Cheney call Bush not because the vice president had overstepped his authority, but as a reminder that they should notify the president.

21. T. Trent Gegax, Arlan Campo-Flores, Alan Zarembo, Gretel Kovach, Evan Thomas, "The Day That Changed America," *Newsweek,* December 31, 2001.
22. Author interview with Josh Bolten, February 27, 2007.
23. Cheney interview, August 9, 2006.
24. Author interview with Condoleezza Rice, August 15, 2006.
25. Ibid.
26. Cheney interview, August 9, 2006.
27. From a transcript of the Air Threat Conference Call, as reported in the 9/11 Commission Report, p. 49.
28. Hastert interview, January 27, 2007.
29. Ibid.
30. James Carney and Judy Keen, White House Pool Report, September 11, 2001.
31. Susan Page, "Crisis Presents Defining Moment for Bush," *USA Today,* September 12, 2001, p. 16A.
32. *Newsweek,* December 31, 2001.
33. Rice interview, August 15, 2006.
34. Author interview with David Addington, n.d.
35. Ari Fleischer, Report to the Pool, 5:30 PM, September 11, 2001.
36. Bob Woodward, *Bush at War.* New York: Simon & Schuster, 2002, pp. 26–27.
37. Ibid., p. 27.
38. McConnell interview, July 31, 2006.
39. Addington interview, n.d.
40. Ibid.
41. Cheney interview, August 9, 2006.

CHAPTER TWELVE: SECURE, UNDISCLOSED

1. *NewsHour with Jim Lehrer,* PBS, October 12, 2001.
2. Author interview with Vice President Dick Cheney, August 9, 2006.
3. Author interview with Condoleezza Rice, August 15, 2006.
4. Ibid.
5. Bob Woodward, *Bush at War.* New York: Simon & Schuster, 2002, p. 91.

6. Rice interview, August 15, 2006.

7. *Meet the Press*, NBC, September 16, 2001.

8. Eric Schmitt, "Out Front or Low Profile, Cheney Keeps Powerful Role," *New York Times*, October 7, 2001. Sec. 1B, p. 4.

9. Mark Schlueb, "Terror Campaign Was Meant to Last Days, Senator Says; Comments Made after CIA Briefing," *Chicago Tribune*, September 18, 2001.

10. Rice interview, August 15, 2006.

11. *Meet the Press*, NBC, September 16, 2001.

12. "The Dollar," *Economist*, September 22, 2001.

13. Steven Pearlstein, "Global Recession Near, Some Economists Say; Confidence, Already Damaged by U.S. Slump, Further Dented by Attacks and War Concerns," *Washington Post*, September 26, 2001.

14. Rice interview, August 15, 2006.

15. "Developments in Terrorist Attacks Investigation," Associated Press, October 13, 2001.

16. George W. Bush, presidential address to the nation, October 7, 2001.

17. *NewsHour with Jim Lehrer*, PBS, October 12, 2001.

18. Rice interview, August 15, 2006.

19. "Is Nation Ready for Botulinum Attack?" Associated Press/CNN, March 25, 2003.

20. *NewsHour with Jim Lehrer*, PBS, October 12, 2001.

21. Dick Cheney, remarks at 56th Annual Alfred E. Smith Memorial Foundation Dinner, New York, October 18, 2001.

22. Charles Krauthammer, "Not Enough Might," *Washington Post*, October 30, 2001, p. A21.

23. R. W. Apple, "A Military Quagmire Remembered: Afghanistan as Vietnam," *New York Times*, October 31, 2001.

24. Mike Allen, *Washington Post*, October 13, 2001.

25. Paul Nelson, "Paul Nelson Farm Experience," www.paulnelsonfarm.com.

26. Author interview with David Bohrer, February 13, 2007.

27. Cheney interview, August 9, 2006.

28. Ibid.

29. Author interview with President George W. Bush, December 11, 2006.

Chapter Thirteen: Back to Baghdad

1. For more see Stephen F. Hayes, "The White House, the CIA, and the Wilsons," *Weekly Standard*, October 24, 2005, pp. 20–27.

2. Interview with Richard Kerr, "The Dark Side," *Frontline*, PBS, January 25, 2006, http://www.pbs.org/wgbh/pages/frontline/darkside/interviews/kerr.html (accessed April 15, 2007).

3. Defense Exhibit 66.2, as reported by Byron York, "Is Everything We Know about Joe Wilson's Trip to Niger Wrong?" *National Review Online*, February 7, 2007.

4. Senate Select Committee on Intelligence, "Report on the U.S. Intelligence Community's Prewar Intelligence Assessments on Iraq," U.S. Senate, July 7, 2004.

5. Michael R. Gordon, "Egypt Vows to Press Iraq on U.N. Arms Inspections, Cheney Is Told," *New York Times,* March 14, 2002, p. A19.

6. "BBC Monitoring Middle East," Iraqi TV, Baghdad, March 15, 2002.

7. Michael R. Gordon and Bernard E. Trainor, *Cobra II: The Inside Story of the Invasion and Occupation of Iraq.* New York: Pantheon, 2006.

8. Author interview with Vice President Dick Cheney, November 21, 2006.

9. David Martin, *CBS Evening News*, CBS, May 15, 2002.

10. *New York Post,* May 16, 2002, p. 1.

11. Hillary Clinton, "Investigate 9/11," 107th Congress, 2nd session, *Congressional Record* (Senate) 148 (63), May 16, 2002, p. 54453.

12. Rep. Richard Gephardt (D-MO), press conference, U.S. Capitol, Washington, D.C., May 16, 2002; see David E. Sanger and Elisabeth Bumiller, "No Hint of Sept. 11 in Report in August, White House Says, but Congress Seeks Inquiry," *New York Times,* May 17, 2002, p. A1.

13. Dick Cheney, address to Conservative Party of New York State, 40th Annual Anniversary Dinner, New York, May 16, 2002.

14. For more, see Stephen F. Hayes, *The Connection: How al Qaeda's Collaboration with Saddam Hussein Has Endangered America.* New York: HarperCollins Publishers, 2004, pp. 117–127.

15. *Meet the Press,* NBC, May 19, 2002.

16. Cheney interview, November 21, 2006.

17. John McLaughlin interview, "The Dark Side," *Frontline,* January 11, 2006.

18. Senate Select Committee on Intelligence, "Report on the U.S. Intelligence Community's Prewar Intelligence Assessments on Iraq," p. 351.

19. Brent Scowcroft, "Don't Attack Saddam," *Wall Street Journal,* August 15, 2002.

20. James A. Baker III, "The Right Way to Change a Regime," *New York Times*, August 25, 2002.

21. Dick Cheney, address, 103rd National Convention, Veterans of Foreign Wars, Nashville, Tenn., August 26, 2002.

22. Ibid.

23. Ibid.

24. Les Aspin, remarks, *This Week with David Brinkley,* ABC, August 4, 1991.

25. *This Week with David Brinkley*, ABC News, August 4, 1991.

26. Cheney, Veterans of Foreign Wars, August 26, 2002.

27. Ibid.

28. Bob Woodward, *Plan of Attack.* New York: Simon and Schuster, 2004, p. 175.

29. Ted Kennedy, speech at the Paul Nitze School for Advanced International Studies, Johns Hopkins University, September 27, 2002.

30. Bill Clinton, interview, *Good Morning America,* September 27, 2002.
31. Author interview with Representative John Murtha (D-PA.), December 4, 2006.
32. Elisabeth Bumiller and Carl Hulse, "Bush Will Use Congress Vote to Press U.N.," *New York Times,* October 12, 2002, p. A1.
33. Commission on the Intelligence Capabilities of the United States Regarding Weapons of Mass Destruction, Report to the President, March 31, 2005, p. 14. As problematic as the NIE of October 2002 was, it was not the community's worst analytic failure on Iraq. Even more misleading was the river of intelligence that flowed from the CIA to top policy makers over long periods of time—in the president's daily brief (PDB) and in its more widely distributed companion, the senior executive intelligence brief (SEIB). These daily reports were, if anything, more alarmist and less nuanced than the NIE. It was not that the intelligence was markedly different. Rather, it was that the PDBs and SEIBs, with their attention-grabbing headlines and drumbeat of repetition, left an impression of many corroborating reports where in fact there were very few sources. And in other instances, intelligence suggesting the existence of weapons programs was conveyed to senior policy makers, but later information casting doubt on the validity of that intelligence was not. In ways both subtle and not so subtle, the daily reports seemed to be "selling" intelligence—in order to keep its customers, or at least the "first customer," interested.
34. Cheney interview, August 22, 2006.
35. Jeffrey Goldberg, "The CIA and the Pentagon Take Another Look at Al Qaeda and Iraq," *New Yorker,* February 10, 2003. "According to a senior Administration official, the CIA itself is split on the question of a Baghdad–Al Qaeda connection: analysts in the agency's Near East–South Asia division discount the notion; the Counterterrorist Center supports it. The senior Administration official told me that Tenet tends to agree with the Counterterrorist Center."
36. Agence France-Presse, November 12, 2002.
37. Woodward, *Plan of Attack,* p. 249.
38. Author interview with Vice President Dick Cheney, August 22, 2006.
39. George W. Bush, State of the Union Address, Washington, D.C., January 28, 2003.
40. Dick Cheney, *Meet the Press,* NBC, March 16, 2003.
41. Woodward, *Plan of Attack,* p. 387.
42. Author interview with President George W. Bush, December 11, 2006.
43. Cheney interview, August 22, 2006.
44. Bush interview, December 11, 2006.
45. President George W. Bush, address to the nation, March 19, 2003.
46. Dick Cheney, address to the Discovery Institute, Seattle, Wa., August 14, 1992.
47. Cheney interview, August 22, 2006.

CHAPTER FOURTEEN: THE WAR OVER THE WAR

1. John H. Cushman and Thom Shanker, "A War Like No Other Uses New 21st-Century Methods to Disable Enemy Forces," *New York Times*, April 10, 2003.

2. Stephen F. Hayes, "Blueprint for the Iraqi Insurgency," *Weekly Standard* 11 (22) (February 20, 2006).

3. President George W. Bush, remarks from USS *Abraham Lincoln*, at sea off the coast of San Diego, Calif., May 1, 2003.

4. Ibid.

5. Gromer Jeffers Jr., "Cheney: I Will Be on Ticket; VP Speaks at SMU, Says Health Fine Despite Laryngitis," *Dallas Morning News*, May 8, 2003, p. 1B.

6. Ibid.

7. Ibid.

8. Ibid.

9. Nicholas Kristof, "Missing in Action: Truth," *New York Times*, May 6, 2003, p. A31.

10. Author interview with Vice President Dick Cheney, August 9, 2006.

11. Ibid.

12. Veronique de Rugy and Tad deHaven, "'Conservative' Bush Spends More Than 'Liberal' Presidents Clinton, Carter," Cato Institute Policy Study, July 31, 2003.

13. Author interview with Senator John McCain, November 17, 2006.

14. Ibid.

15. Ibid.

16. Author interview with Vice President Dick Cheney, August 22, 2006.

17. Author interview with Dennis Hastert, January 27, 2007.

18. Ibid.

19. Ibid.

20. Ibid.

21. Author interview with Bill Thomas, October 27, 2006.

22. Ibid.

23. Author interview with Vice President Dick Cheney, February 15, 2007.

24. Cheney for Congress, campaign brochure, University of Wyoming Photo File, 29542. "One tax cut Dick supports is the one proposed by Congressman Steiger and Senator Hansen, which would cut the maximum tax rate on capital gains to 25%. Such a tax cut would benefit homeowners, farmers, ranchers."

25. Jim Toedtman, *Newsday*, May 26, 2003.

26. Thomas interview, October 27, 2006.

27. Charles Grassley, CNN, May 24, 2003.

28. Alan K. Ota and Martha Angle, "Senate Clears Tax Cut Package for Bush's Signature," *Congressional Quarterly Daily Monitor*, June 2, 2003.

29. See, for example, William Gale, Brookings Institution, comments in *Washington Post* online chat, May 28, 2003. "Members of the Administra-

tion (incl the Treasury Secretary) have been saying that tax cuts (a) will increase economic growth and (b) will increase growth so much that it will raise revenues. The first part (a) is questionable—most estimates suggest that the long-term effect of the tax cut on growth will be zero or worse. But the second part (b) is undoubtedly wrong."

30. Dick Cheney, remarks on growth and jobs package, U.S. Chamber of Commerce, Washington, D.C., January 10, 2003.
31. Letter to the editor, "How to Balance the U.S. Budget by 1991," *New York Times*, October 15, 1985 (signed by Connie Mack, Robert Michel, Richard Cheney, Trent Lott, and Joseph DioGuardi).
32. Author interview with Mitch Daniels, July 24, 2006.
33. De Rugy and deHaven, "'Conservative' Bush."
34. Author interview with Vice President Dick Cheney, June 23, 2004.
35. Daniels interview, July 24, 2006.
36. Ibid.
37. Herbert J. Storing, "The Ford White House," University of Virginia White Burkett Miller Center of Public Affairs, p. 75.
38. Bruce B. Auster, Mark Mazetti, and Edward T. Pound, "Truth and Consequences," *U.S. News and World Report*, June 9, 2003.
39. Walter Pincus and Dana Priest, "Some Analysts Felt Pressure from Cheney Visits," *Washington Post*, June 5, 2003.
40. Walter Pincus, "CIA Did Not Share Doubt on Iraq Data; Bush Used Report of Uranium Bid," *Washington Post*, June 12, 2003.
41. Reference to Libby's notes in David Addington's testimony, January 30, 2007.
42. Spencer Ackerman and John B. Judis, "The First Casualty," *New Republic* 28 (25) (June 30, 2003), p. 23.
43. Joseph Wilson, interview, *Meet the Press*, July 6, 2003.
44. Joseph Wilson IV, "What I Didn't Find in Africa," *New York Times*, July 6, 2003.
45. David Sanger, "Bush Claim on Iraq Had Flawed Origin, White House Says," *New York Times*, July 8, 2003, p. 1.
46. Author interview with Condoleezza Rice, August 15, 2006.
47. David E. Sanger and James Risen, "C.I.A. Chief Takes Blame in Assertion on Iraq: Uranium," *New York Times*, July 12, 2003, p. A1.
48. "In Tenet's Words: I Am Responsible for Review," *New York Times*, July 12, 2003, p. A5.
49. Sanger and Risen, "C.I.A. Chief Takes Blame in Assertion on Iraq: Uranium."
50. "Iraq '04: DNC, Field Suddenly Energized over Uranium," *The Hotline* (*National Journal*), July 11, 2003.
51. Robert Novak, "The Mission to Niger," *Chicago Sun-Times*, July 14, 2003, p. 31.
52. David Corn, "A White House Smear," *The Nation* (Blog: Capital Games), July 16, 2003, http://www.thenation.com/blogs/capitalgames?pid=823.

53. SSIC report, p. 69.
54. Carl Levin, press conference on Iraq: WMD, U.S. Capitol, Washington, D.C., July 23, 2003.
55. SSIC report, p. 69.
56. Ibid.
57. Ibid.
58. Leon D'Souza, "Levin: Bush War Motives All Suspect," *Port Huron Times Herald*, August 7, 2003.
59. Alex Johnson with Andrea Mitchell, "CIA Seeks Probe of White House," MSNBC.com, September 26, 2003, http://www.msnbc.com/id/3087263/.
60. *I. Lewis Libby v. United States*, p. 8. The indictment says that the investigation began "on or about September 26, 2003." Other sources confirm that the order came on that day.
61. Carl Hulse and David E. Sanger, "New Criticism on Prewar Use of Intelligence," *New York Times*, September 29, 2003.
62. Hulse and Sanger, "New Criticism."
63. Robert Novak, "Columnist Wasn't Pawn for Leak," *Chicago Sun-Times*, October 1, 2003, p. 49.
64. Author interview with Kevin Kellems, September 3, 2006.
65. Ibid.
66. James Comey, Department of Justice press conference, Washington, D.C., December 30, 2003.

Chapter Fifteen: War and Politics, Politics and War

1. Dan Shapley, "Cheney Knew about Saddam during Local Visit," *Poughkeepsie Journal*, December 15, 2003, p. 2A.
2. Author interview with Paul Bremer, November 28, 2006.
3. L. Paul Bremer, *My Year in Iraq: The Struggle to Build a Future of Hope*. New York: Simon and Schuster, 2006.
4. Author interview with Vice President Dick Cheney, November 21, 2006.
5. Ibid.
6. Bremer interview, November 28, 2006.
7. David Sanger, "Pakistani Says He Saw North Korean Nuclear Devices," *New York Times*, April 13 2004.
8. *The Today Show*, NBC, May 3, 2004.
9. David Johnston and Richard W. Stevenson, "Former Envoy Talks in Book about Source of C.I.A. Leak," *New York Times*, April 30, 2004, p. A16.
10. *Countdown with Keith Olbermann*, MSNBC, May 4, 2004.
11. Author interview with Kevin Kellems, September 3, 2006.
12. SSIC report, p. 73.
13. Paul Gigot, *Fox News Sunday*, Fox News Channel, June 20, 2004.
14. James P. Gannon, "Cheney Needs to Step Aside for Good of Bush, Party," *USA Today*, June 20, 2004, p. 21A.
15. Debra Saunders, "Bush-McCain," *San Francisco Chronicle*, June 20, 2004.

16. John McCain, *Face the Nation*, CBS, June 20, 2004.
17. Timothy J. Burger and Adam Zagorin, "Did Cheney Okay a Deal?" *Time*, June 7, 2004; Jyoti Thottam, "The Master Builder," *Time*, June 7, 2004.
18. "Democrats Want Cheney-Halliburton Probe," CNN.com, June 1, 2004.
19. "Six Defense Department witnesses at the hearing all said they knew of no Cheney influence. They said the 2002 briefing of the vice president's office was simply a routine notification, not an attempt to win approval." Larry Margasak, "Official: Cheney Not Briefed on Iraq Work," Associated Press, June 16, 2004.
20. Bill Straub, Scripps Howard News Service, June 23, 2004.
21. See FactCheck.org, "Kerry Ad Falsely Accuses Cheney on Halliburton," September 30, 2004, http://www.factcheck.org/article261.html (accessed April 15, 2007).
22. Author interview with Vice President Dick Cheney, February 15, 2007.
23. Ibid.
24. Ibid.
25. Sheryl Gay Stolberg, "Salty Language as Cheney and Senator Clash," *New York Times*, June 25, 2004.
26. Author interview with Merritt Benson, August 10, 2005.
27. James A. Barnes, "Partisanship," *National Journal* 19 (45) (November 7, 1987), p. 2825.
28. Author interview with Ron Lewis, August 11, 2005.
29. Cheney interview, February 15, 2007.
30. Cheney interview, November 21, 2006.
31. Bremer interview, November 28, 2006.
32. Remarks by Thomas Kean, 9/11 Commission Report release press conference, July 22, 2004.
33. Report of the Select Committee on Intelligence on the U.S. Intelligence Community's Prewar Intelligence Assessments on Iraq, July 9, 2004, p. 314.
34. Ibid. p. 315.
35. Chris Hedges and Don McNeil Jr., "New Clue Fails to Explain Iraq Role in Sept. 11 Attack," *New York Times*, December 16, 2001.
36. Cheney interview, November 21, 2006.
37. Senate Intelligence Committee testimony, February 24, 2004.
38. Ibid.
39. Cheney interview, June 23, 2004.
40. Ibid.
41. Author interview with Randy DeCleene, July 18, 2006.
42. *Nightline*, ABC, September 9, 2004.
43. James Gerstenzang, Matea Gold, and Peter Wallsten, "Cheney Warns of Risk If Rivals Win," *Los Angeles Times*, September 8, 2004, p. A1.
44. Laura Meckler, "Cheney Defends Invasion of Iraq," *USA Today*, September 9, 2004.

45. Rick Lyman, "Desperately Seeking Dick Cheney," *New York Times,* September 19, 2004.
46. Author interview with Senator John McCain, November 17, 2006.
47. Billy Cox, "Rush Takes Aim at UFO Politics," *Florida Today*, August 8, 2001.
48. Cheney remarks on the *Diane Rehm Show,* National Public Radio, April 11, 2001.
49. McCain interview, November 17, 2006.
50. Author interview with Rob Portman, August 24, 2006.
51. Cheney interview, August 22, 2006.
52. Warren P. Strobel, Jonathan S. Landay, and John Walcott, "CIA Review Finds No Evidence Saddam Had Ties to Islamic Terrorists," Knight Ridder/Tribune News Service, October 5, 2004.
53. Ibid.
54. Peter Jennings, *Special Report*, ABC News, October 5, 2004.
55. Bush interview, December 11, 2006.
56. *Newsnight,* CNN, October 6, 2004.
57. Author interview with George W. Bush, December 11, 2006.
58. Ibid.
59. *Hardball,* MSNBC, October 13, 2004.
60. Dick Cheney's remarks, *FoxNews.com,* October 16, 2004.
61. Bush interview, December 11, 2006.
62. Ibid.

CHAPTER SIXTEEN: DICK CHENEY: NEW DEMOCRAT?

 1. Author interview with Vice President Dick Cheney, December 7, 2004.
 2. Ibid.
 3. Ibid.
 4. Author interview with Vice President Dick Cheney, August 22, 2006.
 5. Cheney interview, December 7, 2004.
 6. Cheney interview, August 22, 2006.
 7. Ibid.
 8. Author interview with Condoleezza Rice, August 15, 2006.
 9. Author interview with President George W. Bush, December 11, 2006.
10. Ibid.
11. Ibid.
12. *Larry King Live*, CNN, May 31, 2005.
13. *CNN Late Edition*, CNN, June 5, 2005.
14. Author interview with Senator John McCain, November 17, 2006.
15. Jim VandeHei and Peter Baker, "Bush's Optimism on Iraq Debated; Rosy View in Time of Rising Violence Revives Criticism," *Washington Post*, June 5, 2005, p. A1.
16. McCain interview, November 17, 2006.
17. *Wolf Blitzer Reports*, CNN, June 23, 2005.

18. Author interview with Vice President Dick Cheney, November 21, 2006.
19. *New York Times*, "The Sovereign People of Iraq," October 17, 2005.
20. Dick Cheney, remarks, *Meet the Press*, September 16, 2001.
21. Author interview with Dick Cheney, June 23, 2004.
22. McCain interview, November 17, 2006.
23. Ibid.
24. Sheryl Gay Stolberg and Eric Lichtblau, "Senators Thwart Bush Bid to Renew Law on Terrorism," *New York Times*, December 17, 2005, p. A1.
25. Gwyneth K. Shaw, "Congress Republicans Split over Secret Listening by NSA; Bartlett and Specter among Skeptics about Program's Legality," *Baltimore Sun*, December 22, 2005.
26. Ibid.
27. Toni Locy, "Justice Department Defends NSA Spying," December 22, 2005.
28. Author interview with Michael McConnell, November 22, 2006.
29. Ibid.
30. Michael Hayden's remarks, National Press Club, January 23, 2006.
31. Michael Hayden's testimony, House Permanent Select Committee on Intelligence, April 12, 2000.
32. Bernard Gwertzman, "U.S. Aides Divided on Further Raids," *New York Times*, April 27, 1986, p. 1.
33. Transcript, *MacNeil/Lehrer NewsHour*, PBS, April 11, 1986.
34. Ibid.
35. Transcript, *NewsHour with Jim Lehrer*, PBS, February 7, 2006.
36. Sen. Jay Rockefeller (D-WV), letter to Vice President Dick Cheney, July 17, 2003.
37. Author interview with Dennis Hastert, January 27, 2007.
38. Douglas Jehl, "Among Those Told of Program, Few Objected," *New York Times*, December 23, 2005.
39. Dick Cheney, remarks to the traveling press, Air Force Two, en route, Muscat, Oman, December 20, 2005.
40. Dick Cheney, interview, *NewsHour with Jim Lehrer*, PBS, February 7, 2006.
41. McConnell interview, November 22, 2006.
42. Ibid., October 12, 2001.

CHAPTER SEVENTEEN: "THE ALTERNATIVE IS NOT PEACE"

1. Zbigniew Brzezinski, *Washington Post*, March 25, 2007.
2. Author interview with Vice President Dick Cheney, February 15, 2007.
3. Ibid.
4. Nicholas Lemann, "The Quiet Man," *New Yorker*, May 7, 2001.
5. Author interview with President George W. Bush, December 11, 2006.
6. Osprey Travel, *Destinations: The Ponoi River Company*.
7. Cheney interview, February 15, 2007.
8. Ibid.

9. Dick Cheney, interview, *Special Report with Brit Hume*, Fox News Channel, February 15, 2006.
10. Nedra Pickler, "Cheney Accidentally Shoots Fellow Hunter in Texas," Associated Press, February 12, 2006.
11. Cheney interview, *Special Report with Brit Hume*.
12. Ibid.
13. Ibid.
14. Mason, *Houston Chronicle* blog, March 13, 2006, via *Hotline*.
15. Ball, *Las Vegas Review-Journal*, March 13, 2006, via *Hotline*.
16. Author interview with Sean O'Keefe, August 8, 2006.
17. Cheney interview, February 15, 2007.
18. Trip of the Vice President, Pool Report 1, February 23, 2007.
19. Richard Morin, "187.?" *Washington Post*, March 5, 2006, p. B03.
20. Ibid.
21. Author interview with Vice President Dick Cheney, April 27, 2006.
22. Bush interview, December 11, 2006.
23. Tony Hendra, *Huffington Post*, November 23, 2006.
24. Jeffrey Goldberg, "Breaking Ranks," *New Yorker*, October 31, 2005.
25. Author interview with Bob Michel, October 19, 2006.
26. Ibid.
27. Author interview with Lee Hamilton, September 2, 2006.
28. Cheney interview, February 15, 2007.
29. Ibid.
30. Ibid.
31. *60 Minutes*, CBS, October 1, 2006.
32. Ibid.
33. Author interview with Vice President Dick Cheney, November 21, 2006.
34. Author interview with Bob Woodward, March 6, 2007.
35. *Meet the Press*, NBC, October 8, 2006.
36. Erika Bolstad and Lesley Clark, "Shaw: Firing of Rumsfeld Came Too Late," *Miami Herald*, November 11, 2006.
37. Bush interview, December 11, 2006.
38. Cheney interview, November 21, 2006.
39. Dick Cheney, remarks at the Pentagon, Arlington, Va., December 15, 2006.
40. Bush interview, December 11, 2006.
41. Ibid.
42. Ibid.
43. Author interview with Condoleezza Rice, August 15, 2006.
44. Author interview with Michael McConnell, November 22, 2006.
45. Ibid.
46. Ibid.
47. Richard Armitage, comments, *CBS Evening News*, CBS, September 7, 2006.

48. Cheney interview, February 15, 2007.

49. Dan Froomkin, "The Cloud over Cheney," *Washingtonpost.com*,
    February 21, 2007.

50. Michael Duffy, "Cheney in Twilight," *Time* 169 (12) (March 19, 2007).

51. Cheney interview, November 21, 2006.

52. Ibid.

53. Ibid.

54. Ibid.

ACKNOWLEDGMENTS

Writing a book is often said to be a lonely exercise. And that's true. But it is also very much a collaborative effort. The most important of my collaborators were my sources—some of them named, some of them unnamed. Their stories fill this book and provide its shape and texture. Each of those sources chose to cooperate with me, and in doing so gave up their own valuable time to help me tell this story. I am grateful for their willingness to do so.

The staff of the Ford Library was unfailingly polite and helpful during my week there, especially Helmi Raaska, Stacy Davis, Kenneth Hafeli, and David Horrocks. The same is true of the staff at the American Heritage Center at the University of Wyoming. Leigh Farris from CBS News provided me with a tape of Dick Cheney on *Face the Nation* in 1976, his only Sunday show appearance during the Ford administration.

David Hirshey at Harper Collins supported the project through its many different iterations, and his assistant, Kate Hamill, helped make sure that the rush to deadline went as smoothly as possible. Production editor David Koral also deserves a special note of thanks for his gracious indulgence of last-second changes and additions.

Several friends and colleagues read all or part of the manuscript. Matt Labash, David Chase, and Bob Fladung all offered valuable feedback on early versions of individual chapters. Thomas Joscelyn read late drafts and made valuable recommendations on both style and content. Tim Townsend of the *St. Louis Post-Dispatch* provided detailed suggestions on much of the manuscript and even made a special trip to Washington for a long weekend of brainstorming, editing, and a few beers. Tony Mecia of the *Charlotte Observer* offered important guidance throughout the entire project, from broad structural suggestions to specific line edits. Both Tim and Tony endured many hours of my questions and complaints for more than two years. I promise to repay the favor if I ever have the chance, or at least buy them a nice steak.

Several members of Cheney's staff went out of their way to schedule interviews, provide documents, and answer obscure questions about important subjects like Cheney's record on UFOs. Especially helpful were Jennifer Mayfield, Megan McGinn, and Lea Anne McBride, who promptly answered my many calls and e-mails. Jim Steen has worked alongside Cheney on and off since their days together on Capitol Hill in the late 1960s and is known to many in Cheney world as "the keeper of the documents." He spent day after day with me going through old correspondence, papers, photographs, and news clippings.

Eric Simonoff, my agent, surely has had to work harder on this book than on most of his others. He has been an extremely effective advocate as we traveled each of the many circuitous roads that led to the completion of this project. He is both wise and patient beyond his years.

Several other individuals deserve special mention for their contributions. Windsor Mann, my researcher, managed to find obscure articles and reference materials that I never expected he would find. I also began to depend on him for other things: fact-checking, line-editing, notes. He made contributions far greater than I had anticipated when we began to work together.

Nick Trautwein, an extraordinarily gifted editor, agreed to continue working on the book despite the fact that he left HarperCollins for Bloomsbury USA midway through the project. As he did with my first book, Nick worked on very tight deadlines and made sense out of the very rough drafts I sent his way. More than that, however, Nick was a voice of reassurance and sanity over the course of a sometimes trying two years. He answered e-mails containing what must have seemed like very stupid questions and graciously took my late-night phone calls on more than one occasion. I owe him a lifetime of Chong Ching chicken.

My colleagues and editors at the *Weekly Standard* were quick with an encouraging word and understanding about my many obligations. Claudia Anderson and Richard Starr are brilliant editors whose guidance has transformed this below-average writer into an average writer. Fred Barnes provided valuable advice, especially about the Ford years and the current Bush administration. Terry Eastland offered important context to my study of executive power. Bill Kristol read parts of several drafts of the manuscript and spent hours talking about the shape of the book. On several occasions he has given me advice that can only be described as selfless. My leave of absence took far longer than I had anticipated and I never felt any pressure from Fred, Bill, or Terry to do anything other than finish the book and take care of my family. My friends think I'm exaggerating when I say that my job at the *Weekly Standard* is the best job I can imagine. And other than being the general manager of the Green Bay Packers, I'm not.

These people helped shape and improve the book in many ways, but any mistakes that remain are mine alone.

My parents, Stephen and Nancy Hayes, have supported me on this book and in everything I have ever undertaken. They gave me important advice and encouragement during each of the many twists and turns that this process involved. It is not possible to have had a better upbringing or to have had a better example of good parenting to look up to. My siblings—Andy, Julianna, and Dan—always found a way to make me laugh or to give me something exciting to think about as I pressed to finish. I'm the oldest and yet I find myself constantly looking to them as role models. And my grandmother, Grace Forester, now 98, remains an inspiration to all of us.

Most especially, my wife, Carrie, tolerated long stretches of my absence, as the reporting took me again and again across the country and overseas. She took care of the kids and me, doing far more than her fair share for more than two years. Her support and patience, as much as anything, allowed me to take on and complete this project.

INDEX